D1258237

Selected Titles in This Series

36 **Joseph G. Rosenstein, Deborah S. Franzblau, and Fred S. Roberts, Editors,** Discrete Mathematics in the Schools

35 **Dingzhu Du, Jun Gu, and Panos M. Pardalos, Editors,** Satisfiability Problem: Theory and Applications

34 **Nathaniel Dean, Editor,** African Americans in Mathematics

33 **Ravi B. Boppana and James F. Lynch, Editors,** Logic and random structures

32 **Jean-Charles Grégoire, Gerard J. Holzmann, and Doron A. Peled, Editors,** The SPIN verification system

31 **Neil Immerman and Phokion G. Kolaitis, Editors,** Descriptive complexity and finite models

30 **Sandeep N. Bhatt, Editor,** Parallel Algorithms: Third DIMACS Implementation Challenge

29 **Doron A. Peled, Vaughan R. Pratt, and Gerard J. Holzmann, Editors,** Partial Order Methods in Verification

28 **Larry Finkelstein and William M. Kantor, Editors,** Groups and Computation II

27 **Richard J. Lipton and Eric B. Baum, Editors,** DNA Based Computers

26 **David S. Johnson and Michael A. Trick, Editors,** Cliques, Coloring, and Satisfiability: Second DIMACS Implementation Challenge

25 **Gilbert Baumslag, David Epstein, Robert Gilman, Hamish Short, and Charles Sims, Editors,** Geometric and Computational Perspectives on Infinite Groups

24 **Louis J. Billera, Curtis Greene, Rodica Simion, and Richard P. Stanley, Editors,** Formal Power Series and Algebraic Combinatorics/Séries formelles et combinatoire algébrique, 1994

23 **Panos M. Pardalos, David I. Shalloway, and Guoliang Xue, Editors,** Global Minimization of Nonconvex Energy Functions: Molecular Conformation and Protein Folding

22 **Panos M. Pardalos, Mauricio G. C. Resende, and K. G. Ramakrishnan, Editors,** Parallel Processing of Discrete Optimization Problems

21 **D. Frank Hsu, Arnold L. Rosenberg, and Dominique Sotteau, Editors,** Interconnection Networks and Mapping and Scheduling Parallel Computations

20 **William Cook, László Lovász, and Paul Seymour, Editors,** Combinatorial Optimization

19 **Ingemar J. Cox, Pierre Hansen, and Bela Julesz, Editors,** Partitioning Data Sets

18 **Guy E. Blelloch, K. Mani Chandy, and Suresh Jagannathan, Editors,** Specification of Parallel Algorithms

17 **Eric Sven Ristad, Editor,** Language Computations

16 **Panos M. Pardalos and Henry Wolkowicz, Editors,** Quadratic Assignment and Related Problems

15 **Nathaniel Dean and Gregory E. Shannon, Editors,** Computational Support for Discrete Mathematics

14 **Robert Calderbank, G. David Forney, Jr., and Nader Moayeri, Editors,** Coding and Quantization: DIMACS/IEEE Workshop

13 **Jin-Yi Cai, Editor,** Advances in Computational Complexity Theory

12 **David S. Johnson and Catherine C. McGeoch, Editors,** Network Flows and Matching: First DIMACS Implementation Challenge

11 **Larry Finkelstein and William M. Kantor, Editors,** Groups and Computation

10 **Joel Friedman, Editor,** Expanding Graphs

9 **William T. Trotter, Editor,** Planar Graphs

8 **Simon Gindikin, Editor,** Mathematical Methods of Analysis of Biopolymer Sequences

7 **Lyle A. McGeoch and Daniel D. Sleator, Editors,** On-Line Algorithms

6 **Jacob E. Goodman, Richard Pollack, and William Steiger, Editors,** Discrete and Computational Geometry: Papers from the DIMACS Special Year

(*Continued in the back of this publication*)

DIMACS

Series in Discrete Mathematics and Theoretical Computer Science

Volume 36

Discrete Mathematics in the Schools

Joseph G. Rosenstein
Deborah S. Franzblau
Fred S. Roberts
Editors

NSF Science and Technology Center
in Discrete Mathematics and Theoretical Computer Science
A consortium of Rutgers University, Princeton University,
AT&T Labs, Bell Labs, and Bellcore

American Mathematical Society
National Council of Teachers of Mathematics

This DIMACS volume is a collection of articles by experienced educators explaining why and how discrete mathematics can and should be taught in K–12 classrooms. It also discusses how discrete mathematics can be used as a vehicle for achieving the broader goals of the major effort now under way to improve mathematics education. This volume developed from a conference that took place at Rutgers University on October 2–4, 1992.

1991 *Mathematics Subject Classification.* Primary 00A05, 00A35.

Library of Congress Cataloging-in-Publication Data

Discrete mathematics in the schools / Joseph G. Rosenstein, Deborah S. Franzblau, Fred S. Roberts, editors.

p. cm. — (DIMACS series in discrete mathematics and theoretical computer science, ISSN 1052-1798 ; v. 36)

Papers from a conference held at DIMACS at Rutgers University in Oct. 1992.

"NSF Science and Technology Center in Discrete Mathematics and Theoretical Computer Science. A consortium of Rutgers University, Princeton University, AT&T Labs, Bell Labs, and Bellcore."

Includes bibliographical references.

ISBN 0-8218-0448-0 (hardcover : alk. paper)

1. Mathematics—Study and teaching—Congresses. I. Rosenstein, Joseph G. II. Franzblau, Deborah S., 1957– . III. Roberts, Fred S. IV. NSF Science and Technology Center in Discrete Mathematics and Theoretical Computer Science. V. Series.

QA11.A1D57 1997

511′.07′1—dc21

97-23277
CIP

---

10 9 8 7 6 5 4 3 2 1   02 01 00 99 98 97

Contents

Foreword ix

Preface xi

Vision Statement from 1992 Conference xiii

Overview and Abstracts xv

Introduction
Discrete Mathematics in the Schools: An Opportunity to Revitalize School Mathematics
JOSEPH G. ROSENSTEIN xxiii

Section 1. The Value of Discrete Mathematics: Views from the Classroom

The Impact of Discrete Mathematics in My Classroom
BRO. PATRICK CARNEY 3

Three for the Money: An Hour in the Classroom
NANCY CASEY 9

Fibonacci Reflections—It's Elementary!
JANICE C. KOWALCZYK 25

Using Discrete Mathematics to Give Remedial Students a Second Chance
SUSAN H. PICKER 35

What We've Got Here Is a Failure to Cooperate
REUBEN J. SETTERGREN 43

Section 2. The Value of Discrete Mathematics: Achieving Broader Goals

Implementing the Standards: Let's Focus on the First Four
NANCY CASEY AND MICHAEL R. FELLOWS 51

Discrete Mathematics: A Vehicle for Problem Solving and Excitement
MARGARET B. COZZENS 67

Logic and Discrete Mathematics in the Schools
SUSANNA S. EPP 75

Writing Discrete(ly)
ROCHELLE LEIBOWITZ 85

Discrete Mathematics and Public Perceptions of Mathematics
JOSEPH MALKEVITCH 89

Mathematical Modeling and Discrete Mathematics
HENRY O. POLLAK 99

The Role of Applications in Teaching Discrete Mathematics
FRED S. ROBERTS 105

Section 3. What Is Discrete Mathematics: Two Perspectives

What Is Discrete Mathematics? The Many Answers
STEPHEN B. MAURER 121

A Comprehensive View of Discrete Mathematics: Chapter 14 of the New Jersey Mathematics Curriculum Framework
JOSEPH G. ROSENSTEIN 133

Section 4. Integrating Discrete Mathematics into Existing Mathematics Curricula, Grades K–8

Discrete Mathematics in K–2 Classrooms
VALERIE A. DEBELLIS 187

Rhythm and Pattern: Discrete Mathematics with an Artistic Connection for Elementary School Teachers
ROBERT E. JAMISON 203

Discrete Mathematics Activities for Middle School
EVAN MALETSKY 223

Section 5. Integrating Discrete Mathematics into Existing Mathematics Curricula, Grades 9–12

Putting Chaos into Calculus Courses
ROBERT L. DEVANEY 239

Making a Difference with Difference Equations
JOHN A. DOSSEY 255

Discrete Mathematical Modeling in the Secondary Curriculum: Rationale and Examples from The Core-Plus Mathematics Project
ERIC W. HART 265

A Discrete Mathematics Experience with General Mathematics Students
BRET HOYER 281

Algorithms, Algebra, and the Computer Lab
PHILIP G. LEWIS 289

Discrete Mathematics Is Already in the Classroom – But It's Hiding
JOAN REINTHALER 295

Integrating Discrete Mathematics into the Curriculum: An Example
JAMES T. SANDEFUR 301

Section 6. High School Courses on Discrete Mathematics

The Status of Discrete Mathematics in the High Schools
HAROLD F. BAILEY 311

Discrete Mathematics: A Fresh Start for Secondary Students
L. CHARLES BIEHL 317

A Discrete Mathematics Textbook for High Schools
NANCY CRISLER, PATIENCE FISHER, AND GARY FROELICH 323

Section 7. Discrete Mathematics and Computer Science

Computer Science, Problem Solving, and Discrete Mathematics
PETER B. HENDERSON 333

The Role of Computer Science and Discrete Mathematics in the High School Curriculum
VIERA K. PROULX 343

Section 8. Resources for Teachers

Discrete Mathematics Software for K–12 Education
NATHANIEL DEAN AND YANXI LIU 357

Recommended Resources for Teaching Discrete Mathematics
DEBORAH S. FRANZBLAU AND JANICE C. KOWALCZYK 373

The Leadership Program in Discrete Mathematics
JOSEPH G. ROSENSTEIN AND VALERIE A. DEBELLIS 415

Computer Software for the Teaching of Discrete Mathematics in the Schools
MARIO VASSALLO AND ANTHONY RALSTON 433

Foreword

This DIMACS volume on "Discrete Mathematics in the Schools" contains refereed articles on the theme of a conference held at DIMACS at Rutgers University in October 1992. The conference was sponsored by DIMACS with funding from the National Science Foundation.

We would especially like to thank Joseph G. Rosenstein for organizing the conference, and him and Deborah S. Franzblau who together with Fred S. Roberts served as editors of this volume.

Fred S. Roberts, Director
Bernard Chazelle, co-Director
Stephen R. Mahaney, Associate Director

Preface

Discrete mathematics can and should be taught in K-12 classrooms. This volume, a collection of articles by experienced educators, explains why and how, including evidence for "why" and practical guidance on "how". It also discusses how discrete mathematics can be used as a vehicle for achieving the broader goals of the major effort now underway to improve mathematics education.

This volume is intended for several different audiences. Teachers at all grade levels will find here a great deal of valuable material that will help them introduce discrete mathematics in their classrooms, as well as examples of innovative teaching techniques. School and district curriculum leaders will find articles that address their questions of whether and how discrete mathematics can be introduced into their curricula. College faculty will find ideas and topics that can be incorporated into a variety of courses, including mathematics courses for prospective teachers. A description of the organization of this volume and an annotated summary of the articles it contains can be found in the **Overview and Abstracts**.

This volume developed from a conference that took place at Rutgers University on October 2-4, 1992. The conference, entitled "Discrete Mathematics in the Schools: How Do We Make an Impact?" was attended by 33 people, from high schools and colleges, who had played leadership roles in introducing discrete mathematics at precollege levels.[1] The conference was sponsored by the Center for Discrete Mathematics and Theoretical Computer Science (DIMACS)[2] and funded by the National Science Foundation (NSF).

The invitation to the conference noted that "Although primarily a re-

[1]A list of conference participants and an abbreviated conference program appear as appendices to the **Introduction**.

[2]DIMACS is an NSF-funded Science and Technology Center which was founded in 1989 as a consortium of Rutgers and Princeton Universities, AT&T Bell Laboratories, and Bellcore (Bell Communications Research). With the reorganization of AT&T Bell Laboratories in 1996, it was replaced in the DIMACS consortium by AT&T Labs and Bell Labs (part of Lucent Technologies). DIMACS is also funded by the New Jersey Commission on Science and Technology, its partner organizations, and numerous other agencies.

search center, DIMACS is committed to educational programs involving discrete mathematics ... as discrete mathematics activities at K-12 levels increase, it is appropriate for a national center in discrete mathematics to bring together those associated with such activities for an opportunity to reflect on how all of our activities can make an impact on mathematics education nationally." The rationale for the conference is further described in the **Introduction**, and the **Vision Statement** concerning discrete mathematics in the schools that emerged from the conference appears directly after this **Preface**.

This volume was originally conceived as the proceedings of the conference. However, as we began receiving and reviewing articles, we realized that an expanded and more comprehensive book would have greater value and impact. Accordingly, we solicited additional articles from appropriate authors; approximately two-thirds of the articles are based on conference presentations, and the remainder were written independently. All of the authors received comments and suggestions from both anonymous referees and the editors, and revised their articles accordingly; this lengthened considerably the time to produce the volume, but greatly enhanced its quality.

The editors wish to thank the authors for their cooperation and patience, as well as for their contributions. We also thank the referees for their assistance, Reuben Settergren for many hours spent in editorial work, typesetting, and creating figures, Pat Pravato for her able secretarial help, and NSF for a supplementary grant that enabled us to complete the volume.

Compiling a volume like this, involving 34 articles from different authors, is not an easy task, and we are quite pleased that this task has now been completed.

Joseph G. Rosenstein
Deborah S. Franzblau
Fred S. Roberts

Vision Statement from 1992 Conference[1]

A major reform effort is now underway in mathematics education. The goals of this reform are to enable us to educate informed citizens who are better able to function in our increasingly technological society; have better reasoning power and problem-solving skills; are aware of the importance of mathematics in our society; and are prepared for future careers which will require new and more sophisticated analytical and technical tools.

We feel that *discrete mathematics is an exciting and appropriate vehicle for working toward and achieving these goals.* It is an excellent tool for improving reasoning and problem-solving skills. It lends itself well to the evolving consensus on effective instructional strategies expressed in the *Curriculum and Evaluation Standards for School Mathematics* of the National Council of Teachers of Mathematics (NCTM). Discrete mathematics has many practical applications that are useful for solving some of the problems of our society and that are meaningful to our students. Its problems make mathematics come alive for students, and help them see the relevance of mathematics to the real world. Discrete mathematics does not have extensive prerequisites, yet poses challenges to all students. It is fun to do, is often geometrically based, and stimulates an interest in mathematics on the part of students at all levels and of all abilities.

At the same time, we feel that *discrete mathematics needs to be introduced into the K-12 curriculum for its own sake.* During the past 30 years, discrete mathematics has grown rapidly and has become a significant area of mathematics. Increasingly, discrete mathematics is the mathematics that is being used by decision-makers in business and government; by workers in fields such as telecommunications and computing that depend upon information transmission; and by those in many rapidly changing professions involving health care, biology, chemistry, automated manufacturing, transportation, etc. Increasingly, discrete mathematics is the language of a large body of science and underlies decisions that individuals will have to make

[1]An initial draft of this "vision statement" was developed during the October 1992 conference, reflecting the goals of the conference and the consensus of its participants. The statement was revised at a meeting of a designated committee of conference participants the following January.

in their own lives, in their professions, and as citizens.

It should be stressed, however, that *we are not advocating any specific set of topics in discrete mathematics that should be taught*; discrete mathematics includes many different areas, each of which is valuable. Rather, we feel it is important that students be able to speak the language of discrete mathematics and be exposed to the ways of thinking and reasoning that are inherent in modern discrete mathematics; all students should know and be able to apply discrete mathematics concepts and skills in a variety of contexts. And it is *especially* important for teachers to become excited about their own experiences with discrete mathematics and to share that excitement with their students.

Overview and Abstracts

As noted in the **Preface**, this volume makes the case that discrete mathematics should be included in K–12 classrooms and curricula, and provides practical assistance and guidance on how this can be accomplished. The organization of this volume parallels these two goals. After the **Introduction** the articles are arranged in the following eight clusters:

Section 1. The Value of Discrete Mathematics: Views from the Classroom

Section 2. The Value of Discrete Mathematics: Achieving Broader Goals

Section 3. What is Discrete Mathematics: Two Perspectives

Section 4. Integrating Discrete Mathematics into Existing Mathematics Curricula, Grades K–8

Section 5. Integrating Discrete Mathematics into Existing Mathematics Curricula, Grades 9–12

Section 6. High School Courses on Discrete Mathematics

Section 7. Discrete Mathematics and Computer Science

Section 8. Resources for Teachers

Everyone's first question is of course, "What is discrete mathematics?" Everyone's second question is, "Why should I use discrete mathematics?" Explicit discussion of the first question is delayed until Section 3, and the focus of the **Introduction** and Sections 1–2 is the second question. These sections make the case for discrete mathematics — from the perspective of teachers in the classroom, and from the perspective of researchers involved in improving mathematics education. These articles encompass a variety of agendas — implementing the four NCTM process standards (problem-solving, reasoning, communicating mathematical ideas, and making connections), improving the public's perception of mathematics, conveying the usefulness of mathematics, and providing a new start for students, teachers, and curricula.

Everyone's third question is, "How can I use discrete mathematics in my classroom?" This question is addressed in Sections 4–7. One set of responses involves incorporating discrete mathematics into existing curricula; these responses appear in Sections 4 and 5, arranged by grade level. Another set of responses involves introducing new courses, typically at the high school level, and these are addressed in Section 6. Section 7 addresses the role of computer science in the high school curriculum, as well as the role of discrete mathematics in the teaching of computer science.

Section 8 describes resources available to teachers who decide to enrich their classrooms with discrete mathematics.

Following are abstracts of the articles in this volume, prepared by the editors. The abstracts are arranged by section, and within each section are presented alphabetically, as are the articles in the volume.

Introduction

Joseph G. Rosenstein's article **Discrete Mathematics in the Schools: An Opportunity to Revitalize School Mathematics** serves as an introduction to this volume and describes why discrete mathematics can be a useful vehicle for improving mathematics education and revitalizing school mathematics. He provides rationales for introducing discrete mathematics in the schools, noting that discrete mathematics is applicable, accessible, attractive, and appropriate, and argues that discrete mathematics offers a "new start" in mathematics for students. This article is based on a concept document distributed to participants prior to the October 1992 conference, and on the opening presentation of the conference.

Section 1. The Value of Discrete Mathematics: Views from the Classroom

Bro. Patrick Carney's article **The Impact of Discrete Mathematics in My Classroom** describes anecdotally how the author aroused in his students an interest in mathematics, and developed in his students a more "positive attitude toward mathematics and their ability to do it".

Nancy Casey's article **Three for the Money: An Hour in the Classroom** describes the excitement generated in a class of high school students, participating in a special summer program, when they are presented with an unsolved mathematical problem, and the mathematical journeys that they take to learn what the problem is and to try to solve it. It also provides a vivid description of how the teacher's role in the classroom changes when the class embarks on an uncharted adventure of mathematical discovery.

Janice C. Kowalczyk's article **Fibonacci Reflections: It's Elementary!** is an account of her experiences giving a workshop on the Fibonacci sequence (1, 1, 2, 3, 5, 8, ...) to a fourth-grade class. She gives a detailed description of the workshop activities, including student investigations of the classical rabbit population problem that leads to the sequence, and spiral-counting in pinecones, sunflowers, shells, and other objects whose growth patterns exhibit the sequence. The article illustrates how using a topic with a strong visual appeal, along with a focus on student exploration, can bring out the strengths in many students who have had difficulties in the traditional elementary mathematics curriculum.

Susan H. Picker's article **Using Discrete Mathematics to Give Remedial Students a Second Chance** is an account of her experiences introducing discrete mathematics to a class of remedial tenth-grade students in Manhattan, and their success in solving complex graph-coloring problems. More than that, it is an account of the impact that this course had on the students' perceptions of mathematics and their own abilities, as well as on their subsequent school careers. The author learned from this experience the extent to which students' dislike of arithmetic serves as an obstacle to their progress and success in mathematics.

Reuben J. Settergren's article **"What We've Got Here is a Failure to Cooperate"** describes a cooperative game, based on the classical Prisoner's Dilemma, that the author played with twelve-year-old students in a summer program. The game gave students insight into why individuals are sometimes motivated to behave in a way that harms the larger community, providing an opportunity to discuss moral and social issues in a mathematics class.

Section 2. The Value of Discrete Mathematics: Achieving Broader Goals

Nancy Casey and Michael R. Fellows' article **Implementing the Standards: Let's Focus on the First Four** argues that in order to properly address the NCTM process standards — reasoning, problem-solving, communications, and connections — in the elementary school classroom, new content must be introduced into the K–4 mathematics curriculum. The authors show by example how elementary versions of problem situations that arise in computer science and discrete mathematics make it possible to realize the goals of the process standards. They describe their approach to teaching mathematics as parallel to the "whole language" approach to teaching reading.

Margaret B. Cozzens' article **Discrete Mathematics: A Vehicle for Problem Solving and Excitement** provides examples of discrete mathematics activities from several curriculum development projects funded by the NSF division that the author heads. The author argues that discrete mathematics can motivate students to think mathematically, to become better problem solvers, and to increase their interest in mathematics.

Susanna S. Epp's article **Logic and Discrete Mathematics in the Schools** argues that logical reasoning should be a component of the discrete mathematics that is discussed at all grade levels. Students should not have to wait until they are college students to explore the reasoning involved in "and", "or", and "if-then" statements, or to understand how quantifiers are used. This need not be done formally (e.g., through truth tables) but through concrete activities which ultimately will support the students' transition to abstract mathematical thinking. The author illustrates the value of explicit discussion of logic with experiences from a discrete mathematics course she has taught at DePaul University.

Rochelle Leibowitz' article **Writing Discrete(ly)** argues that discrete mathematics serves as an excellent vehicle for teaching students to communicate mathematically. Through describing carefully simple proofs and algorithms (e.g., instructions for building a Lego model), students acquire technical writing skills that will be useful in a variety of career and life situations.

Joseph Malkevitch's article **Discrete Mathematics and Public Perceptions of Mathematics** contrasts the kinds of problems typically discussed in high school mathematics classes, usually involving extensive manipulation of symbols, with the kinds of problems that manifest the ways in which mathematics influences daily life. Malkevitch argues that the negative perceptions that the general public has about mathematics arise in part from an unbalanced mathematical diet — too much of the former, too little of the latter — and notes that problems from discrete mathematics can play an important role in changing these perceptions.

Henry O. Pollak's article **Mathematical Modeling and Discrete Mathematics** discusses mathematical modeling in general, noting that "applied mathematics", "problem solving", and "word problems" all start with an idealized version of a real world problem, and so normally omit the initial and final parts of the modeling process. The author notes that in discrete mathematics situations, however, it is often possible to introduce the entire mathematical modeling process into the classroom; he provides five examples of modeling situations which lead to discrete mathematics and which can be made accessible to high school students.

Fred S. Roberts' article **The Role of Applications in Teaching Discrete Mathematics** notes that "one of the major reasons for the great increase in interest in discrete mathematics is its importance in solving practical problems." The author introduces several "rules of thumb" about the role of applications in teaching discrete mathematics, and illustrates those by providing many applications of the Traveling Salesman Problem, graph coloring, and Euler paths.

Section 3. What is Discrete Mathematics: Two Perspectives

Stephen B. Maurer's article **"What is Discrete Mathematics?" The Many Answers** provides and discusses a variety of proposed definitions and descriptions of discrete mathematics, along with several proposed goals and benefits for including discrete mathematics in the schools. The article concludes with a set of goals and topics for discrete mathematics in the schools on which the author thinks there might be general agreement.

Joseph G. Rosenstein's article **A Comprehensive View of Discrete Mathematics: Chapter 14 of the New Jersey Mathematics Curriculum Framework** contains a comprehensive discussion of topics of discrete mathematics appropriate for each of the K–2, 3–4, 5–6, 7–8, and 9–12 grade levels. The author spearheaded the development of the Framework in his role as Director of the New Jersey Mathematics Coalition. Grade-level overviews are accompanied by several hundred activities appropriate for the various grade levels. The material reflects the experiences of teachers in the Leadership Program in Discrete Mathematics, discussed in a separate article in Section 8.

Section 4. Integrating Discrete Mathematics into Existing Mathematics Curricula, Grades K–8

Valerie A. DeBellis' article **Discrete Mathematics in K–2 Classrooms** describes the author's visits to several classrooms and what she learned about the reasoning and problem-solving skills exhibited by young children who are introduced to situations involving discrete mathematics. It also describes how topics in discrete mathematics can be reformulated for children at early elementary levels.

Robert E. Jamison's article **Rhythm and Pattern: Discrete Mathematics with an Artistic Connection for Elementary School Teachers** describes the material that the author has used in programs for both inservice and preservice elementary school teachers. It focuses on how elementary school teachers can use geometric activities involving drawing polygons and planar representations of polyhedra, moving in geometric patterns, and using modular arithmetic in movement and music — to provide their students with foundational experiences for future study of mathematics.

Evan Maletsky's article **Discrete Mathematics Activities in Middle School** provides a wealth of activities that are appropriate at the middle school level; these involve counting (e.g., finding the triangular numbers when you count rectangles on a folded piece of paper), graphs, and iteration (e.g., generating Sierpinski triangles). The author discusses how these can be incorporated into the activities that are already taking place in the classroom.

Section 5. Integrating Discrete Mathematics into Existing Mathematics Curricula, Grades 9–12

Robert L. Devaney's article **Putting Chaos into Calculus Courses** describes how fundamental ideas of dynamical systems, including iteration, attracting and repelling points, and chaos, can be introduced in a beginning calculus class, through an in-depth investigation of the behavior of Newton's Method, using a computer or graphing calculator. The author's approach integrates discrete with continuous mathematics and provides a connection from calculus to the fascinating world of fractals and chaos.

John A. Dossey's article **Making a Difference with Difference Equations** shows how difference equations can be used to model change in a number of real-world settings. The author recommends the use of difference equations to provide a unified development of standard sequences studied in mathematics, such as arithmetic, geometric, and Fibonacci sequences.

Eric W. Hart's article **Discrete Mathematical Modeling in the Secondary Curriculum: Rationale and Examples from the Core-Plus Mathematics Project (CPMP)** discusses the questions of what discrete mathematics belongs in the secondary curriculum, and how it should be incorporated, from the perspective of the curriculum developer. The article presents examples adapted from CPMP materials which illustrate the CPMP approach — that discrete mathematics should be woven into an overall integrated mathematics curriculum, and that the emphasis should be on discrete mathematical modeling.

Bret Hoyer's article **A Discrete Mathematics Experience with General Mathematics Students** describes how the author introduced topics in discrete mathematics first into intermediate algebra and geometry classes, and then, as a result of the students' positive experiences, into other classes as well — including general mathematics and consumer mathematics courses. The article focuses on the "Street Networks" unit on Euler paths and circuits that was woven into these courses.

Philip G. Lewis' article **Algorithms, Algebra, and the Computer Lab** describes how the author's high school students used the LOGO computer environment to explore and develop concepts in linear algebra. These explorations, which took place in a computer lab, enabled students to view linear algebra algorithmically and to learn how to construct and analyze algorithms.

Joan Reinthaler's article **Discrete Mathematics is Already in the Classroom — But It's Hiding** argues that many problems in high school courses are discussed as problems with continuous domains when a discrete perspective would be more realistic, and would lead to different investigations and solutions. Several examples are given involving standard textbook problems in algebra.

James T. Sandefur's article **Integrating Discrete Mathematics into the Curriculum: An Example** describes how he uses the handshake problem to review with his precalculus class the notions of function, domain and range, and graphing quadratic functions. The author argues that "this approach integrates discrete mathematics into the existing curriculum, results in deeper student understanding, and can be accomplished in about the same amount of time as is presently devoted to these topics."

Section 6. High School Courses on Discrete Mathematics

Harold F. Bailey's article **The Status of Discrete Mathematics in the High Schools** reports on a survey that the author did to ascertain how many high schools offer courses in discrete mathematics, what those courses contain, and the goals of the schools in offering such courses.

L. Charles Biehl's article **Discrete Mathematics: A Fresh Start for Secondary Students** describes a project-based discrete mathematics course developed by the author for juniors and seniors of average ability. The students explored a variety of mathematical topics in real-world settings; moreover, since many topics in discrete mathematics have few prerequisites, these students were able to become successful problem solvers and to develop more positive attitudes to mathematics. The article includes an outline of the course.

Nancy Crisler, Patience Fisher, and Gary Froelich's article **A Discrete Mathematics Textbook for High Schools** describes the textbook they have co-authored, providing a discussion of its origins and development. The organization and content of the book is based on the NCTM report, *Discrete Mathematics and the Secondary Mathematics Curriculum*; it addresses five broad areas (social decision making, graph theory, counting techniques, matrix models, and the mathematics of iteration) and interweaves six unifying themes (modeling, use of technology, algorithmic thinking, recursive thinking, decision making, and mathematical induction). The article includes summaries of and examples drawn from each chapter of the book.

Section 7: Discrete Mathematics and Computer Science

Peter B. Henderson's article **Computer Science, Problem Solving, and Discrete Mathematics** addresses the role of discrete mathematics in a first course in computer science, based on the author's experience in developing a "Fundamentals of Computer Science" course at SUNY Stony Brook. Although the course described was developed originally for students planning a career in computer science, it has drawn students with a wide variety of goals. The author notes that "With its emphasis on logical reasoning and problem analysis and solution, discrete mathematics provides a catalyst for general thinking and problem-solving skills ... ," making such a course valuable for teaching computer science to high school students as well.

Viera K. Proulx' article **The Role of Computer Science and Discrete Mathematics in the High School Curriculum** identifies six key themes in computer science that the author argues should be taught to all high school students, and sketches activities for students to explore these themes. The ideas in the article grew out of the author's participation in the Association for Computing Machinery (ACM) Task Force on the High School Curriculum, which produced a "Model High School Computer Science Curriculum" in 1993.

Section 8. Resources for Teachers

Nathaniel Dean and Yanxi Liu's article **Discrete Mathematics Software for K–12 Education** describes two workshops involving teachers and software developers in which teachers solved problems using software developed for research, and shared their reflections on the features that would make such software useful in their classrooms. In the first workshop, teachers used NETPAD, written by Dean when he was at Bellcore; in the second workshop, teachers used Combinatorica, written by Steven Skiena of SUNY Stony Brook. The article also provides an annotated list of other software packages that are potentially useful to teachers.

Deborah S. Franzblau and Janice C. Kowalczyk's article **Recommended Resources for Teaching Discrete Mathematics** identifies outstanding resources, including books, modules, periodicals, literature, Internet sites, software, and videos for the K–12 mathematics teacher or supervisor building a core resource library for teaching topics in discrete mathematics. There are extensive reviews of four popular textbooks; other resources are accompanied by briefer descriptions. The list of resources, which is indexed by topic and grade level, and which includes publisher information, was developed from recommendations by participants and instructors in the DIMACS Leadership Program in Discrete Mathematics.

Joseph G. Rosenstein and Valerie A. DeBellis' article **The Leadership Program in Discrete Mathematics** describes the DIMACS-sponsored programs for K–12 teachers that have taken place for the past nine years at Rutgers University, the development and implementation of the program's goals, and how the program is serving as a continuous resource for the dissemination of discrete mathematics to K–12 schools.

Mario Vassallo and Anthony Ralston's article **Computer Software for the Teaching of Discrete Mathematics in the Schools** provides a number of criteria for judging the suitability of computer software for educational use, and then describes and evaluates three software systems (Mathematica/Combinatorica, GraphPack, and SetPlayer) against these criteria.

Introduction

Discrete Mathematics in the Schools: An Opportunity to Revitalize School Mathematics

Joseph G. Rosenstein

This article serves as an introduction in four different but overlapping ways:

- As an introduction to a volume advocating discrete mathematics in the schools, it outlines the case for this position.
- As an introduction to a collection of thirty-four diverse articles, it provides some context for those articles.
- As an introduction to the 1992 conference which led to this volume, it provides information about the conference and its themes.
- As an introduction to my perspective as conference organizer, author, and editor, it summarizes the main reasons for my involvement in this enterprise.

The author's perspective

Starting at the end, which is of course the beginning, there are two major reasons for my ongoing efforts to promote discrete mathematics in the schools — that in two major ways, discrete mathematics offers an opportunity to revitalize school mathematics.

- Discrete mathematics offers a new start for students. For the student who has been unsuccessful with mathematics, it offers the possibility for success. For the talented student who has lost interest in mathematics, it offers the possibility of challenge.
- Discrete mathematics provides an opportunity to focus on *how* mathematics is taught, on giving teachers new ways of looking at mathematics and new ways of making it accessible to their students. From this perspective, teaching discrete mathematics in the schools is not an end in itself, but a tool for reforming mathematics education.

These two themes first appeared in a concept document that I developed in January 1991 and that grew out of the first two years of my experience directing the Leadership Program in Discrete Mathematics, an NSF-funded teacher enhancement program for high school teachers, at Rutgers University.[1] Participants reported changes in their classrooms, in their students, and in themselves. Their successes taught us that discrete mathematics was not just another piece of the curriculum. Many participants reported success with a variety of students at a variety of levels, demonstrated a new enthusiasm for teaching in new ways, and proselytized among their colleagues and administrators.

These two themes are discussed further in this article in sections entitled **Discrete mathematics: A new start for students** and **Discrete mathematics: A vehicle for improving mathematics education.**

The October 1992 Conference

These two views of discrete mathematics — as a new start for students and as a vehicle for improving mathematics education — seemed to me to establish an agenda for those interested in both discrete mathematics and mathematics education. If discrete mathematics could have a significant impact on mathematics education, how can that impact be actualized? This question led to a conference entitled "Discrete Mathematics in the Schools: How Do We Make an Impact?"

The Conference took place on October 2-4, 1992 at Rutgers University and was sponsored by the Center for Discrete Mathematics and Theoretical Computer Science (DIMACS), an NSF-funded Science and Technology Center. It brought together thirty-three educators who had been involved in a variety of ways in introducing discrete mathematics in the schools; see Appendix A for a list of conference participants. The concept document containing the two themes described above was distributed in advance of the conference and was reflected in the opening presentation at which I welcomed and challenged the conference participants.

The conference program was designed to inform the participants about various perspectives of discrete mathematics and its role in K–12 education, and about all of the various activities taking place that promoted discrete mathematics in the schools. An abbreviated version of the program, showing presentations and session titles, appears in Appendix B. Presentations were followed by extended discussions.

[1]The NSF-funded Leadership Program in Discrete Mathematics is co-sponsored by the Center for Discrete Mathematics and Theoretical Computer Science (DIMACS) and the Rutgers Center for Mathematics, Science, and Computer Science Education (CMSCE). Although originally (in 1989-1991) for high school teachers, the Leadership Program subsequently (beginning in 1992) also enrolled middle school teachers, and now (since 1995) focuses on K–8 teachers. See the article by Rosenstein and DeBellis in this volume for further information about the Leadership Program.

One outcome of the discussions at the conference was the **Vision Statement** which appears at the beginning of this volume. Two major points of the Vision Statement were that "discrete mathematics is an exciting and appropriate vehicle for working toward and achieving these goals" (referring to the goals of those striving to improve mathematics education), and that "discrete mathematics needs to be introduced into the curriculum for its own sake" because of the increasing importance and prevalence of its applications.

What is discrete mathematics?

It is, of course, natural for K–12 teachers and administrators, as well as parents and the press, to ask this question. Unfortunately, it is not an easy question to answer. The problem is that the phrase "discrete mathematics" does not refer to a well-defined branch of mathematics — like algebra, geometry, trigonometry, or calculus — but rather encompasses a variety of loosely-connected concepts and techniques. Moreover, it is not a branch of mathematics which is generally familiar to the public. At the dedication ceremony of DIMACS as a Center in 1989, then-Governor Thomas Kean (NJ) quipped that, before participating in this ceremony, his impression was that discrete mathematics was what accountants did behind closed doors. That may be a common initial impression of discrete mathematics.

I have found that one effective way of answering the question is by giving lots of examples of the kinds of situations where the mathematics that is used is "discrete". Though not actually defining discrete mathematics, the examples give a flavor of what comprises discrete mathematics, and also helps to demystify the phrase. Here is the list that we are currently using in one of the brochures of the Leadership Program in Discrete Mathematics; this list contains examples that we anticipate will make sense to the teachers that we hope to attract to the program.

- What is the quickest way to sort a list of names alphabetically?
- Which way of connecting a number of sites into a telephone network requires the least amount of cable?
- Which version of a lottery gives the best odds?
- If each voter ranks the candidates for President in order of preference, how can a consensus ranking of the candidates be obtained?
- What is the best way for a robot to pick up items stored in an automated warehouse?
- How does a CD player interpret the codes on a CD correctly even if the CD is scratched?
- How can an estate be divided fairly?
- How can ice cream stands be placed at various street corners in a town so that at any corner there is a stand which is at most one block away?
- How can representatives be apportioned fairly among the states using current census information?

These problems — and many others from different areas within discrete mathematics — share several important characteristics. They are easily understood and discussed, readily seen as dealing with real-world situations, and can be explored without extensive background in school mathematics. This is discussed in more detail in the following section.

Although I have used this "definition-by-examples" of discrete mathematics for a number of years, in the spring of 1996, as the New Jersey Department of Education was preparing to present its recommendations for mathematics standards to the State Board of Education, I was told that I had to provide a "real definition" for the document. So here is discrete mathematics as it appears in New Jersey's *Core Curriculum Content Standards*:

> Discrete mathematics is the branch of mathematics that deals with arrangements of discrete objects. It includes a wide variety of topics and techniques that arise in everyday life, such as how to find the best route from one city to another, where the objects are cities arranged on a map. It also includes how to count the number of different combinations of toppings for pizzas, how best to schedule a list of tasks to be done, and how computers store and retrieve arrangements of information on a screen. Discrete mathematics is the mathematics used by decision-makers in our society, from workers in government to those in health care, transportation, and telecommunications. Its various applications help students see the relevance of mathematics in the real world.

In This Volume. Two articles in Section 3 of this volume address directly the question, "What is discrete mathematics?" Stephen Maurer's article explores a number of possible charactizations of discrete mathematics, none of which proves to be fully satisfactory. Joseph Rosenstein's article provides an extended elaboration of the description above, as it appears in the *New Jersey Mathematics Curriculum Framework.*

Why introduce discrete mathematics into the curriculum?

A number of different arguments have been presented for including discrete mathematics in the school curriculum; these arguments can each be viewed against the backdrop of the problems posed above. Discrete mathematics is:

Applicable: In recent years, topics in discrete mathematics have become valuable tools and provide powerful models in a number of different areas.

Accessible: In order to understand many of these applications, arithmetic is often sufficient, and many others are accessible with only elementary algebra.

Attractive: Though easily stated, many problems are challenging, can interest and attract students, and lend themselves to exploration and discovery.

Appropriate: Both for students who are accustomed to success and are already contemplating scientific careers, and for students who are accustomed to failure and perhaps need a fresh start in mathematics.

In This Volume. A number of articles in this volume illustrate and elaborate on these reasons for incorporating discrete mathematics into the curriculum. Several articles that particularly address each of the above themes are provided below.

Applicable: The articles by Henry Pollak, Fred Roberts, John Dossey, and Eric Hart address the applications of discrete mathematics and how it provides models for real-world situations.

Accessible: The articles by Janice Kowalczyk, Susan Picker, Nancy Casey and Michael Fellows, Joseph Rosenstein, Valerie DeBellis, Robert Jamison, and Evan Maletsky show, for example, how discrete mathematics can be used in elementary and middle school grades.

Attractive: The articles by Patrick Carney, Nancy Casey, Reuben Settergren, and Margaret Cozzens discuss how discrete mathematics excites student interest.

Appropriate: The articles by Nancy Casey, Susan Picker, Bret Hoyer, and L. Charles Biehl discuss how discrete mathematics is appropriate for students who need a fresh start in mathematics. Other articles in this volume discuss how discrete mathematics can be combined with and enhance existing topics like algebra (Bret Hoyer, Philip Lewis), precalculus (John Dossey, Joan Reinthaler, James Sandefur), calculus (Robert Devaney), and computer science (Peter Henderson, Vera Proulx).

Discrete mathematics: A new start for students

The traditional topics of school mathematics — arithmetic, algebra, geometry, etc. — are of course important; without a good grounding in these topics, students will be seriously disadvantaged in career options. And the nation will continue to have a serious shortfall in technically skilled personnel.

However, many students find school mathematics to be a serious stumbling block, and ultimately give up. The most frequently prescribed remedy for students who have failed in school mathematics appears, unfortunately, to be more of the same. And "more of the same" usually means not only repetition of content, but also repetition of method. Thus, many students come to see school mathematics only as a set of unintelligible procedures, which is not surprising since they were never given an opportunity to explore concepts meaningfully and apply them in new situations.

At the other end of the spectrum, many talented students also find school mathematics to be uninteresting and irrelevant, and thus opt for other careers. For these students, who are looking for a spark of life and challenge in mathematics, a frequent response is "wait until you get to calculus"; but many have lost interest by the time they get to calculus.

Discrete mathematics offers a new start. For the student who has been unsuccessful in mathematics, discrete mathematics offers the possibility of success. Students who have encountered mathematics which they can do successfully are encouraged to take another look at the mathematics at which they have failed. Students who have found that they can solve meaningful problems gain a sense of empowerment. Teachers in the Leadership Program have reported that, for students who have a history of failure in mathematics, being able to use terminology and solve problems in areas with which other school personnel — teachers and guidance counselors, as well as students — are unfamiliar is a very heady experience.

The ranks of students who have been unsuccessful in mathematics contain a disproportionate number of minorities and women. Such students, who have given up hope of ever learning school mathematics, can become interested in and can learn discrete mathematics since they do not associate it at the outset with routine school mathematics. Teachers in the Leadership Program in Discrete Mathematics have used discrete mathematics successfully with these students in all types of schools, including those in urban areas.

For the talented student who has lost interest in mathematics, discrete mathematics offers the possibility of challenge. Discrete mathematics serves as a natural context for many of the puzzle-like questions that intrigue the talented student, offers open-ended problems which quickly lead to the frontiers of knowledge, and provides easy access to applications which mathematicians are now making in a variety of real-life situations. One can imagine students engaged in discrete mathematics saying "This is how I would like to spend my professional life", as well as "This is fun".

In This Volume. See the articles cited under "accessible", "attractive", and "appropriate" in the previous section.

Discrete mathematics: A vehicle for improving mathematics education.

The introduction of new material into the curriculum affords a particular opportunity to infuse new instructional techniques at the same time. When there is no specific body of material that districts and teachers feel obligated to "cover", there is clearly "time" for experimentation — with computers, with group learning, with problem solving. When the problems are new to the teachers, and close to the cutting edge of knowledge, there is greater acceptance of a classroom open to discussion, to reasoning together, and to the excitement of discovering new solutions which are not "in the book".

Moreover, as teachers become familiar with these techniques and see that they work with their students in their own classrooms, they will adapt them for use in their other classes. Those teachers who have taken the time from traditional teacher-oriented instruction to try these learner-oriented techniques know that the time is well spent. The difficulty is in getting them to try.

Discrete mathematics offers a wealth of new material and, more important in this context, consists of many topics which lend themselves readily to approaches to learning that are recommended in the national reports: discovery learning, experimentation, problem solving, cooperative learning, use of technology. With discrete mathematics, students can easily become involved in the doing of mathematics, can see themselves as "mathematicians" rather than as followers of routine instructions.

In This Volume. Nancy Casey and Michael Fellows argue in their article that only if they use discrete mathematics will K–4 teachers have sufficiently rich mathematical content to properly address the process standards of "reasoning, problem-solving, communications, and connections" stressed in the NCTM Standards.[2] Other articles focus on how discrete mathematics can help teachers achieve educational objectives such as teaching students mathematical communication (Rochelle Leibowitz), reasoning (Susanna Epp), and problem-solving (Margaret Cozzens, Peter Henderson), and change public perceptions of mathematics (Joseph Malkevitch). The article by Joseph Rosenstein and Valerie DeBellis discusses the impact of the Leadership Program in Discrete Mathematics on the activities of its participants.

Resources for introducing discrete mathematics in the schools

At the time of the conference, there were relatively few resources available to teachers interested in including discrete mathematics in their classrooms and curricula. Increasingly in recent years, in part because discrete mathematics is addressed in the NCTM Standards, more effort has been placed both on developing materials related to discrete mathematics and to incorporating discrete mathematics activities in textbooks. As a result of the efforts of the Leadership Program in Discrete Mathematics and the "Implementation of the NCTM Standard in Discrete Mathematics Project" program directed by Margaret Kenney at Boston College and other sites across the country, there are now nearly 2000 teachers who have had extensive exposure to discrete mathematics; many of them have been taking leadership roles, developing curriculum materials and making presentations at conferences.

In This Volume. The article by Deborah Franzblau and Janice Kowalczyk, based on recommendations of teachers in the Leadership Program in Discrete Mathematics, provides an extensive review of available print and

[2] *Curriculum and Evaluation Standards for School Mathematics*, National Council of Teachers of Mathematics, 1989, Reston, VA.

video resources. Two articles, one by Eric Hart and the other by Nancy Crisler, Patience Fisher, and Gary Froelich, discuss texts for high school students which include discrete mathematics. Two articles, one by Nate Dean and Yanxi Liu, and the other by Mario Vassallo and Anthony Ralston, discuss discrete mathematics software. Two articles, by Harold Bailey and L. Charles Biehl, discuss high school courses in discrete mathematics. And the article by Joseph Rosenstein and Valerie DeBellis discusses the Leadership Program in Discrete Mathematics.

Conclusion

Speaking for the editors, the conference participants, and the authors, we hope that this volume will be a major contribution both to facilitating the use of discrete mathematics in K–12 schools and to demonstrating the potential of discrete mathematics as a vehicle to improve mathematics education and revitalize school mathematics.

DEPARTMENT OF MATHEMATICS, RUTGERS UNIVERSITY
E-mail address: joer@dimacs.rutgers.edu

Appendix A.

Discrete Mathematics in the Schools: How Do We Make an Impact? October 2–4, 1992

Conference Participants

NAME	STATE	AFFILIATION (at time of conference)
Bailey, Harold F.	NY	College of Mount Saint Vincent
Biehl, L. Charles	DE	McKean HS, Wilmington
Carrs, Marjorie		University of Queensland, Brisbane, Australia
Crisler, Nancy	MO	Pattonville School Dist., St. Louis County
Dance, Rosalie	MD	Ballou Science/Math HS, Takoma Park
DeBellis, Valerie	NJ	Rutgers University
Dean, Nathaniel	NJ	Bellcore
Epp, Susanna	IL	DePaul University
Fellows, Michael		University of Victoria, British Columbia, Canada
Froelich, Gary	ND	Bismarck HS
Hart, Eric	IA	Maharishi International University
Henderson, Peter	NY	SUNY Stony Brook
Hoover, Mark	NJ	Educational Testing Service
Hoyer, Bret	IA	John F. Kennedy HS, Cedar Rapids
Kenney, Margaret	MA	Boston College
Kowalczyk, Janice	RI	Teacher Education and Computer Center
Lacampagne, Carol B.	DC	U.S. Department of Education
Leibowitz, Rochelle	MA	Wheaton College
Lewis, Philip G.	MA	Lincoln Sudbury Regional HS
Malkevitch, Joseph	NY	York College (CUNY)
Maltas, James	IA	Malcolm Price Laboratory School, University of Northern Iowa
Maurer, Stephen	PA	Swarthmore College
McGraw, Sue Ann	OR	Lake Oswego HS
Piccolino, Anthony	NJ	Montclair State College
Picker, Susan	NY	Office of the Superintendent, Manhattan Public Schools
Pollak, Henry	NJ	Columbia University
Proulx, Viera	MA	Northeastern University
Reinthaler, Joan	DC	The Sidwell Friends School
Roberts, Fred	NJ	Rutgers University
Rosenstein, Joseph G.	NJ	Rutgers University
Saks, Michael	NJ	Rutgers University
Vassallo, Mario	NY	SUNY Fredonia
Yunker, Lee	IL	Community HS Dist. 94, West Chicago

Appendix B.

Discrete Mathematics in the Schools: How Do We Make an Impact? October 2–4, 1992

Conference Program (Abbreviated)

Friday October 2
- Presentation: Joseph G. Rosenstein
  - **"Discrete mathematics as a new start for students and teachers"**
- Classroom Perspectives, Experiences, and Models — Session 1
  - L. Charles Biehl — **"Discrete mathematics for students of average ability"**
  - Susan Picker — **"Discrete mathematics: Giving remedial students a second chance"**
- Presentation: Stephen Maurer
  - **"What is discrete mathematics: The many answers"**
- Classroom Perspectives, Experiences, and Models — Session 2
  - Gary Froelich — **"A semester discrete mathematics course at the high school level"**
  - James Maltas — **"Implementing a discrete mathematics course for non-math students"**
  - Nancy Crisler — **"My experiences as a teacher and math coordinator"**
  - Philip Lewis — **"Using a computer lab: Algorithms, algebra, and axioms"**
- Presentation: Joseph Malkevitch
  - **"Discrete mathematics and the public's perception of mathematics"**
- Classroom Perspectives, Experiences, and Models — Session 3
  - Rosalie Dance - **"Integrating discrete and continuous approaches in secondary math"**
  - Lee Yunker - **"Current and future trends on discrete mathematics in the curriculum"**
- Presentation: Eric Hart
  - **"Curriculum materials for discrete mathematics in the schools"**
- An overview and a taste of ...
  - For All Practical Purposes — Joe Malkevitch and Tony Piccolino
  - COMAP Project — Nancy Crisler and Gary Froelich
  - UCSMP materials — Susanna Epp
  - CORE-PLUS — Eric Hart
  - Several textbooks — Lee Yunker

Saturday October 3
- Programs for teachers
  - Georgetown project — Rosalie Dance and Joan Reinthaler
  - NCTM project — Peg Kenney and others
  - Iowa Project — Eric Hart and others
  - Rutgers Project — Joe Rosenstein and others

Classroom Perspectives, Experiences, and Models — Session 4
Joan Reinthaler — "**Teaching modeling to weak math students**"
Sue Ann McGraw — "**Integrating discrete mathematics into traditional math courses**"
Bret Hoyer — "**A discrete mathematics course using For All Practical Purposes**"
Perspectives
Susanna Epp — "**Strengthening thinking skills using discrete mathematics**"
Rochelle Leibowitz — "**Strengthening writing skills using discrete mathematics**"
Anthony Piccolino — "**Discrete mathematics: Making math accessible to all**"
Presentation: Henry Pollak
"**The role of modeling in teaching discrete mathematics**"
Presentation: Fred Roberts
"**The role of applications in teaching discrete mathematics**"
Presentation: Mario Vassallo
"**Computer software for teaching discrete mathematics in the schools**"
Presentation: Nate Dean
"**What computer software is currently being developed?**"
Presentation: Michael Fellows
"**Discrete mathematics and computer science in the elementary schools**"
"**How Do We Make an Impact?**"
Organizing our suggestions
Structuring Sunday's discussions
Sunday October 4
Perspectives
Viera Proulx — "**Computer science in high school**"
Peter Henderson — "**Computer science, discrete mathematics, and problem solving**"
Mark Hoover — "**Assessment and discrete mathematics**"
Harold Bailey — "**Assessing current practice in discrete mathematics**"
"**How Do We Make an Impact?**"
Work sessions in smaller groups
Reports from groups
The next steps

Section 1

The Value of Discrete Mathematics: Views from the Classroom

The Impact of Discrete Mathematics in My Classroom
BRO. PATRICK CARNEY
Page 3

Three for the Money: An Hour in the Classroom
NANCY CASEY
Page 9

Fibonacci Reflections—It's Elementary!
JANICE C. KOWALCZYK
Page 25

Using Discrete Mathematics to Give Remedial Students a Second Chance
SUSAN H. PICKER
Page 35

What We've Got Here Is a Failure to Cooperate
REUBEN J. SETTERGREN
Page 43

DIMACS Series in Discrete Mathematics
and Theoretical Computer Science
Volume **36**, 1997

The Impact of Discrete Mathematics in My Classroom

Bro. Patrick Carney

Early in the year, our school has a "Back to School Night" when teachers meet with parents to explain the courses and answer questions. One parent started the evening session by asking what I was doing that resulted in her daughter's looking at the bar codes on envelopes and on commercial products at home. It would be unfair to describe her as hostile, but she was certainly questioning what we were doing in the math class. I briefly explained the various topics that could be learned through the study of codes, check digits, and the like, and showed how they led students to review the four basic arithmetical operations, remainders, position value, etc. She agreed that such review was probably necessary, and that it was not a routine that students would take to with any degree of enthusiasm. I then pointed out that the mere fact that she was asking me about the class indicated that the young woman in question had enough interest in the material on codes to bring it home and use it. The student's mother finally agreed that this gave a fresh approach to learning necessary skills; from my vantage point in the front of the room, I could see the looks on the faces of the other parents and nods of agreement. In fact, one of the most enthusiastic seemed to be a gentleman who himself is a math teacher in another school.

I begin with this anecdote to make the point that even though discrete mathematics is often stressed for its practical value, I believe that an even more important aspect is that it captures the imagination of the students in a way that routine drill can never do. Many things are practical but do not create enthusiasm for learning. In my opinion, the real impact of discrete mathematics on the curriculum is its ability to instill an interest in students who might not otherwise find mathematics as exciting as we teachers do.

When I first studied in the Leadership Program for Discrete Mathematics at Rutgers University [**2**] in the summer of 1991, I enjoyed what we did very much and thought about how best to use it. I teach in a small school (a little more than 300 students in grades 6 - 12) and had just finished one year

1991 *Mathematics Subject Classification.* Primary 00A05, 00A35.

of teaching a Problem Solving course to 7th and 8th graders. It met once a week, so the class had to be carefully planned to be self-contained. While some things I had done worked well, I was not satisfied with it. Some of the discrete math activities seemed perfectly suited to the course and I decided to work them into my classes as replacements for the topics that did not work well.

The more of these items I tried, the more enthusiastic I became. Even when one was a disaster such as my first try at the Parker Brothers' game *Instant Insanity*, I realized that the problem was not the material, but my approach. (In this case, I had approached it too abstractly. When I revised the activity, it became very popular.) At the end of each year, I asked the students what topics were the most interesting, useful, boring, difficult, etc. That year, although we did many different problems, the discrete math problems clearly were the most popular.

In fact, our study of fair division ranked first. That was a topic I had barely heard of the previous year and I certainly had never thought of including it in the program. It is interesting to note that initially, the students were very frustrated with the fair division problem. I had prepared them for it a few weeks in advance by discussing the strategy of two people sharing a candy bar by having one child break it in "half" and the other select the piece he or she wants. They all grasped that. Then I said "what would be a fair way to divide it among three people?" At the end of each of four weekly classes, I posed that question and they offered solutions until the bell rang. I promised them that a lesson was coming. In the beginning, they were very creative (e.g., successive dividing by 2 hoping to get a multiple of 3, having each person do a different part of the process, one cut, the other two select first, etc.) but as each of their solutions was knocked down because of some flaw (usually found by a peer), they became frustrated.

When one boy finally suggested "kill one and then divide it between the two remaining," I thought it was about time to have that lesson. We looked at the idea of Persons A and B breaking the whole into two equal parts. Then they each broke their piece into what they considered three equal parts. Person C then selected one of the "thirds" from each of the other two people. We went through a couple of other methods which reviewed fractions (at least halves, thirds, and sixths) and they seemed to agree that that was an answer they could understand and perhaps could have figured out. But we did not stop there. One of my colleagues had given us a computer program for the "Moving Knife."[1] Here was a totally different approach. A "knife" moves across the "candy bar" from left to right. When any person feels the amount to the left of the knife is a fair share, he or she hits a key to stop the knife and the computer tells what percent was selected. The moving knife then continues until another person is satified with the

[1]The *moving knife* software I used was written by the class of Jim Lorentz (a fellow 1991 alumnus of the Leadership Program in Discrete Mathematics.) Similar software is distributed with [**1**].

amount displayed. All of their original approaches had been really modeled on dividing the candy bar in two pieces (except for the totally bizarre ones). Here was another view.

We decided to use the software to hold a tournament. We broke the class up into groups of three. We tried the simple rectangular and oval shaped "candy bars" for practice, but the contest used random shapes. Now, from my point of view, we were also reinforcing hand-eye coordination and, more importantly, estimation skills. I have now done this for three years with students of different ages, and I have observed that most students become much better at estimating fractions after playing it a few times.

We ran the contest so that the winner was the individual getting the "largest" piece. When I presented this in a follow-up workshop for teachers at Rutgers, they came up with a great improvement. They said why not have the students compete in teams, and let the winner be the team which ends up with the closest to a fair division. Now the groups would have to cooperate among themselves to compete against others.

Competition has proved to be a great motivator in another way. Last year, we offered a Discrete Mathematics course to our high school students. Generally, it was taken by students who would probably not elect mathematics in their senior year. Certainly some had had great trouble and had even failed previous math classes. One of the problems we studied was the the Traveling Salesperson Problem (TSP). The problem is to find the best route that a salesperson could take if he or she would begin at the home base, visit each customer, and return to the home base ("best" was defined as minimizing total distance). I explained to them that nobody had ever come up with an algorithm to solve the problem which could run in a reasonable amount of time. When we had an exam, I added a TSP problem for extra credit. Whoever had the first, second, and third shortest routes would be awarded extra points. In addition, I would try the problem myself at the same time and hang my answer outside of the door. There were extra points to be gained for beating my answer. Of course there were enough cities that they could never try all cases in the allotted time. Two girls tied for first, and when they went outside, they saw that their answers matched mine. They came back into the classroom all excited about it. It may have been the first time they ever found themselves on a par with the teacher. I cannot say our study of graph theory was perfect, for there were areas which I never did communicate to my satisfaction, but this one aspect really inspired their interest. They were very proud of themselves.

Later in that course, we were building Sierpinski's Pyramid, similar to the one inside which Valerie DeBellis (Associate Director of the Leadership Program in Discrete Mathematics at Rutgers University) is standing in Figure 1.

I had heard many ways of constructing the Pyramid, and remembered there was one which hit me as most suitable for a person with my lack of construction skills, but I had forgotten the details. I posted a note to

FIGURE 1. Valerie DeBellis in Sierpinski's Pyramid

discrete math teachers via email and got back about 14 answers. One was what I wanted, and it came from Evan Maletsky (Professor of Mathematics at Montclair State University and a staff member of the Rutgers program). He closed with the sentence "Send me a picture when you finish – whoops, it's a fractal, you'll never finish." I printed it and brought it to class. We were using Evan's book and I showed them his picture in it. They were most impressed that a famous author would take time to write to our class and also that he would joke with them.

We started the project slowly, but eventually the students caught on. Only the many snow holidays kept us from getting to the height of the classroom. Building the Pyramid turned out to be a great cooperative project. My students were very proud when other students, who were taking the more "advanced" mathematics courses, would come into the room and ask what this was (eventually it became hard to miss) and I would have one of my students explain it. I think it did much for their self-confidence that there were areas of mathematics in which they were the "experts." This may well have been the first time in their lives that this had occurred.

When we studied codes in class, I assigned students to research a code that we did not study in class and give an oral presentation on it to the class. One young man in particular (who had failed my class as a freshman and dropped out of our school for a year) did an outstanding job. I was so impressed that I offered his handout on Vehicle Identification Numbers to those in our discrete math email group. But I posted it so that requests would be sent to the student's address. The first request he received came from a university professor. He carried that letter around in his book with the return address sticking out very noticeably for about a week. He later got more involved in using the Internet and now I am pleased to say he is enrolled in the local community college. No doubt he had matured over time, but there was also a vast difference between his approach to the discrete math course and to the more traditional course in which I had taught him previously.

I think that these anecdotes help illustrate the way that discrete mathematics involves individual people of both sexes, of all ages (well, at least grades 7-12), and of varying abilities. The impact might best be summed up by relating a discussion I had with the students when the course ended last year. I asked if the course turned out to be what they expected when they signed up for it. There was a unanimous "No!" Then individuals added comments such as "I didn't think I was going to like it," or, "I thought it would be boring like the other math classes I've had" (I hate to admit that I was her teacher the year before), and "You should have told us what it was like and more people would have taken it." I do not for a minute claim that all of these students will become great mathematicians, but I do think they have a new and far more positive view of mathematics and their ability to do it. If nothing else, it is my hope that they will not pass on to their children the "I was never any good at math and never found any use for it so why should you try" syndrome which haunts so many students in our schools today.

References

[1] Bennett, Sandi et. al., *Fair Divisions: Getting Your Fair Share,* HiMAP Module #9, Consortium for Mathematics and Its Applications (COMAP), 1987.

[2] Rosenstein, Joseph G., and Valerie A. DeBellis, "The Leadership Program in Discrete Mathematics", this volume.

BISHOP WALSH MIDDLE/HIGH SCHOOL, CUMBERLAND MD
E-mail address: pcarney@dimacs.rutgers.edu

DIMACS Series in Discrete Mathematics
and Theoretical Computer Science
Volume **36**, 1997

Three for the Money: An Hour in the Classroom

Nancy Casey

I am the teacher in a crowded, windowless, cubicle of a classroom, winging it somewhat, as always. The chairs are arranged in a circle, although at the very most, a quarter of the students are sitting down. On the floor in front of us is the largest known 3-regular planar graph of diameter 3. (See Figure 1. If these terms are unfamiliar to the reader, their meaning is explained below.) It is drawn with masking tape and its 12 vertices are large enough for a person to stand in. A jumble of conversations fills the room. Students interrupt one another; occasionally someone asks me a question. Mostly I am watching. What I see is so exciting that it is hard for me not to demand that they all sit down and be quiet so that I can give a lecture about what I see. For the first time since the class period began, I relax. Teaching, learning, the classroom—it is all going the way it is supposed to.

The 20 or so young people in the classroom are high school students participating in the Idaho Science Camp.[1] This is our fourth one-hour session together; we are not quite the strangers that we were to one another a few days earlier. I have been trying to expand their understanding of what it means to do mathematics, and also put into practice some things that I understand theoretically—that students can learn much more in a rich and stimulating environment, that hierarchical exposition isn't always the best way to convey information, and that students learn a lot by talking to each other about their ideas. I am trying to understand what it means to have a learning-centered classroom instead of a teacher-centered one. It is the uncertainty of giving up tight control of the program that has made me tense; seeing what good things can happen when I do is what makes me relax.

Before the students arrived, I had drawn the graph on the floor with masking tape. I know a lot about this particular graph and about graphs in general. Trying to organize that information into a coherent outline seemed

1991 *Mathematics Subject Classification.* Primary 00A05, 00A35.

Research supported by the U.S. Department of Energy Los Alamos National Laboratory Megamath Project and Department of Computer Science, University of Idaho, Moscow, ID.

[1]The Idaho Science Camp is sponsored by the University of Idaho College of Engineering and the U.S. Department of Energy.

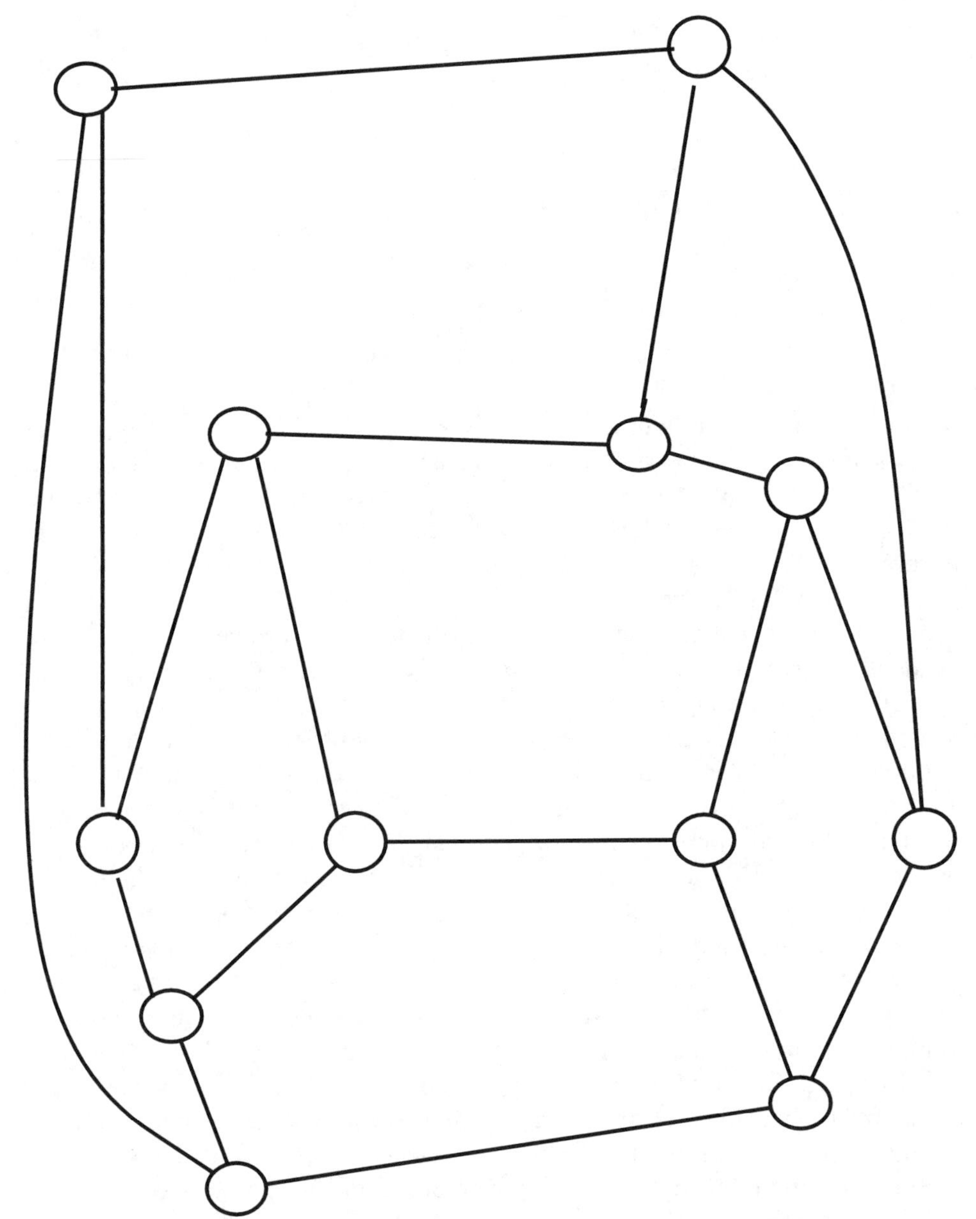

FIGURE 1. *This is the graph that was drawn on the classroom floor with masking tape. No two lines cross. Every vertex of this planar graph has degree 3. The maximum distance between two vertices of the graph is 3. It is not known whether it is possible to draw a graph with more vertices which preserves these properties.*

impossible. The structure of the information in my own mind more closely resembled the drawing on the floor than an ordered list with indentations. I decided to take a risk and come to the classroom equipped primarily with my understanding of graphs. I had in mind a brief introduction with which I would begin. Otherwise, my only plans were to draw the graph in Figure 1 on the floor ahead of time, to take my cues from the students, to be alert, and to try to think on my feet. I was more than a little bit nervous when it was time to get started. I asked:

> *Do you remember I told you about my friend Mike*[2]*—the Computer Scientist?*

"The guy from San Diego who invented the Orange Game![3]" someone called out.

> *Yes, he's the one. Well, I have learned a lot from him—a lot of mathematics, and also a lot about what mathematicians* actually *do. One thing that drives him crazy is that people have the mistaken idea that mathematics is primarily about numbers. Really, there is much more to mathematics than that. One day when he was particularly excited, he plucked a drawing from a jumble of papers on the table and exclaimed, 'An object like this is every bit as important as a number—and just as useful, too. Why, this* is *a number! This is a* psychedelic *number!'*
>
> *Well, this drawing is the one he was referring to. What he called a psychedelic number that day is what most mathematicians call a graph. It is made of dots (or circles) and lines. You'll notice right away, I hope, that even though it's called a graph, it doesn't have anything to do with the kind of graph you've probably studied in school—you know, bar graphs, line graphs and such. This is a a whole different kind of a graph, simply made of dots and lines. This might look like an arbitrary configuration of dots and lines to you, but Mike has found it to be very interesting, so much so that this past year he offered students a cash prize if they could solve a certain problem that involves this graph. The problem sounds simple enough: Is it possible to draw a graph that is larger than this one, yet preserves three critical properties?*

There was a flurry of excitement—how much money, how soon do we get it, and quickly someone called out, "I can draw one bigger, just gimme the masking tape!" I clarify:

---

[2]Mike Fellows, Department of Computer Science, University of Victoria, Victoria BC, Canada, `mfellows@csr.uvic.ca`

[3]The Orange Game illustrates routing and deadlock in networks. Any number of people can play. All players label two oranges with their names, then the oranges are mixed up, a single orange is removed, and the players each pick up two oranges which are not their own. (One person will have only one.) Players then stand in a circle. A player may place his or her orange into the *empty* hand of an *adjacent* player only. The game ends when all players have their own oranges back.

> *Bigger means having more nodes or circles; in graph theory they are called vertices. Also, you still need to know what the three critical properties are that you have to preserve. Size is a property. The size of this graph is 12 because it has 12 vertices. Obviously that's not the property that you have to preserve.*

"Well, what are the properties then?" At that moment the idea occurred to me to turn the problem inside-out and make them guess what the properties were. I also decided to use precise mathematical terms whenever the opportunity arose, and not to define them unless someone asked me what they meant. In retrospect, I would see that these were two wonderful ideas. Scrutinizing and manipulating a mathematical object so as to understand its properties is so much more like what mathematicians do than being presented with a list of the properties that some object has (and being asked to memorize them). Because I used the special terminology of graph theory right off, the students heard the words and began using them themselves, long before they were absolutely sure what they meant. These words were with us from the beginning, and as the class period progressed, students attached meanings to them that they clarified and shared. I answered:

> *You have to guess what the properties are. Name all the properties that you can think of and each time you name one of the three critical ones, I will tell you that you have guessed one.*

Students began talking all at once, one of the many chaotic eruptions that this class period would have. I doubted that the students had more than a foggy notion of what I meant by the word *property*, but since this didn't dampen their enthusiasm, I wasn't going to interrupt and tell them that. I also didn't think that they would be able to guess the three properties without help. I waited until the excited conversation died down before I continued.

> *The prize-winning graph will be of size greater than 12 and also be* planar, 3-regular, *and have a* diameter *of 3. That's a hint. All you have to do is figure out what that means and you will know what the three properties are.*

A few people wrote those words own. Quite a few students were copying the graph. Others were excited about winning the money.

Next time I had everyone's attention, I told them about Paul Erdös, an elderly Hungarian mathematician who spends most of his time on the road, finding it far more interesting to travel and visit mathematicians all over the world, to work with them and get them excited about problems that interest him, than to stay home and be famous. I held up a photocopied article about him [**5**].

> *He has a great mathematical mind, and would be famous and world-renowned by simply working alone. But he prefers*

working with other people, planting seeds, sharing ideas and questions, and offering prizes for solutions of unsolved problems. As a result, university mathematicians love to have him visit and give talks. These visits often result in a collaboration between Paul Erdös and a professor or student, and together they share the credit for some breakthrough.

So Mike's offer of a cash prize for solving this problem is an imitation of Paul Erdös. The prize adds a little excitement to the problem, and Mike is always trying to find ways to get other people excited about mathematics. Here is another interesting thing that Mike told me: he thinks that the reason why no one has found the solution for this problem is that no one has fiddled with it enough. It seems like you should be able to figure it out if you spent enough time understanding how graphs that have the same critical properties as these are constructed. All you have to do here is either draw a larger graph, or explain why it is impossible to do it.

Think about it. Imagine someone offered you a prize if you could find two odd numbers that made an odd number when you added them together ...

Immediately a voice called out, "You can't do that!" Other voices agreed.

Right. You would never say that someone who couldn't add two odd numbers and get an odd number needed to try harder. You would be sure they couldn't do it. With some careful thinking, you can come up with an explanation why no one will be able to add two odds and get an odd.

Some students began explaining to one another why this is so, but I continued speaking.

But this problem is different. No one has been able to come up with a good logical argument for why it would be impossible to draw a graph larger than this one that preserves the three properties. So if there's no reason that anyone knows of why there can't be a bigger one, it's likely enough that there is one, only no one has discovered it. If that's the case, it doesn't take any special advanced mathematics to just draw it. It just takes a pencil, and a lot of time, patience, and inclination to fiddle with it. Mike shows this problem to grade school kids whenever he can because he thinks one of them might just stumble on the solution.

"Tell us what the properties are, then!" The room became quiet.

OK, I'll give you some hints: Planar, 3-regular, and diameter 3. That's a lot of 3's. That's a hint. The number three. What's with the number 3?

Someone noticed that there are three edges (lines) touching every vertex. This was explained to others. They checked all the vertices. Indeed it was true. I told them that the number of edges that touch a vertex is called the *degree* of that vertex. In this graph, every vertex is of degree three. When every vertex of a graph has the same degree, then the graph is *regular*. So this graph is *3-regular*. When you draw the graph with more vertices, it has to be 3-regular.

Immediately some students began trying to draw 3-regular graphs of size larger than 12. Soon the chaos was back. A short while passed before enough students realized that they didn't have all the information to solve the problem. A few more moments passed before they could shush the others so I could tell them more.

> *I'll give you hints by showing you some games that you can play on this graph. The games will help you understand the structure of the graph better. No doubt that the person who actually solves this problem will know many more things about the structure of this graph than just the three properties that are preserved. We'll play some games that will help you see lots of properties. When someone guesses a property that is one of the three critical ones, I will tell you.*

I asked a couple of students to label the vertices with little strips of masking tape with the numbers from 1–12, reminding everyone that the numbers weren't part of the graph. The numbers would serve as names for the vertices and make it easier to talk about them.

Two boys had begun to play cards. I asked them to loan me the clubs. Surprised, they shrugged and obliged. I removed the King, handed the remaining clubs to a student, and asked him to go stand in any circle. Then I explained a game.

> *Here's the rule: You can travel along the lines and walk from vertex to vertex. Every time you come to a vertex, you must lay a card down in it before you leave. You cannot return to a vertex that already has a card in it. Can you deal out all the cards?*

He set out. It is not very hard to do. Soon all the cards were laid in circles on the floor. "It's traversible!" cried a voice from the crowd. "We had that last year."

> *Right.* Traversible *is a word you can use to describe this graph. You can find a path through the graph that allows you to travel to all of the vertices without touching any of them twice. When you can do that, you can say that the graph is traversible, or that it has a Hamiltonian Path, named for the Irish mathematician William Rowan Hamilton who was very interested in graphs that have this property. This graph is indeed Hamiltonian. It's not one of the critical properties*

that has to be preserved, but it is a property. Do you think it will be preserved in the larger one?

Some yes's, some no's. They wanted more properties. I announced another game, walked to the center of the room, picked up a card from the graph, and called for volunteers.

We need 11 volunteers. One person for each vertex where there is a card.

Students came eagerly forward and scrambled for places. When they were set, I gave them directions.

OK, pick up the cards. Look at the number that you have. Jack is 11, Queen is 12, Ace is 1.

"Oh no!" groaned a voice. "The Orange Game!" We had played the Orange Game a few days earlier. This was indeed a variation.

You're absolutely right! Routing and deadlock in networks. Let's give it a try. Everyone has to get to the vertex with their number on it, but the only way you can move is from the place where you are standing and along a line into an empty vertex.

The fun begins. There is only one empty vertex. This requires a lot of maneuvering, planning and working together. It is much harder than the Orange Game. Once I realize this, I am not even sure it is possible, and wonder if I haven't made an awful mistake by asking them to do it.

Several leaders emerge. They argue. Some students are confused, but follow orders willingly. Some refuse to do what others tell them. No one gives up. A few students in the audience add their suggestions too. Others continue to pore over graphs as they have for the last 20 minutes. A girl has gone to the board to demonstrate to two friends why you can't draw a 3-regular graph with 13 vertices. ("With 39 ends-of-lines, one line would just have to hang there loose, with no vertex on its other end," she says.) When a student struggling with the puzzle on the floor is too frustrated and wants to quit, someone jumps up and offers to take his place. The student who quits stands aside, continues to watch, and is soon making suggestions.

With the students moving about the graph in front of me, once again the series of questions related to those that leaped to my mind when I watched them play the Orange Game comes back. What is the fewest number of moves it will take to get everyone back to their places when there is just one empty slot? How is it different when there are 2, 3, or 4 empty slots? If you chart these numbers, will you see a pattern? Won't that depend on how "mixed up" everyone was in the first place? What does "mixed up" mean? How did we know the numbers were "good and mixed up" when we started? Is it possible to start out so mixed up that everyone can't get back to their places? How would the game be different if you didn't know what numbers other people had? What if you couldn't know the number of a vertex until you stood in it?

Another observer might find this classroom disorderly and unfocused. It *is* disorderly and unfocused. This, however, is the moment at which I relax and feel that I have done my job as teacher well. The room is filled with excitement and enthusiasm about mathematics. As the students explore this problem, they are participating in a large collaboration set up by one of the greatest mathematicians of our time. Even though there is a monetary prize for a "right answer", this is not the sole focus of the room. In fact, none of the students understands the question yet! They have lost themselves in the preliminary exploration. I pat myself on the back for the lucky on-the-spot inspiration that had me use the technical terms for the hints. Those big, unfamiliar words have indeed grown meanings as they practiced saying them and tried to figure out what they mean.

I regret that I will say goodbye to these students for the last time tomorrow. When will be the next time that I will have 20 enthused and clever people to think about problems like this with me so that we can act them out together? Next week, my learning of graph theory will slow to its usual solitary, plodding pace as I sit at a table and move labeled bottle caps over graphs drawn on sheets of paper, trying to explore some of the questions raised for me during these sessions.

At last the students have sorted themselves on the network. They stand triumphantly in the vertices whose numbers correspond with their cards. A student who is in his seat looks up from his notebook with an expression that is both confused and intense. He asks, "What are those properties again?"

> *You are trying to find a planar graph of size larger than 12 that is 3-regular, and has a diameter of 3.*

I count to 3 on my fingers as I say the words *planar*, *3-regular*, and *diameter of 3*. Voices join and say the words along with me. I am ready with another proposal.

> *Do you want to play a game that is easier than that last one? When you figure the game out, you will know what the diameter business is.*

Of course they do. I held up an illustrated storybook[4] I had made.

> *This is a game that I invented to teach this problem to a group of 3-year-olds ...*
>
> *Now, with children that age, it takes a lot of effort just to get them to stay on the edges when they walk around the graph. So you can imagine how teaching them about properties takes a lot of doing. I invented this story and the game to go with it after watching them play dress-up one day. The*

---

[4]The text of this and other stories with Gertrude, Superperson, and the Monster that illustrate properties and problems on graphs through games are available on the World Wide Web `http://www.c3.lanl.gov/mega-math/` and in Chapter 3 of [**1**], Games on Graphs.

characters were ones that they had made up for their game: Gertrude (she's a goose), a Monster and Superperson. We'll need three volunteers to play.

When the three-year-olds play, they spend the first 10 minutes dressing up in their costumes and capes. But you know that we're doing this so that you can find out what diameter *means, so we can get right to the point.*

In this game, the circles of the graph are ponds. The lines that connect them are the flyways that you can use to fly from one pond to the next. Superperson has the most flying power, Gertrude has the least, and the Monster is in the middle. The Monster is frightening and ugly, but harmless. He wants to play with Gertrude, but Gertrude is afraid of him. Since Gertrude is going to try to escape from the Monster, and since the Monster has more traveling power, it's more fair to let the Monster choose where he is going to start first. Then Gertrude can pick a good place to hide from him. Superperson will pick her place last.

The students who have volunteered to play the roles of each of the characters choose places to stand on the graph, and the game is ready to start.

The game begins because Gertrude is bored, this pond is boring and she wants to be somewhere else. So she flies off to another pond. She's not in very good shape, though, so by the time she gets to the next pond, she crash lands in it and has to rest.

The person playing Gertrude acts out her part.

Now the monster wants to play with Gertrude. He is so ridiculously large that it takes a huge amount of effort just for him to get off the ground. He is still going up *when he passes the first pond. But soon he is tired and he crashes into the second one with an enormous splash.*

The good-natured Monster acts out his ungainly flight. This time, anyway, he doesn't reach the pond where Gertrude is hiding from him.

Now Superperson flies. She has the most power of all. She can fly three ponds before she gets tired, zip-zip-zoom! She flies over two ponds and lands in the third.

Superperson does this.

Now it's Gertrude's turn. All three of them will keep flying around—first Gertrude one, then the Monster two, and Superperson three—until the Monster catches up with Gertrude.

It doesn't take long before the Monster lands in the same pond as Gertrude, and the two students perform the ensuing action.

> *Now the Monster isn't going to hurt Gertrude, but Gertrude doesn't know that, so she attacks the Monster who has to cry to Superperson for help. C'mon Gertrude! Attack! Monster, yell for help!*
>
> *Can Superperson come and save the Monster before Gertrude drowns him in mis-directed self-defense?*

Of course she can. The diameter of the graph is three. But it takes a few times before the students figure this out. Soon they complain that the the game is boring and kind of silly. The Monster always gains a step on Gertrude and will always catch up with her, and Superperson always gets there in time to save him.

> *Yes! The game is horribly dull because of the properties of the graph that is being used for a game board. Superperson always saves the day, and Superperson always saves the day because the* diameter of the graph is 3. *So what does that mean? Think about that statement: the graph has a diameter of 3?*

It takes a little bit of arguing and some experimentation, but soon everyone realizes that you can get from one vertex (no matter which) to any other in three or fewer steps.

Now the students are divided roughly into two groups: those who want to keep playing the Gertrude game and invent more rules so it is less predictable, harder, and more interesting, and those who want to work on the prizewinning problem. I insist that everyone sit down and listen one last time.

> *Planar. That's the other property. What could that mean?*

The students are silent. There is an uncomfortable shifting of feet. They avoid my gaze.

> *It has something to do with a plane. Like points, and lines and planes that you may have learned about in geometry.*

There are still no guesses. They need a hint. What hint can I give that's not tantamount to telling them? On the spur of the moment I can only come up with one idea.

> *OK, here's a hint. Imagine that this is the map of a city. The edges are roads, the vertices are intersections. What is something that is expensive to build that this city doesn't have any of?*

What a stupid, misleading hint, I think, and am surprised when a chorus of voices shouts, "Bridges!" Soon everyone knows that *planar* means you can draw the graph on a plane or flat surface in such a way that the edges only touch at vertices. I tell them that planarity is especially important when people are trying to design electronic circuits that can fit on one side of a silicon chip. Distance and degree are also important properties in the graphs that computer scientists use to model the chips that they design.

About ten minutes remain in the class period. On the one hand, it has taken over 45 minutes simply to state a problem. On the other hand, it is mind-boggling to consider the amount of things we have done in that short time. This group is heterogeneous, not "gifted". In two days the Science Camp will be over. They are winding down, have been staying up too late, far more interested in getting in the last days of socializing than the last moments of mathematics. I did not anticipate this level of enthusiasm and tenacity. I never could have planned this. Had I not been there, I would argue that it is not possible to "cover" so much in one class period. And yet, it has happened.

Alone, in pairs or in groups some students are thoughtfully drawing graphs. Others want to keep playing games and I turn my attention to them. I rip up five sheets of paper that are different colors, and randomly place strips of all five colors on the vertices.

> *You want to make it so that when you walk along the edges and leave from a vertex of one color, the next vertex you come to will not be the same color. You might need these.*

I give a handful of the colored strips to each student. It's not hard to do a vertex-coloring of this graph with five colors. When they finish, I ask them to try to color the vertices in the same way and use fewer than five colors. They begin experimenting.

I turn to the students who have queued up to show me graphs they have drawn, trying to win the prize for a planar, 3-regular graph, with maximum distance 3 and size larger than 12. I systematically check out the properties, showing everyone my thought-processes out loud. Soon they are checking their own and each others. It is more efficient to help each other than to wait for my undivided attention.

When the game-playing students have found a vertex-coloring that works for three colors, I ask them why they can't do it with two colors, and they don't have a hard time showing me why two colors just can't work. I scoop up the colored strips from the vertices and place them on the edges and ask them what they think they are supposed to do now. They don't seem to hear me. They are already working, trying to find a way to arrange the colors so that no two edges of the same color touch.

I interrupt and ask them how many colors they think it will take. Three, someone says. Agreement is muttered; they all seem sure. I ask how they know that, but no one answers. When they finish the edge-coloring, there is still time to color the regions.

A few students who have been drawing graphs that solve the original problem are convinced that it is impossible to do.

"Tell that Orange Game guy you can't do it," someone says. I explain that if they can demonstrate why it is not possible, they can probably win the prize. They are excited—until I insist that a valid demonstration is not

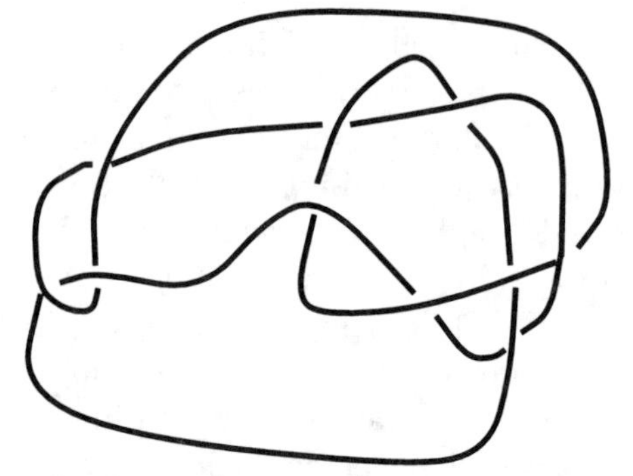

FIGURE 2. *The story of mathematics is much more the story of long and fruitless searches for solutions to problems, than it is the story of a discipline full of quick answers that are easily verified as right and wrong—even though the latter is the more common experience of many students. In 1889 after several years of study and experimentation, mathematician P.G. Tait published a table of knots. He believed it was not possible to twist, pull, or otherwise deform (without cutting) any of the knots in the table so that they looked like any of the others. 75 years later, Kenneth Perko, a New York attorney and amateur mathematician demonstrated that these two knots from Tait's tables were, in fact, the same knot. Make these knots out of rope and try for yourself to deform one so that it looks like the other. Chances are it won't take you 75 years!*

simply saying that you tried and tried and couldn't find it. How do you know if you have tried long and hard enough? I happen to have a book [**3**] with me that has a picture of two knots which were thought to be different for 75 years. (See Figure 2.) Then after all that time, someone showed that they were really the same. I warn them:

> *Even if you try for a long time and can't do it, that doesn't mean that someone else won't get lucky and figure it out.*

We review the reason why we are sure that you can't draw a graph with 13 (or any odd number) vertices that has these properties. Maybe it won't work for 14, but how can you be sure it doesn't work for 50, or 246? And how do you *know* it doesn't work for 14? How can you be sure you have tried every possible combination? Is there a systematic way to list the different things you try to that you can tell if you have left any possibility out? They want that prize. They are talking about how they will spend the money. They keep working.

At the end of the class period, I shoo the students away so they will be on time for their next class. Later in the afternoon and the next day, the final day of classes, students will bring me graphs that they have drawn and

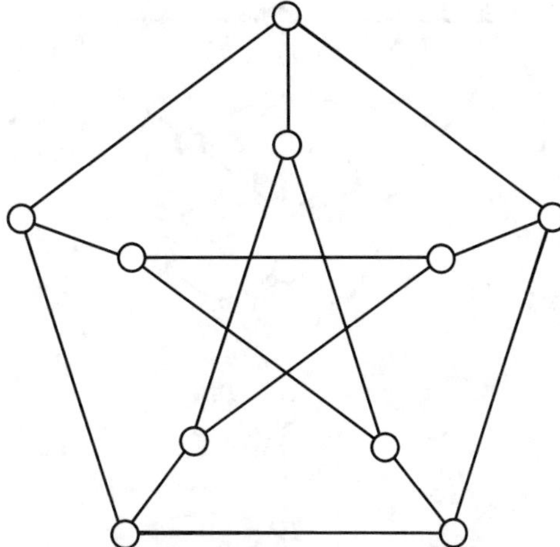

FIGURE 3. *The Peterson Graph is one of the most famous graphs in all of graph theory. Its renown stems from the unusual list of properties it both possesses and fails to possess. It is an interesting graph for students to examine to see what properties it has and does not have.*

we will go over them carefully. No one solves the problem, but they have tried hard. I have to examine them carefully to find the flaws.

"What will you do if I find it?" someone asks.

"We will get on the phone right away and call Mike."

I ask the students to write down what they have learned about mathematics this week. One girl writes, "My teachers try to make me think abstractly, and I refuse. You tricked me into doing it."

How I wish I had been able to teach for *both* weeks of the Science Camp. This session would have been in the first week, and on the weekend I could have accompanied them on their Saturday trip to picnic and swim at the Snake River. They would show me even more graphs, we could look more closely at the games and talk about other things that came up during the week, such as logic puzzles, and what it takes, in mathematics, to say that something is true. Perhaps there would be a student or two tenacious enough to pore over Mike's paper on dense planar networks [**2**] with me and decipher it. Certain elements of his methods for drawing graphs with minimal degree and diameter are accessible and might help them in their efforts to find a prizewinning graph.

If we were together a second week, we would surely gather with a different graph drawn on the floor in front of us. The Tutte Graph perhaps, or the Peterson Graph—they are pretty and symmetric and have different properties to discover and talk about. (See Figures 3 & 4.) We could even draw it in colored chalk on the sidewalk in front of the library or somewhere downtown; or we could use lawn paint to spray paint it on the grass.

If we had more time, we could return to Mike's idea of a graph as a psychedelic number, and try to figure out what he meant. I would be able

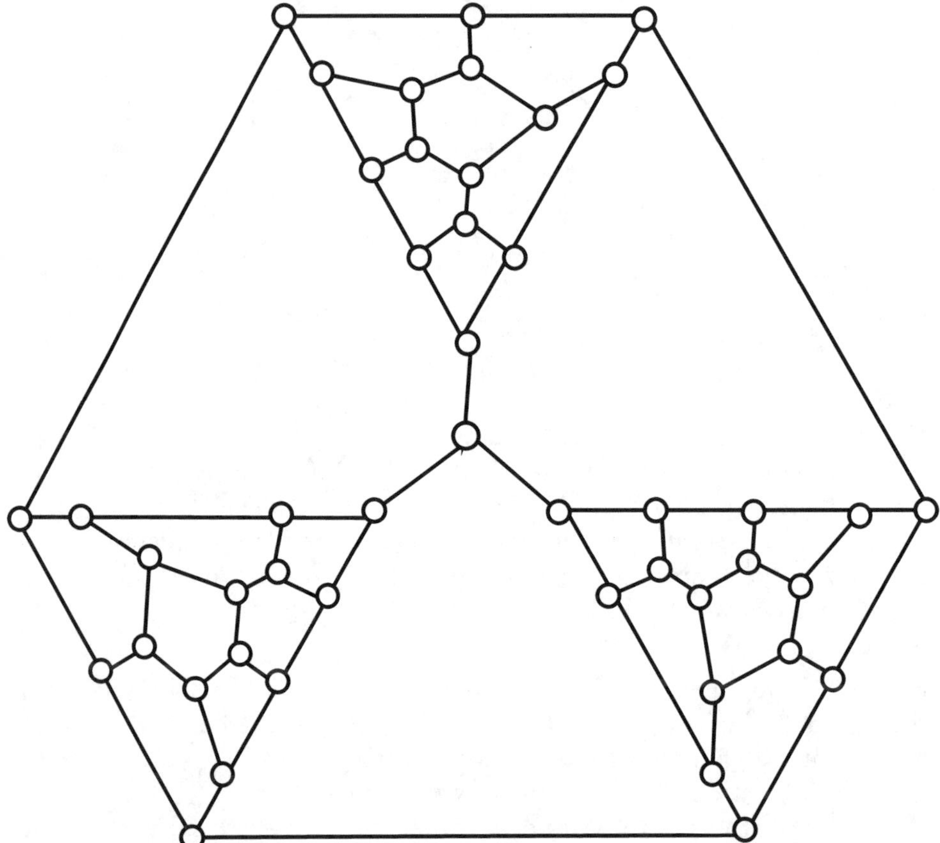

FIGURE 4. *When Canadian mathematician William T. Tutte drew this graph in 1946, a question which had been open for over 60 years was solved. The Tutte graph is 3-regular and has the property of being* 3-connected, *which means that in order to break the graph into two disconnected components, you must delete at least 3 vertices, along with the edges that are connected to them.. In 1880, the English mathematician P.G. Tait (see also Figure 2) conjectured that every 3-connected, planar, 3-regular graph is Hamiltonian. (A graph is Hamiltonian if it is possible to find a path that allows you to walk from vertex to vertex, touching each vertex exactly once and return to the vertex where you began.) The Tutte graph is not Hamiltonian, so it provides the counterexample which disproved Tait's conjecture—even though Tait's conjecture seemed to be true for all those years.*

to talk about mathematical objects, their properties and operations on them. I could watch these terms acquire meaning for the students as we looked for properties of objects such as graphs and numbers, knots and maps and then discussed what it meant to do operations—such as add, subtract, multiply and divide—on them.

No matter how engaging my presentations, I would expect them to stop listening to my teacherly lectures and begin to experiment and talk about properties that they suspect or discover. They may adapt versions of the games we have played to the mathematical object that is at hand. They will get in each others' way, interrupt one another and make each other angry. Some students may try to take charge. Others will ignore them and work furiously in their seats on problems they may or may not understand. Some students will stand aside and appear to be doing nothing at all. This, I am learning, is often a symptom of careful reflection.

I will remind myself how hoarse I get whenever I raise my voice in the classroom. As I (and my agenda) slip from the center of their attention, I can be confident that their attention will come back. Perhaps someone on the other side of the room will call out a question to me. I will pantomime my inability to talk loud enough to be heard. Perhaps I will adopt a mock professorial attitude and say, "Ahem, I thought I was asked to impart some valuable information here."

Eventually someone who likes to give orders will want to hear what I have to say and shout, "Hey! Psssh! Hey you guys, *shut up!* Can't you see she's trying to *tell* you something?!"

I will begin talking in a quiet voice, but before I have finished a sentence, all the other conversations in the room will resume. I will stop talking, and the students who have accepted the responsibility of making the others be quiet and listen will again take the floor and make it quiet so that I can talk. Eventually, they will be ready to listen, and I will draw on what has happened to pick up the thread of the story that has woven throughout all of our class sessions—the story of what it means to do mathematics.

I will point out that whenever I try to tell them about a good math problem, if it *is* a good problem, it is so rich and many-tentacled that immediately, when I begin talking about it, their minds dart to interesting and compelling places. They notice things, they want to experiment, they make guesses and argue; they want to tell someone else what they are thinking. This happens because they are doing mathematics: churning up ideas, filling the room with creative thoughts and guesses that eventually they can pin down, refine, and either keep or reject, by marching carefully through their thoughts with rigor, logic, and proof. They are *doing* mathematics, and that is far more stimulating than sitting still and listening to a teacher talk about it. I hope I will have to shoo them out the door to be on time for their next class. I will leave them, as I always do, excited about learning, about mathematics and about questions of my own, because their enthusiasm and excitement are so very, very contagious.

References

[1] Casey, Nancy and Michael Fellows, *This is MegaMathematics: Stories and Activities for Mathematical Thinking and Problem Solving*, Los Alamos National Laboratory,Los Alamos NM, 1993. Available via anonymous ftp from `ftp.cs.uidaho.edu` in the directory `pub/mega-math/workbk`.

[2] Fellows, M., P. Hell, and K. Seyffarth, "Constructions of Dense Planar Networks", Unpublished paper, University of Victoria, 1993. Available from Mike Fellows, Department of Computer Science, University of Victoria, Victoria, BC, Canada V8W 2Y2 or `mfellows@csr.uvic.ca`

[3] Peterson, Ivars, *Islands of Truth: A Mathematical Mystery Cruise*, W. H. Freeman, New York, 1990, p. 55.

[4] Tait, P.G., "On Knots I, II, and III." *Scientific Papers*, Cambridge University Press, 1990.

[5] Tierney, John, "Paul Erdös is in Town. His Brain is Open." *Science '84*, October 1984, pp. 40-47.

[6] Trudeau, Richard J., *Dots and Lines*, Kent State University Press, Kent OH, 1976. (For information about other accessible graph theory problems and their applications)

Department of Computer Science, University of Idaho, Moscow, ID 83843
E-mail address: `casey931@cs.uidaho.edu`, `http://www.cs.uidaho.edu/~casey931`

DIMACS Series in Discrete Mathematics
and Theoretical Computer Science
Volume **36**, 1997

Fibonacci Reflections—It's Elementary!

Janice C. Kowalczyk

Background

While I am presently in the position of the assessment coordinator for the Leadership Program in Discrete Mathematics, I have set aside some time to practice the mathematical ideas and teaching methodologies that I gleaned from being a participant in the first program for middle and elementary school teachers in 1992. During recent years, I have conducted a number of teacher workshops and sessions with students. The following is an account of a recent experience in a fourth-grade class.

Connections

In December of 1993 I conducted a series of one-hour sessions on the Fibonacci numbers with my daughter's fourth-grade class at the Forest Avenue School in Middletown, Rhode Island. My daughter created the opportunity when she recognized the Fibonacci numbers in a class science unit when they were studying monocots and dicots. Her Fibonacci comments caught the attention of her curious teacher and before long a connection was made, and a date set for this series.

The Fibonacci sequence begins with the numbers 1 and 1.
Each number that follows is the sum of the previous two numbers, hence:
1, 1, 2, 3, 5, 8, 13, 21, 34, 55, 89, 144, ...
are the first 12 terms of this infinite sequence.

The Beginning—Some Thoughts on Mathematicians

On our first day together we investigated the classic rabbit question; however, before we began that activity, we took some time to discuss what mathematicians look like and what mathematicians do. None of the students had any clear images of a mathematician; however, many of the students felt

1991 *Mathematics Subject Classification.* 00A35, 00A05.

that mathematicians add, subtract, multiply, and divide for a living. One student did reply that mathematicians solve problems using numbers.

Some Thoughts on Doing Math

Next, we pursued a number of real mathematical questions and looked at the process associated with them. For example: If I gave you 10 cents every day to put in your bank, how much money would you have at the end of a week? What information did you have to know first? What would you need to know if I asked you how much money you would have after a year (with no interest)? I was surprised at this point, to find out that most of the students did not know how many days were in a year; however, they were clever enough to question how much money was in the bank when the problem began. I gave them the bank question for a homework challenge and added that they had to tell me how they got the information they needed as well as how they got their answer.

When Am I Ever Going to Use This?

Next, we took a look at the classic rabbit question below. I teased them with the idea that the rabbit question did not appear to be a significant question with any "real world" application when it was first posed by Leonardo of Pisa (better known by his nickname Fibonacci). Then I assured them that we would see that the exploration of this question has led to an incredible number of important "real world" connections in a number of other fields.

The Rabbit Question

> *If a pair of rabbits were put into a walled enclosure to breed, how many pairs of rabbits would there be after one year if it is assumed that every month each pair produces a new pair, which, in turn, begins to bear young two months after its own birth?* [**1**]

With the overhead projector, the class helped me develop a rabbit population diagram, as in Figure 1, month by month, from January through April. To make the experience more concrete, I created 12 envelopes containing 18 cardboard pairs each of baby rabbits, mature rabbits, and "married" producing rabbits; and I gave one envelope to each pair of students in the class. We then broke the class up into pairs to recreate the growth of the rabbit population through the months of May and June. Each pair of students used the envelopes of rabbits to build a model of the problem. Rabbits were taped or glued onto large sheets of paper while labels and arrows were drawn to clarify the growth pattern of the rabbits.

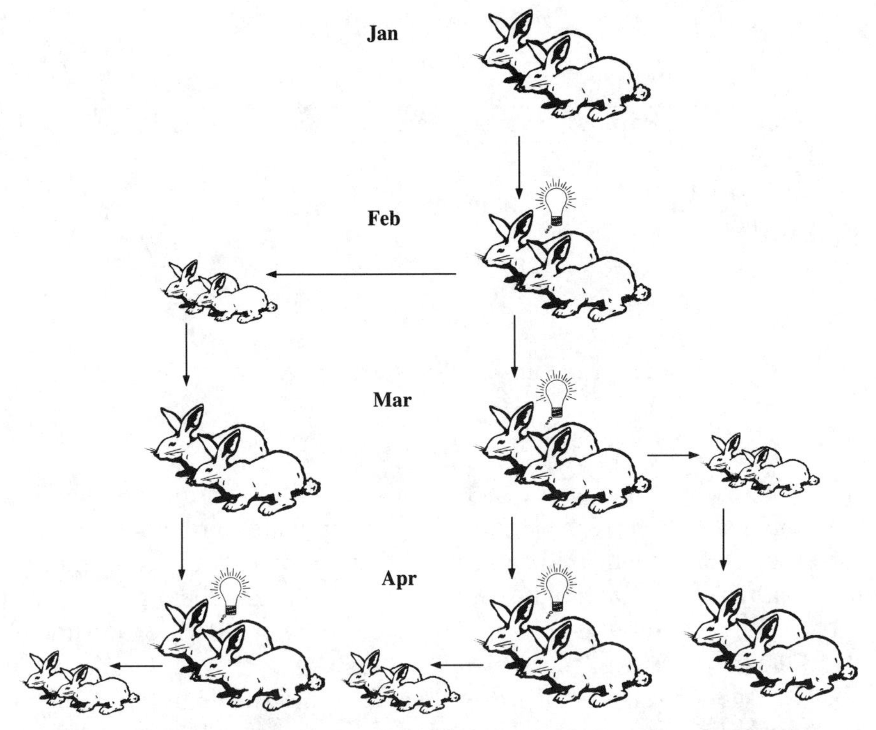

FIGURE 1. The Fibonacci rabbit-breeding model, January through April.

Building a Model

About half of the groups did extremely well and were ready to help others trace the growth of the rabbits through the next month, while one group seemed to miss the idea of tracing the growth of the population from one month to the next. The students that went off track had a tendency to trace the growth of only one or two rabbits through a number of months and lost track of the other rabbits. These students were able to understand readily how each kind of rabbit progressed in their development but were not able to organize themselves around the idea that in each month the rabbit population changed. However, after about 20 minutes and help from either me, their teacher, or their peers, all of the class members were able to complete their models. On our chart on the overhead projector, we continued to trace and draw the growth of the rabbit population together. This time we continued solving the problem through the months of May and June. A few students had already successfully completed the problem through June with their cardboard rabbits and had commented that July's rabbits would probably not fit onto their large pieces of paper. Next, as a group, we discussed ways to answer the original question, "How many rabbits would there be after a year?"

Thinking Ahead

Month	Pairs of Babies	Pairs of Adults	Total Pairs
January 1	0	1	1
February 1	1	1	2
March 1	1	2	3
April 1	2	3	5
May 1			
June 1			

TABLE 1. Chart with numbers of pairs of rabbits by month.

We organized the numbers of each kind of rabbits each month on charts that were handed out, similar to Table 1. The students were quick to recognize patterns. One student realized that she could predict the next number of pairs of baby rabbits by looking at the sequence of numbers produced by the pairs of all the rabbits. Another student commented that he thought that the difference between successive numbers was 1, 2, 3, 4, 5, and so on. We looked at the series to see if we could verify this and found out that we could not get a difference of 4 in this expected series; but this idea did lead the class to discover that the difference between successive terms was also a Fibonacci sequence. Interest was climbing! Eventually, one of the girls in the class realized that each term was determined by the sum of the 2 previous terms. She expressed this as, "You can get the new number by adding the last two numbers together." The groups worked together to verify her idea and reported back that it seemed to be true. The homework challenge was to use this method or any other method that they could justify to try to answer the rabbit question through the first year.

"Thirty Days Hath September ..."

The next day we warmed up by revisiting the first homework challenge to explain the answer and the process to the "10 cents every day for a year" question. Most students had asked others how many days were in a year. A few had looked at calendars. I commented that while there was nothing wrong with getting the answer from others, many times we have the knowledge to find an answer within ourselves and do not even realize it. I told them I expected that this was the case with the question, "How many days are in a year?" and reminded them of the verse, "Thirty days hath September". We spent a few minutes, reciting and recording the numbers from the verse and then put together all the numbers to get 365. While all of the students knew the verse, no one had thought to call upon it as a resource.

Revisiting the Rabbits

Next, we revisited and retraced the rabbit problem to the end of the year using the charts I had handed out to keep track of the number patterns. Many students came to class with the correct answer of 377, but a few had numbers like 381 or 367. I asked the class to guess why some answers might be off by a little. I asked them, "What might have happened in the process to cause answers such as 381, etc?" Some students concluded that a small addition error could have caused the discrepancy. The students with answers around 377 were asked to try to retrace their process and report back if the "addition error" theory was correct. In all cases the students were pleased to respond that this theory had been correct.

Spiraling In

The main activity for the second day was to investigate the items that had been set up on the first day in a display area. The purpose for this activity was to help students make connections to Fibonacci numbers in nature and to help them recognize a connection with spirals [**5**]. These items included: The poster, "Fibonacci Numbers in Nature" [**2**]; five different varieties of pinecones; an artichoke; a cactus; a pineapple; a large sunflower; the book, *Fascinating Fibonaccis* [**1**]; and some sea shells. Many of these items were marked with either white-out or colored Elmer's glue where spirals could be seen. While passing the items around, we tried to determine what they all had in common. After some discussion, spirals were agreed upon. The word *helix* was introduced and students were asked to think of other things in nature that display this shape. Tornadoes, whirlpools, and seahorse tails were three of the responses.

Seeing is believing

Students were put together for a "pair share" activity, an arrangement in which each pair of students explore a problem and then two pairs are put together to compare, discuss, or verify each other's answers. Each pair of students was asked to count the number of clockwise and counterclockwise helices on at least two varieties of pinecone (see Figure 2). Some of the students could see the helices, but had difficulty counting them, others had difficulty seeing the helices when they were given the unmarked pine cones. The teacher, myself, and my daughter were kept very busy rotating from group to group helping students visualize, mark, and count the helices. Successful groups were employed to help others. Eventually, everyone seemed to be able to accomplish the task and, in the process, found over and over again the presence of the Fibonacci numbers. I left the students with the challenge of counting the helices on the large sunflower.

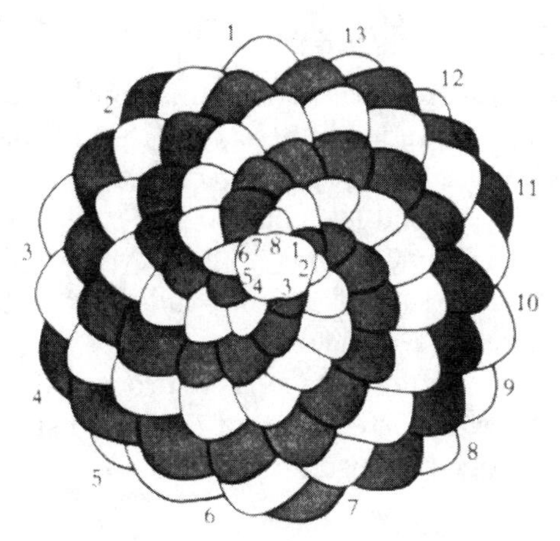

FIGURE 2. Eight clockwise helices on a pinecone, colored alternately for clarity. There is a set of thirteen helices winding counter-clockwise, which is harder to see.

From: FOR ALL PRACTICAL PURPOSES 3rd Edition by COMAP (p. 662). Copyright 1994 by COMAP, Inc. Used with permission of W. H. Freeman and Company.

Is All Fair In Love and War?

For a closing, cardboard bookmarks were passed out. These markers were decorated with flowers that contain petal counts with Fibonacci numbers from 3 to 89 [**5**]. Examples on the bookmark include the Oxeye daisy with 34 petals and the African daisy with either 55 or 89 petals. We ended the session by thinking about even and odd in the "Loves me, Loves me not" activity that is classically played with daisies. We discussed whether the activity was fair. (If you have never thought about this before, now is the time to do so.)

Fibonaccis are Very Prolific

On the third day, we reviewed the Fibonacci sequence. One student asked how long it would take for the rabbit population to grow to one million. Other students showed their interest in this question immediately, and as I quickly discovered, these fourth-grade students were in love with the concept of "a million". I asked students to write down their time estimates. On the overhead I uncovered the population numbers slowly, month by month. At the end of the second year, students were given the opportunity to revise their estimates. Shortly after, they were amazed to find out that the walled enclosure contained over 1 million rabbits after only 31 months and that their estimates had been too high. To get an additional sense of 31 months, students were asked to convert their answers to the number of days and then to years and to share their answers and their thinking with the class.

Fleas, Bees, Trees, and Piano Keys

I followed this activity with a presentation intended to quickly make many other Fibonacci connections. In the presentation, I proceeded to:

- trace the male honeybee family tree to the black keys of the piano;
- look at three common musical scales that use 5, 8, and 13 keys;
- examine a scale of the keyboard with 13 keys including 8 white keys and 5 black keys with the black keys arranged in 2s and 3s;
- clap out the rhythm of a limerick about fleas with beats of 2s and 3s and a total of 13 beats completed in 5 lines;
- demonstrate the Fibonacci (fractal) growth pattern of some trees and root systems;
- try to explain the connection with the "Elliot Wave principle" and the Dow Jones averages.

In the 1930's, Ralph Elliot studied patterns in the stock market, realizing that business swings are a result of human patterns of optimism and pessimism. His observations lead to the prediction that the stock market follows Fibonacci patterns. I carefully connected this idea to some issues that students could connect to their lives. We talked about optimism and pessimism in spending and related this to the recession in Rhode Island. I explained that when the economy is in a downswing people tend to hold on to their money, and when it is in an upswing people are more apt to spend. I related this to the school bond issue which had been defeated twice in the last few years, but in the recent elections it had been passed by the voters. I explained that the economy was beginning to improve and therefore the voters were finally agreeing to spend. Since this particular school is one of the two schools that will be most affected by this bond issue, students were keenly aware of the election and therefore able to make the connection. By this point in the presentation, students were fascinated and literally on the edge of their chairs. After some discussion, I asked students if they thought that Fibonacci knew the significance of this seemingly silly question about rabbits, when he set out to explore and write about it. While some did not respond, many doubted that Fibonacci had these applications in mind.

The Finale

For the final activity, we organized ourselves into groups to explore one last Fibonacci connection. The students were told that they would be directed to use building blocks whose dimensions were Fibonacci numbers to create puzzle pieces that could be reassembled into a picture of something that might be found in Jurassic Park. I also said that I had a fossil in my pocket and that it was 350 million years old. They were told that the check of their work would be comparing their completed puzzle to the fossil in my pocket. After discussing the idea of 350 million for a few minutes, each pair of students measured and cut out: two 1-unit squares; one 2-unit square; one 3-unit square; one 5-unit square; and one 8-unit square. They used compasses to swing an arc joining diagonally opposite corners of each square and then cut out all six squares. They were then asked to assemble these six puzzle pieces into an picture. I also told them that the completed puzzle would be rectangular. Group by group, they verified their completed puzzles with the Ammonite fossil that I had in my pocket. This is similar to a Nautilus shell, shown in Figure 3. Once verified, each group was given a handout with a picture and story about Ammonites and the creation of fossils. (This activity could be a springboard for explorations with the Golden Ratio.) This puzzle activity became the connecting and culminating activity of the three days, in that it drew together the themes of Fibonacci numbers and helices discussed earlier.

FIGURE 3. Cross-section of a Nautilus shell, similar to that of an Ammonite.

From: Tannenbaum/Arnold, EXCURSIONS IN MODERN MATHEMATICS (p. 285), Copyright 1992. Reprinted by permission of Prentice-Hall, Inc., NJ.

The Beat Goes On

We closed with some discussion about helices in nature and the world around us, and about the occurrence of the Fibonacci numbers that we had seen over the last three days. I showed some spiral images that I had gathered from books and an invitation was extended to students to continue to make Fibonacci discoveries and to share any new findings. The Fibonacci display was left up with a dinosaur book and the Ammonite fossil for about a week (the students were able to cut up and enjoy the pineapple).

Afterwords

I am not sure what my expectations of fourth-grade students were when I began, but I know now that they can understand and connect with the Fibonacci numbers. The fourth-grade teacher, who is a very honest and open person, made two particular comments that caught my attention. First she noticed that students who usually do not do well in math, were doing well with this series of activities and that some of these same "weak" math students were particularly strong in their visualization of the helices (something which she herself found challenging). Her second comment had to do with her teaching of math. She said that while she feels very comfortable teaching science and writing, she does not feel comfortable helping students who have difficulties in math, because she can not seem to comprehend why they don't get it. She said her teaching of mathematics has been rather cut-and-dried, since she has always seen mathematics as either right or wrong. She felt that I had given her a new view of teaching math. She said she noticed how much emphasis I put on the process of getting the answer rather than the answer itself and that she had never thought to go about teaching mathematics that way. She said that the sessions were inspiring to her and that she learned a lot from having me visit her classroom.

The Call of the Classroom

Personally I found these three days valuable and enjoyable and once again hear the "call of the classroom". Teaching mathematics around a theme is a great deal of fun. I expect that my approach to teaching mathematics in this fourth-grade classroom would have been somewhat different if I had been the permanent teacher. Because I was invited to this class to introduce the Fibonacci numbers, and because the time limit was predetermined, my approach was much more guided than I would expect it to be in an ongoing self-contained classroom. In a year-long classroom experience, I imagine that Fibonacci connections could be woven into the classroom experience many times and in many places. I have come to believe that mathematics is part of our lives and I hope that through my enthusiasm for richly connected topics such as the Fibonacci numbers this belief will also be contagious to my students. As I guided our three-hour exploration, I allowed some mathematical side trips as student interests and questions sometimes exposed their "need to know". Creating and guiding this "need to know" will be for me the most exciting task when I return to the classroom.

The Impact of the Leadership Program in Discrete Mathematics

My own curiosity and awareness of the Fibonacci numbers was developed a few years ago when I first heard about them in an article by Michael Tempel in the *Logo Exchange*. I can't say whether I fully understood the article at the time, but it did get me to open my ears and eyes to this sequence. At the Leadership Program in Discrete Mathematics at Rutgers [**4**], I became aware of the connections of this number sequence to science and nature. These connections sparked my interest in the topic. I spent the school year 1992-93 learning more about the Fibonacci numbers and in the summer of 1993, I collaborated with other Leadership Program participants to put together some classroom materials and a staff development workshop on this topic. The classroom activities in this article are drawn from these materials. They are designed to make connections, encourage concrete exploration and foster communication. The inspiration for this kind of mathematics teaching is drawn from the Leadership Program.

References

[1] Garland, Trudi Hammel, *Fascinating Fibonaccis: Mystery and Magic in Numbers*, Dale Seymour Publications, Palo Alto CA, 1987

[2] Garland, Trudi Hammel, and Edith Algood, Poster: "Fibonacci Numbers in Nature", Dale Seymour Publications, Palo Alto CA, 1988.

[3] Kappraff, Jay, *Connections*, McGraw Hill, New York, 1990.

[4] Rosenstein, Joseph G., and Valerie A. DeBellis, "The Leadership Program in Discrete Mathematics", this volume.

[5] Tannenbaum, Peter, and Robert Arnold, *Excursions in Modern Mathematics*, Prentice Hall, 1992.

[6] Wahl, Mark, *A Mathematical Mystery Tour: Higher-Thinking Math Tasks*, Zephyr Press Learning Materials, Tucson AZ, 1988.

Rhode Island School of the Future, P.O. Box 4692, Middletown, RI 02842
E-mail address: kowalcjn@ride.ri.net

DIMACS Series in Discrete Mathematics
and Theoretical Computer Science
Volume **36**, 1997

Using Discrete Mathematics to Give Remedial Students a Second Chance

Susan H. Picker

In the summer of 1990, as a participant in the Leadership Program in Discrete Mathematics at Rutgers University [**2**], I first came to study discrete mathematics. When I returned to my classroom at Murry Bergtraum High School in lower Manhattan that fall, I was excited and enthusiastic, even though I would only be teaching discrete topics to the 10th grade remedial classes. They were often referred to as the "classes from hell". In fact, the "classes from hell" gave me the most rewarding term I'd ever experienced.

As the term began, I defined some goals for myself. They are basic questions which helped me to clarify what I hoped to accomplish, and they were putting discrete mathematics to a test. The first question was: can students be encouraged to come to class regularly? In New York City, where I teach, we hear stories of students who will literally climb out the windows to get out of class. And this attitude was widespread among the remedial students. I felt and hoped that with the material I was bringing back to them, students would want to come to class on a regular basis.

Next, I wondered, can students be encouraged to believe that there's more to mathematics than arithmetic computation? At that point I'd had nine years of hearing students proclaim: "I hate math!" But I knew that this was going to change because at Rutgers that summer I had come to understand finally that when students said they hated mathematics they were really saying they hated arithmetic. They didn't know anything about mathematics; during their entire student careers most of them had only seen the four operations of arithmetic, over and over and over. And I knew that discrete mathematics was real mathematics and not some watered-down version of mathematics.

My third question was more specific: Can the students be encouraged to like at least some topics which they clearly knew to be mathematics? I saw this as a key to changing their minds —having them reconsider their attitude to mathematics as a whole.

1991 *Mathematics Subject Classification.* Primary 00A35, 00A05, 05C15.

The last question I posed was based on their need to be able to take some additional mathematics as preparation for college —a goal in my school and in New York City: Can students be encouraged to study mathematics further —in this case to study algebra?

The results were beyond anything I could have imagined. I found in time that I could answer yes to all these questions. I found myself enjoying being in the classroom each day, and so did the students.

I started slowly. I didn't have a lot of materials so I began buying or getting publishers to send me everything that had the word "discrete" in the title. Eventually I began developing my own materials —such as taking the map of New York City or the map of Paris and creating a problem which was a variation of the famous Königsberg Bridge problem. But at a certain point I needed even more materials, so I found myself going to college textbooks and even graduate-level texts. Because of the nature of discrete mathematics and the students' increasing interest, I found that it didn't matter that these were intended as college-level problems. The students could understand them and solve them.

An example of the type of problem students came to be able to solve by December of that term, comes from [**1**]. The problem (Figure 1), which is sophisticated for high school students, gives a matrix indexed by chemicals, many of which are unsafe to transport together, and sets up a situation where a train has to be assembled with one chemical per car. Because of the incompatibility of some chemicals with each other, their cars can only travel next to each other if two open gondolas of sand separate them. The challenge is to find the minimum number of gondolas of sand that can be used in setting up this train with twelve different chemicals.

As preparation for understanding the problem, we talked first in class about grocery products travelling together to the supermarket and I asked students what they would not want travelling together say, in the same truck. Students said such things as meat and bleach, strawberries and onions–they saw that these would not affect each other well. Next we talked about the chemicals. Students commented on what they knew of the chemicals with which they were familiar, like acetone, which some of the students recognized from nail polish remover bottles. Students noticed that in the problem nitrogen travels safely together with everything and they were curious about why that was. We also discussed various chemical spills which had been in the news; trains which had derailed carrying chemicals and how that had affected the environment. It became an interdisciplinary discussion about a real-world situation.

In studying the topic of graph coloring, students had already solved many smaller and simpler problems involving scheduling conflicts but none as large or as complicated as this. They now knew that when there is a possible scheduling conflict as, say, subcommittees of a legislature having members in common and therefore not able to meet at the same time, that the way to begin an efficient scheduling is to create a conflict graph of vertices and edges. The vertices represent those things which could be in conflict, like the subcommittees, or in this case, chemicals. An edge is drawn between two vertices if and only if there is a conflict between them. In this case each "U" in the matrix indicated to students that those chemicals would

> *A manufacturer of chemicals ships its products by railroad tank cars. To reduce the danger that might occur through accidental spills of chemicals, the company specifies that the train must be made up in segments in such a way that*
>
> **a:** *no two chemicals in the cars of each segment react dangerously with each other,*
>
> **b:** *two open gondola carloads of sand must precede each segment to separate dangerously reactive chemicals in case of a derailment or other emergency,*
>
> **c:** *two open gondolas of sand must separate the last segment of chemical cars from the caboose.*
>
> *Determine the smallest possible number of gondolas of sand needed to make up a train that carries one tank car of each of the following 12 chemicals. Show your work and reasoning.*
>
> *Chemicals*
>
> **1.** *Toluene* **7.** *Dimethyl hydrazine*
> **2.** *Acetone* **8.** *Dinitrogen tetroxide*
> **3.** *Phosphoric acid* **9.** *Chromic anhydride*
> **4.** *Sulfuric acid* **10.** *Nitrogen*
> **5.** *Potassium cyanide* **11.** *Chlorine*
> **6.** *Sodium hydroxide* **12.** *Potassium dichromate*

	1	**2**	**3**	**4**	**5**	**6**	**7**	**8**	**9**	**10**	**11**	**12**
1	-	S	S	U	S	S	S	U	U	S	U	U
2	S	-	U	U	U	U	U	U	U	S	U	U
3	S	U	-	S	U	U	U	S	S	S	S	S
4	U	U	S	-	U	U	U	S	S	S	U	S
5	S	U	U	U	-	S	S	U	U	S	U	U
6	S	U	U	U	S	-	S	U	U	S	U	S
7	S	U	U	U	S	S	-	U	U	S	U	U
8	U	U	S	S	U	U	U	-	U	S	U	U
9	U	U	S	S	U	U	U	U	-	S	U	S
10	S	S	S	S	S	S	S	S	S	S	S	S
11	U	U	S	U	U	U	U	U	U	S	-	U
12	U	U	S	S	U	S	U	U	S	S	U	-

Reaction Table: S (relatively safe when mixed); U (unsafe when mixed)

FIGURE 1. Chemical transportation problem [**1**]

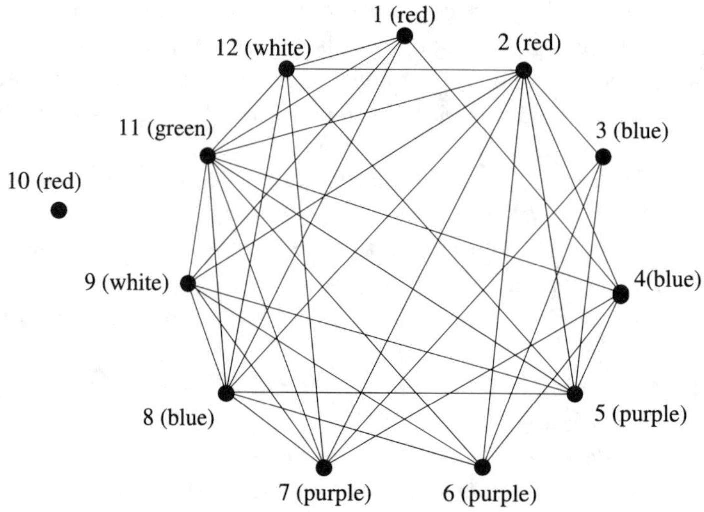

FIGURE 2. The conflict graph, with a 5-coloring

react unsafely and would therefore be in conflict with each other, so students rightly concluded that chemicals with a "U" between them were to be joined by an edge in the graph.

After creating the graph (Figure 2), students colored the vertices using the fewest colors possible to determine the chromatic number. Students were aware that since the edges joined conflicting things, those vertices joined by an edge would have to have different colors. Therefore those vertices which had the same color were compatible and could be scheduled or grouped together. Through discussion, students saw that in this problem, the coloring would separate the chemicals into groups which could travel safely together. The chromatic number told them how many groups there would be. It was these groupings —the students determined there were 5 of them —which then had to be separated by the two open gondolas of sand. In the graph in Figure 2, which accompanied models students drew of the train (see Figure 3, for example), students indicated the chromatic number of 5 with the Greek character χ. They became enthusiastic about using mathematical symbols which they hadn't seen before.

Students at this point in the term had become used to working in groups, for problems in discrete mathematics often are solved more successfully by a group than by a student working alone. A number of students got so involved that they started coming to school regularly, especially when we began the unit on graph coloring. My chairman was pleased.

This chemicals-carrying-train problem was the culminating project involving graph coloring and it took a week for the students to do it. Though the problem might have taken other students a much shorter time to solve, the important thing is that these students understood what the problem demanded and what the constraints were and were able to work methodically to arrive at a solution. For nearly all of them it was in this class that they did real problem-solving in a mathematics class for the first time.

What they were able to accomplish surpassed what I had believed students who were apparently so low-level could produce. And yet these students weren't really so low-level anymore. There were new concepts in mathematics and approaches to problem-solving that they were coming to understand and utilize.

FIGURE 3. "The Picker Express" (with the author as engineer) was drawn by a student in the class to illustrate the solution to the chemical train problem. Freight cars piled high with sand are situated between safe combinations of chemicals. Each car is labeled with its cargo, and only the first ten cars are pictured here.

The month after this project, in January, as the term was ending, I was invited to guest teach for a week in a couple of the calculus classes. I planned to do as much as I could of graph theory and coloring. These were classes with very high registration —maybe 34 students each, and I realized that I would not be able to be everywhere at once. It occurred to me that perhaps it would serve a dual purpose to bring four different discrete mathematics students with me each day to assist and answer questions. When I suggested this to them, my students were terrified. "How can we go into classes with kids who are taking calculus? We're just tenth graders."

It was not easy for these students to understand that they knew things of which the twelfth graders had no idea —that they had never seen before and could not do. To see themselves as the "experts" and others as learning from them was a completely new way of thinking for them, but in the end they came with me to the calculus classes. It turned into a great experience for them. As the hands began going up, the tenth graders started going over to the calculus students and answering their questions. My students were amazed that they could do such a thing. And it gave them a confidence that they had never had before.

About half of the class, eleven students, went from my discrete mathematics class into algebra the following term. These were students who were not in the "academic math track" and it is unlikely that they would have left high school having studied any algebra, had they not been encouraged in this term to see that they could succeed in a mathematics class. I followed up with these students the best I could during the term that followed. Meeting them in the halls, I asked them how they were doing in their first term of algebra, and the responses were generally favorable. I also heard about their progress from other teachers who now had them in their classes. But it was hard to continue to keep track of the students and continue to encourage them because I left the school at the end of that year, in June.

The following fall I was periodically in the school and I happened to see a videotape of one of the second-term algebra classes. There were four of my former discrete mathematics students actively participating and clearly doing well. I ran into one of the students and told her that I had seen the tape. "Your class made all the difference," she said.

At the beginning of the term I always give students a questionnaire about their attitudes. I always ask the students "Do you think you'll ever be a mathematician?", because I am concerned that their image of mathematicians lacks a clear understanding of what a mathematician is and does. Students always write, "NO!" —big letters; exclamation points. "I don't think so!"

At the end of that term a few days before our last meeting I asked my students if they remembered their answers to that question on the questionnaire. And students said —"You know...we like this, but we don't think we're going to be mathematicians!" And I said well, I have to tell you, I've learned that you are mathematicians because you've been doing mathematics. And a mathematician is a person who does mathematics. What happened next took me completely by surprise: the students spontaneously burst into applause. I was never moved in a mathematics class before. And it moves me still when I think of it —that these students had come to have such a different sense of themselves and such a different attitude in class.

As I look back on that term I can see that I had changed, too. Discrete mathematics had given me a wider view of mathematics as a live and growing subject with new areas for exploration. Perhaps because of this I was also open to see new things in my students including strengths I had previously overlooked. Many were very visual, some more visually adept than the calculus students, and they had an easier time with graph theory and other topics because of it. They showed a great creativity in their approaches to problems, and in presenting solutions to those problems, as in the train problem. They got very involved in topics which I hadn't thought would interest them, like chromatic polynomials, which they were able to work with without the algebraic notation. I came to respect my students more through this, and to believe even more in their ability to learn. It had never been clearer to me that a huge obstacle to these students' progress and success was their dislike of mathematics, and this dislike is what I saw lessen and change.

References

[1] Finkbeiner, D.T. II, and Lindstrom, W.D., *A Primer of Discrete Mathematics*, W. H. Freeman and Company, 1987 (problem used with permission).

[2] Rosenstein, Joseph G., and Valerie A. DeBellis, "The Leadership Program in Discrete Mathematics", this volume.

Office of the Superintendent, Manhattan High Schools, 122 Amsterdam Avenue, New York, NY 10023

E-mail address: spicker@dimacs.rutgers.edu

DIMACS Series in Discrete Mathematics
and Theoretical Computer Science
Volume **36**, 1997

What We've Got Here is a Failure to Cooperate

Reuben J. Settergren

1. The Setup

At twelve years old, you feel terribly mature and independent living in a real college dorm. At the moment, in the sunny Los Angeles afternoon, at the Johns Hopkins Center for Talented Youth (CTY), you are stretching your mind and body, trying to master the strategy and technique of Ultimate Frisbee. Suddenly, from out of nowhere, your Residential Advisor (RA) grabs you and stuffs an envelope into your hands, saying, "Tell no one of this." How intriguing; the RA usually hands out student mail at the nightly hall meetings. At the first opportunity, the next break between activities, you tear open the envelope (marked only with your name and CTY's logo), and read the following letter:[1]

> *Dear _______,*
>
> *I am sending this letter out via Special Delivery to fifteen of 'you' (namely, various friends of mine at CTY). I am proposing to all of you a game, the payoffs to be in* real money *(provided by me). It's very simple. Here is how it goes.*
>
> *Each of you is to give me a single letter: 'C' or 'D', standing for 'cooperate' or 'defect'. This will be used as your move in a game against* each *of the other players. Here is the payoff scheme for each of the two-player games. If both players cooperate, each gets 5 cents. If both defect, each gets 1 cent. If one cooperates and one defects, though, the defector gets 9 cents, while the cooperator gets nothing.*
>
> *Thus, if everyone sends in 'C', everyone will get* $14 \times 5 = 70$ *cents, while if everyone sends in 'D', everyone will get* $14 \times 1 = 14$ *cents. You can't lose! And of course, anyone who sends*

1991 *Mathematics Subject Classification.* Primary 00A05, 00A35.

An earlier version of this article appeared as "A Classroom Dilemma" in [**5**].

[1]The text of this letter is taken from [**2**], and was slightly modified to fit my audience and financial resources.

in 'D' will get at least as much as everyone else will. If, for example, 8 people send in 'C' and 7 send in 'D', then the 8 C-ers will get 5 cents apiece from each of the other C-ers (making 35 cents), and zero from the D-ers. The D-ers, by contrast, will pick up 9 cents from each of the C-ers, making 72 cents, and 1 cent from each of the other D-ers, making 6 cents, for a grand total of 78 cents. No matter what the distribution is, D-ers always do better than C-ers. Of course, the more C-ers there are, the better everyone *will do!*

By the way, I should make it clear that in making your choice, you should not aim to be the winner, but simply to get as much money *for yourself as possible. Thus you should be happier to get 35 cents (say, as a result of saying 'C' along with 7 others, even though the 7 D-sayers get more than you) than to get 14 cents (by saying 'D' along with everybody else, so nobody 'beats' you). Furthermore, you are not supposed to think at some subsequent time you will meet with and be able to share the goods with your co-participants. You are not aiming at maximizing the total amount of money I shell out, only at maximizing the amount that comes to* you*!*

Of course, your hope is to be the unique *defector, thus really cleaning up: with 14 C-ers, you'll get $1.26, and they'll each get 13 times 5 cents, namely $0.65! But why am I doing the multiplication or any of this figuring for you? You're very bright. So are all of you! All about equally bright, I'd say, in fact. So all you need to do is tell me your choice.*

It is to be understood (it almost *goes without saying, but not quite) that you are not to try to get in touch with and consult with others who you guess have been asked to participate. In fact, please consult with no one at all. The purpose is to see what people will do on their own, in isolation. Finally, I would very much appreciate a short statement to go along with your choice, telling me* why *you made this particular choice.*

Sincerely,

Hank D. Bank

"Hank D. Bank?" you think, "Gimme a break! How stupid does Reuben think I am?" What a transparent guise for a classroom exercise. Obviously, the right thing to do, the nice thing to do, is to cooperate. But you're no sheep; why not try for the big bucks? But what if everybody else defects too? You mentally enumerate your classmates, and decide that enough other players will cooperate that you will get quite a sizable payoff. Later, you chuckle to yourself as you embellish the big, dark "D" on your paper, knowing that many of your classmates will foolishly count on you to cooperate, because of your generally meek demeanor. Ha, ha, ha ....

2. The Setting

Every summer, the Johns Hopkins University Center for Talented Youth (CTY) offers two three-week sessions of residential instruction at college campuses up and down the east coast, in Los Angeles, and even in Europe. Their courses include not only mathematical and physical sciences, but also a large array of innovative humanities courses, from archaeology to political science to etymology. In 1994, CTY offered a new class called Applications of Contemporary Mathematics (ACOM). ACOM was intended for CTY's youngest, least experienced students, those entering eighth grade—possibly without even pre-algebra under their belts.

I eagerly accepted CTY's offer to teach two sessions of ACOM in Los Angeles. CTY had chosen a textbook for ACOM, *For All Practical Purposes* [**1**], but also allowed latitude in developing a syllabus. So, while waiting for my copy of the text, I brainstormed and searched for activities. I ran across an article [**2**] I had read years before by Douglas Hofstadter (author of *Gödel, Escher, Bach: an Eternal Golden Braid*) in his collection of *Scientific American* articles [**4**]. Hofstadter described how he devised the cooperation game above, and discussed the results of playing it with twenty of his friends. The results were so compelling that I decided to use the game to teach my students about the Prisoner's Dilemma.

In this famous and fundamental game theory situation, two partners in crime are detained by the police and held incommunicado. The police tell each suspect "If you give a full confession, and your confession leads to the conviction of your partner, then you can go free." Each partner knows that the evidence against them is slim, and silence by both would mean small sentences, but can each partner trust the other not to confess and stick him with a long sentence? By having my students play a game simulating the Prisoner's Dilemma, I could force them to wrestle with the dilemma themselves, not just listen to me talk about it.

I timed this activity to coincide with classroom material on Game Theory (Chapter 15 in [**1**]); I distributed the letters on Tuesday and Wednesday of the second week, and announced the results, gave out money, and held discussion on Friday afternoon. I distributed the letters individually, outside of the classroom, and as mysteriously as possible, in order to establish the atmosphere of secrecy and isolation necessary for proper play of the game. In the second session, when there were two sections of ACOM and another instructor who wanted her class to play as well, I divided the larger number of players into three games, with game groups cutting across class divisions.

3. The Settlement

On Friday, just before the weekend, we gathered to reveal and discuss the outcome of the game. Not surprisingly, in all cases there were dismally few cooperators. Each of the second session games had two or three cooperators each (out of fifteen or sixteen players), and in the first session, there was one lone cooperator among fifteen (who of course received nothing)!

Student response was varied. Not only cooperators, but also some defectors were shocked that their classmates were so greedy. Some defectors expected that most of their classmates would defect, but lemming-like, defected themselves — hoping to squeeze the most out of the few sucker cooperators, and fearful of being counted as one of them. One defector, who had obviously agonized over his decision, responded to the results by lamenting, "I feel crummy!", and became my most ardent advocate for cooperation in the ensuing discussion. Another student, more Machiavellian, broke the game's premise of secrecy to form a pact of cooperation with two henchmen — in order to sweeten his defection! During the discussion, however, he repented, and also embraced cooperation. Of course, this was just during the discussion. I'm sure that he would be among the wily students that would be quick to take advantage of the lessons other students learned, if the game were to be played a second time.[2]

So far in this article, I have tried only feebly to conceal that the lesson to be learned is that cooperation is the "right" answer to this game; that in this type of situation, the players must think not individually, but collectively. Hofstadter explained it best: if all the players are equally and perfectly rational (and thus their moves are "right"), they will all make the same move. Since each and all do better when the collective move is "C" than when it is "D", each player will cooperate.

Even if my students didn't see the game in exactly these terms, most of them instinctively understood this logic. Ironically, this was the seed of their downfall. Since most knew that everybody "should" cooperate, few could resist the temptation to try to reap big profits by relying on the cooperation of their classmates.

After initial reactions and handing out of money, I passed out copies of Hofstadter's article, and we went through it, discussing Hofstadter's reasoning, and comparing the responses of the original players to those of the class. Hofstadter also provides a number of variations on the game, with more extreme payoffs, and even penalties. With more at stake, students have to rethink their moves, because it is simultaneously more important, and yet more dangerous to cooperate.

The simplest extrapolation is to consider what the students would do if they were offered Hofstadter's original game, which used dollars instead of cents? Or how about millions of dollars instead of dollars? This leads to an excellent demonstration of the nonlinearity of value. Psychologically, while

[2]This brings up a separate issue of repeated play of the Prisoner's Dilemma, about which Hoftsadter has written another article [**3**] (which also appears in [**4**]). In a computer-run contest, the consistently best strategy (called TIT FOR TAT) was to cooperate the first time, and thereafter repeat your opponent's last move. Incidentally, the administrator of the computer tournament unhesitatingly played 'D' himself in Hofstadter's one-time game.

three dollars might be three times as valuable as one, three million dollars are not three times as valuable as one million. Another way to say this is, a player would be more willing to have their payoff cut from three million dollars to one million than to have it cut from three dollars to one. Thus, it is likely that more players would cooperate in a million dollar game than in a penny game, where the most significant part of the payoff is in status and ego reinforcement.

There are many nearly equivalent social situations that students can probably discover themselves, if sparked with an example or two.

- During the onset of rush hour traffic, fast-moving drivers notice an accident at the side of the road. "I don't have to slow down very much to get a good look at the carnage," they each think. Pretty soon, frustrated drivers in bumper-to-bumper traffic are thinking instead, "Why does everybody have to stop and stare? ... But since I've been waiting so long, when I get up to the accident, I might as well take a look, since traffic is moving so slowly anyway".
- A local museum is exhibiting a painting by Van Gogh, and you are enthralled by the texture of the paint; so chunky, so three-dimensional, so—"Hey, I wonder what it feels like? It will be all right if I'm the only one that touches it."
- It's a beautiful day, and you're out for a drive with your family, even though gas prices are sky-high, and the air could be a lot cleaner. You sure do look good driving that powerful, eight-cylinder Lincoln Continental, though.
- It's a Tuesday in November, and you're curled up for the evening in your La-Z-Boy waiting through commercials for your favorite sitcom, and you see that the local news is pumping their election coverage, which will begin in two hours, when the polls close. "My one vote wouldn't matter," you think, as you put up the footrest and dip into a bag of cheese snacks.

There are many other examples which your students can devise. You can hope that after they have played this game and discussed real-world applications, they will be able to recognize situations where this kind of cooperation can make a difference, and decide more rationally how to respond.

References

[1] COMAP, *For All Practical Purposes: Introduction to Contemporary Mathematics*, 3rd ed., W. H. Freeman, New York, 1994.

[2] Hofstadter, Douglas, "Dilemmas for Superrational Thinkers", and "The Tale of Happiton", *Scientific American*, June 1983.

[3] Hofstadter, Douglas, "The Prisoner's Dilemma and the Evolution of Cooperation," *Scientific American*, May 1983.

[4] Hofstadter, Douglas, *Metamagical Themas*, Basic Books, New York, 1985, ch. 29, 30, 32.

[5] Settergren, Reuben, "A Classroom Dilemma," *In Discrete Mathematics: Using Discrete Mathematics in the Classroom*, #6, Spring/Summer 1995, p. 2.

Rutgers Center for Operations Research, P.O. Box 5062, New Brunswick, NJ 08903

E-mail address: `reuben@rutcor.rutgers.edu`

Section 2

The Value of Discrete Mathematics: Achieving Broader Goals

Implementing the Standards: Let's Focus on the First Four
NANCY CASEY AND MICHAEL R. FELLOWS
Page 51

Discrete Mathematics: A Vehicle for Problem Solving and Excitement
MARGARET B. COZZENS
Page 67

Logic and Discrete Mathematics in the Schools
SUSANNA S. EPP
Page 75

Writing Discrete(ly)
ROCHELLE LEIBOWITZ
Page 85

Discrete Mathematics and Public Perceptions of Mathematics
JOSEPH MALKEVITCH
Page 89

Mathematical Modeling and Discrete Mathematics
HENRY O. POLLAK
Page 99

The Role of Applications in Teaching Discrete Mathematics
FRED S. ROBERTS
Page 105

DIMACS Series in Discrete Mathematics
and Theoretical Computer Science
Volume **36**, 1997

Implementing the Standards: Let's Focus on the First Four

Nancy Casey and Michael R. Fellows

1. Introduction

The *Curriculum and Evaluation Standards for School Mathematics* of the National Council of Teachers of Mathematics [**8**] can be viewed as an attempt to shift attention in the mathematics curriculum to high-level cognitive issues, and away from the traditional focus on the accumulation of low-level rote computational skills (tasks that increasingly ubiquitous machines do quite well). At all levels of mathematics education, and in many different ways throughout modern culture, we see this same general shift to higher cognitive issues and skills. A consensus has rightly emerged that one of the principal goals of mathematics education is mathematical literacy and confidence in mathematical modes of thinking. The purpose of this paper is primarily to discuss the role of mathematical content in achieving this goal.

At every grade level, the following four standards appear at or near the head of the list:

Standard 1: Mathematics as Problem Solving.
Standard 2: Mathematics as Communication.
Standard 3: Mathematics as Reasoning.
Standard 4: Mathematical Connections.

We will call these *the First Four*. No doubt they appear at each grade level because they address directly what it means to *do* mathematics. The items that follow the First Four in the various Standards lists by grade level describe, for the most part, new approaches to old content with a minimal amount of new content. We argue that more needs to be done in terms of content — particularly in grades K–4.

1991 *Mathematics Subject Classification.* Primary 00A05, 00A35, 05C15.

Research supported by the U.S. Dept. of Energy, Los Alamos National Laboratory, MegaMath Project.

Research supported by the National Science and Engineering Research Council of Canada, and by the MegaMath Project of the Los Alamos U.S. National Laboratories.

For example, what about the following content possibilities (and we will argue, necessities) in the early grades: proof, infinity, variable, logic, induction, recursion and computational complexity?

And what about the following mathematical experiences?

- The experience of a surprising mathematical truth that contradicts intuition.
- The experience of understanding a simply-stated mathematical problem with no known solution.
- The experience of logical paradox.
- The experience of wrestling with the idea of a limit.
- The experience of mathematical exploration.

In the following we will describe ways in which these and other mathematical experiences and concepts that are typically considered advanced, can engage children in grades K-4 (ages 5–9), and why they should be introduced to this age group. Many of the topics by which these ideas and experiences are conveyed are relatively new as mathematics — many are part of computer science and its discrete mathematical roots.

The main points of our argument are summarized as follows:

- The First Four curriculum standards cannot be meaningfully implemented except in the context of a significantly enriched mathematics content agenda. They are not independent of content issues.
- There is natural compatibility between the First Four curriculum standards and the goals and methods of effective mathematics popularization.
- Literature and literacy provide useful metaphors for understanding many of the important issues in mathematics education.
- Discrete mathematics and computer science have an important role to play as sources of content enrichment for the elementary grades.

We hypothesize that all of the problems with mathematics education at all levels are abundantly represented in the first five years of school, and for that reason, draw our comments from our experiences with children in classrooms at these grade levels.

By the end of even the first year many, if not most, of the children we have met have already formed a dismal impression of mathematics, considering it a boring and intimidating discipline devoted primarily to speedy and accurate manipulations of numbers. By the end of the fourth year they have typically had an abundance of the traditional experiences of school mathematics: the meaningless seat work, the rote memorization of procedures, the stilted word problems and pointless obscure vocabulary, the anxiety of parents and teachers, and the testing that separates the winners from the losers. They have already experienced "mathematics as crowd control"[1] where the reward for mastering a drill sheet is — another drill sheet.

[1] See *the paranoid theory of mathematics education* in [**5**].

One of the ironies in this age group is that their playground culture is rich with combinatorial games, with riddles and word-play, with informal discussions of infinity, space-time, and the Liar Paradox. They are busy with topological and dynamic amusements such as tether ball, jump rope, cat's cradle and braids. These activities and puzzlements are in many ways closer to the spirit of mathematics as it known by mathematicians than what is presented as "mathematics" in the classroom.

2. The First Four: What do they really mean?

Mathematicians understand that making connections, communication, problem-solving and reasoning are at the heart of their discipline. As measurable skills, however, these are nebulous — far more difficult to teach and track than the ability to count, compare, or compute. One of the things that is commonly happening in practice as school districts and curriculum developers wrestle with the Standards is that the First Four are in many cases being split off and treated differently from the rest. In particular, they are in many cases being interpreted merely as process standards having no particular connection to any kind of mathematical content.

One well-meaning principal of an elementary school in British Columbia, which has been used as a model for curriculum reform, put it this way. "These four standards are really important — we handle them elsewhere in the curriculum!" By this was meant that communication skills are practiced in creative writing, problem-solving skills are practiced in designing art projects, etc.

Our central argument here is that the First Four cannot be realized without an expanded agenda of interesting mathematics and mathematical experiences to reason, communicate and problem-solve about. We simply cannot realize these standards by means of classroom discussions about our ideas for doing long division or naming triangles. If the current impoverished K-4 mathematics agenda is not capable of supporting any meaningful realization of the First Four, we must look to all of mathematics for expanding the range of ideas that are brought to the K–4 classroom.

We believe the K–4 content curriculum should include anything and everything suitable for a Mathematical Sciences Museum and thus the project of realizing the First Four for K-4 is naturally allied with the vital project of mathematics popularization for all ages. In these first years, an enduring sense should be formed of what mathematical science is about and how it feels to participate in this adventure of the human spirit, central as it is to all of modern science and technology.[2]

In fact, science popularization is inherently concerned with the K-4 audience because science museum exhibits are, more or less, designed for the

[2]Notice that if in this sentence the words "mathematical science" are replaced by "print literacy," then the result is a common-place. Not only do children routinely master the decoding of print in K-4, but they engage exciting poetry and stories (including their own) and form a basic sense of why one would want to read and write.

4th grade audience in order to be just about right for children, grandparents and everyone else in between.

For scientists and mathematicians, the K-4 audience is a delight. They are full of vibrant curiosity and enthusiasm. They are endowed with natural tendencies to abstract representation and the play of ideas. (Think of the odd bits of wood they have asked you to regard as a space laser.) In the next section, we describe some of the "advanced" mathematical ideas that can be engaged by this age group.

3. Some Content Examples for the First Four

The purpose of this section is to describe some examples of how advanced mathematical ideas can be engaged by children in grades K-4. One of the most fundamental appreciations that one can have of mathematics is a sense of the power, the sheer variety and the marvelous interconnections of mathematical models of things in the world. We may begin with the K–4 audience by exhibiting and engaging a rich collection of examples of mathematical models.

3.1. Example 1: Map and Graph Coloring. The basic **Map Coloring Problem** is that of trying to discover the minimum number of colors needed to properly color a map. A map is properly colored if no two countries sharing a border have the same color. (See Figure 1.)

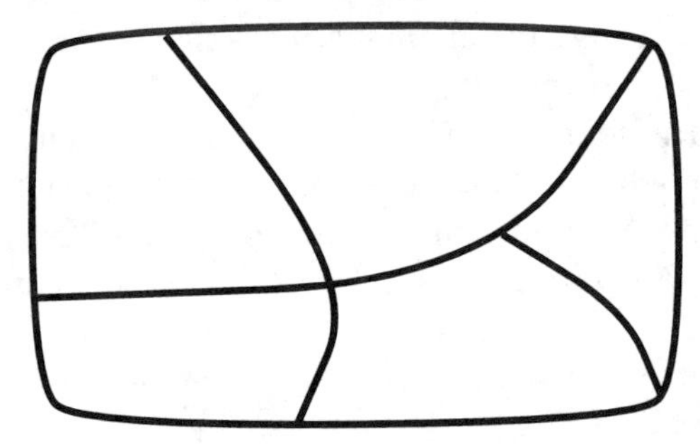

FIGURE 1. A map which will require 3 colors to be colored correctly.

This problem (like any of hundreds of such combinatorial optimization problems) can be presented in a classroom setting by doing the following.[3]

1. Beforehand, make up and photocopy 3 or 4 maps of varying sizes, such as with 5, 10 and 20 regions. (Do not "solve" them.)
2. In class, discuss how maps are ordinarily colored, how regions on the map that share a boundary are colored different colors so that they are not easily confused. Discuss also, how it would make sense commercially to color maps with as few colors as possible, due to the cost of ink, the complexity of printing many colors, etc.

[3]Detailed instructions for classroom use of this and other problems along with sample handouts, ideas for discussion, and explanations of their relationship to the whole of mathematics can be found in [**3**].

3. Pass out the maps and invite students to find ways to color them with as few colors as possible, working individually or in groups, as they prefer.
4. Be an attentive listener and facilitator. Encourage the children to describe their ideas for solving this problem and to explain what they are doing to each other.
5. Afterwards, have the children write about their ideas, draw maps of their own to color, and/or share the activity with a different group of children.

The coloring problem is one of the great gems of discrete mathematical modeling. Some of its applications include: the assignment of non-interfering frequencies to radio stations, the timing of traffic lights, the scheduling of meetings and machines, and the scheduling of garbage truck routes. Coloring is also an activity to which the K-4 audience is already natively inclined. There is some satisfaction in connecting this ordinary childhood artistic activity to the deep and important mathematics that concerns it. Surprising to most people is the fact that coloring problems (of various kinds) remain a subject of vigorous mathematical investigation. They are important to all kinds of discrete mathematical modeling, including, for example, the analysis of DNA sequences.

3.2. Coloring: What's In It? In K–4 classrooms, where children are puzzling over finding the minimum number of colors for various maps, all kinds of interesting and deep mathematical issues naturally arise. We are concerned that if these ideas are left off of the content agenda, teachers will lack an adequate reference framework to appreciate, stimulate and support the problem-solving strategies that the children will invent. The following is an unsystematic inventory of various fragments of our classroom experiences with the coloring problem, pointing to various "advanced" mathematics content that emerged.

3.2.1. *"Two is not enough!"* What typically happens with a hypothetical map M, like the one above, with chromatic number 3 (that is, where 3 colors are required) is that someone first colors it with 7 colors, and then someone colors M with 5 colors, ... the number gradually improves. But eventually we are left wondering (publicly, as we celebrate this progress) whether we can do it with 2 colors. Inevitably, some child will figure out that (and explain energetically why) two is not enough for M, typically by finding three regions each of which borders the other two. The moment when a child gives that excited shout needs to be appreciated as a "teachable moment" for the fundamental topic of **mathematical proof**. A teacher not equipped with the idea of the importance of mathematical proof, and expecting to encounter and develop this concept, is not equipped to fully appreciate and empower the problem-solving going on.

3.2.2. *"I did it with two!"* Consider the same scenario with a different map M' having chromatic number 2 (Figure 2.). We have the same gradual improvements. First there is a solution with 5 colors, then one with 4 or 3 colors. Finally someone shouts that they have done it with 2 colors.

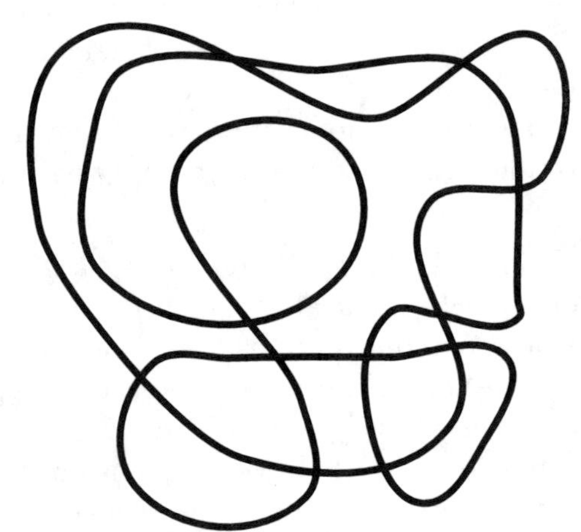

FIGURE 2. Maps drawn as closed curves can always be colored with two colors.

If we interview the children who have found a 2-coloring for M', asking about their method and their ideas, we usually find that they have hit upon the following systematic approach. They first color a region, with, say, red. And then choose another color, say, blue for those neighboring regions that are forced to be blue because they share a border with the first region. And then (conservatively) they proceed by coloring red those further regions that share a border with the newly-colored blue ones, and thus are forced to be red, and so on.

This is a very interesting strategy! In fact, it is an **algorithm** that can serve to determine for any map whether two colors is enough, and do so with great **algorithmic efficiency**. It can be compared to a different (but also interesting and respectable) **greedy algorithm** that many other students will discover: pick up a crayon and (at random) use it to color regions until it can't be used anymore, then pick up a new crayon and repeat the process until all regions are colored. A teacher (and a curriculum agenda) not equipped with the idea of an **algorithm** is not equipped to appreciate the problem-solving going on here, the ideas that are emerging, and their substantial ultimate significance in mathematics education.

Here is also an opportunity to point out to the children one of the most important **unsolved problems** in all of mathematics and computer science. This is that while there is a simple and efficient algorithm to determine whether two colors is enough (sketched above), no one knows whether there is a fast way to find out whether 3 colors are enough. It seems a good thing not only to share with children significant problem-solving situations that have no single right answer, but also situations where no one, not even the adults, presently knows the answer.

3.2.3. *"I want to try to do it with 3!"* When the children are working with crayons, the natural thing to do at first is to dive in and color as well as you can. However, once students want to experiment so as to truly minimize their colorings, a certain weakness in using crayons becomes apparent — it is impossible to back up! Once a region has been colored red, it's messy (if not impossible) to try to change it to green. On their own, or with minimal encouragement, some children will switch to using colored tokens to mark the colors that they assign to the regions. This provides a far more powerful means to try to achieve an optimal coloring.

In one classroom, a teacher observing the children moving the colored markers around on the maps remarked, "That's a higher level of abstraction." The teacher obviously (and rightly) felt the need to appreciate this more powerful problem-solving approach in some way. Rather than rely on psychological concepts for this, we can appreciate what's going on in a straightforward mathematical way: the regions are now functioning (manipulatively) as **variables** that can be conveniently instantiated to a color value by a marker. This is precisely why this is such a powerful problem-solving strategy, and a good demonstration of why the concept of a variable is so fundamental in mathematics. If **variable** is not on the content agenda, then teachers are left to ad hoc psychological appreciations, with no sound connection to the enduring and important mathematical ideas that are emerging in the children's activity.

3.2.4. *"These maps can always be done with 2!"* If you place your pen on a piece of paper and draw any sort of intersecting continuous curve, eventually returning your pen to its starting point without lifting it from the paper, you will have drawn a map that is 2-colorable! (See Figure 2.) Try it out. Colored, it looks like the kind of "psychedelic checkerboard" that Salvador Dali might have preferred. It is generally regarded as somewhat surprising that these kinds of maps are always 2-colorable. How can we be convinced that this is true?

One way to explore being convinced is to make a loop of string. Imagine that it is black string, imitating the black ink of a pen that would draw such a map. Surely you will agree that you could lay the loop of string right on top of the curve that you drew. If it were just the string, lying like that on the white paper, we might think, "What a mess!" We might decide to gradually, very slowly, one step at a time, move the loops apart. We might in this way obtain a very boring situation: the string is now just lying in a loop that does not intersect itself. If this were a map, it would just be one island and the sea surrounding it — of course we have no trouble 2-coloring this!

Now let's slowly go backwards, gradually putting things back the way they were. At each step of the way we will notice that we make one of a few kinds of moves, and in each case, the property of the map being 2-colorable is preserved! (See Figure 3.) Now we see why all these maps are 2-colorable: they are all just mixed-up forms of the single Two-Colorable Island (and the

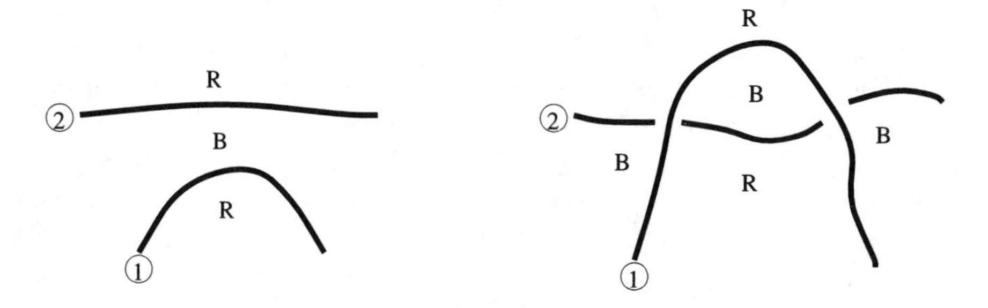

FIGURE 3. If the map contaiing the left diagram is 2-colorable, then the map containing the right diagram, where string 1 is pulled over string 2, is also 2-colorable.

mixing up doesn't hurt anything). There is a lot of fun and contemplation in this for young children. It is really the essence of a proof by **induction** of this surprising theorem. (No matter how old you are, you really should try this out with a piece of string and two kinds of colored markers, and see that induction is, after all, really quite suitable for 7 year-olds.)

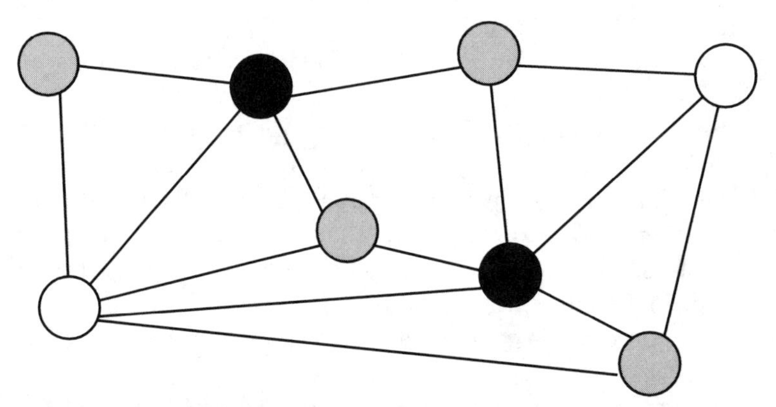

FIGURE 4. This graph is colored correctly with 3 colors.

3.2.5. *"This one can be done with 3! See if you can find how!"* A problem closely related to map coloring is graph coloring. A graph is a network of dots (called vertices) connected by lines (called edges). The vertices of a graph are properly colored when no two vertices joined by an edge receive the same color. (See Figure 4.)

Several children in one of our second-grade encounters spontaneously decided that the 3-coloring puzzles for graphs that were passed out were so much fun, they would have **Graph 3-Coloring** as an activity at their birthday parties!

Here is a little mystery for further exploration. How was it possible to announce to them, "It is easy to draw a graph that can be colored with 3 colors so that you know exactly how to color it, but other people will have a hard time figuring out how." The puzzle is solved with a very entertaining activity: begin by making a polka-dot pattern of 3 kinds of colored dots. On top of this lay a fresh piece of paper, and tracing through, make a circle around each dot. Now add edges between these circles, but only between circles that surround differently colored dots! In this way you have created a graph for which you know a secret 3-coloring, but it might be pretty hard for someone else to find one. This is a kind of combinatorial **one-way function**, a topic of profound importance in modern mathematical cryptography (for further explorations beginning from this point and involving **polynomials** as encryptions of public-key messages see [**6**]).

4. Other topics of interest for the early grades

We next describe (a bit more telegraphically) a few more topics that have proved fruitful in exploring the first four standards in the early grades.

4.1. Minimum Weight Spanning Trees. We have come to call this the "Muddy City Problem". (See Figure 5.) The scenario is a city with unpaved roads in which transportation becomes impossible when it rains. The vertices of the graph represent houses and the edges are roads. The labels on the edges of the graph are the costs of paving each segment of road. The question becomes: What is the least expensive way to pave roads so that everyone can get to everyone else's house when it rains (even if it is by a circuitous route)?

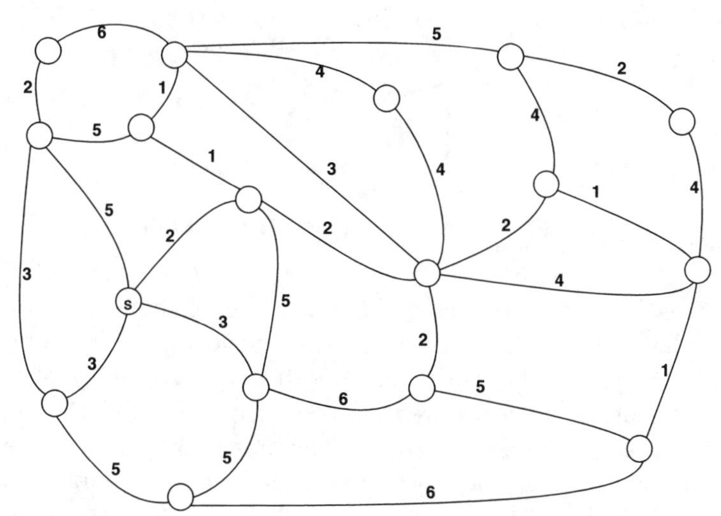

FIGURE 5. A map that can be used for the Muddy City Problem.

In the attempts to find an optimal solution for a given weighted graph, a typical classroom experience invokes a hurricane of arithmetic as children work to create ever better solutions, and to match the best that have been found so far. The are many interesting nontrivial ideas and observations that children will typically make and be prepared to explain and argue: such as the fact that an optimal solution has no cycles. Here again we have the vital content of **mathematical proof** arising. The fact that there is a (surprising and elegant) fast algorithm for the minimum weight spanning tree problem raises the issues of **algorithm** and of **algorithmic efficiency**.

Someone will notice that all the best solutions for a given graph (optimal solutions are generally not unique) involve the same number of edges. We have here again an opportunity for a simple, visually presented argument by **induction** to explain this.

4.2. Knot Theory. The Canadian Navy donated to us a number of large ropes, and we have had wonderful experiences presenting some of the rudiments of knot theory in elementary classrooms. This is an excellent topic for mathematics popularization for several reasons. First of all, it is first-rate mathematics that has recently moved center-stage in the research world in a very exciting way. Secondly, everyone uses knots, and almost no one is aware that they are an object of mathematical investigation. To share the fact that there is a *mathematics* of knots is a powerful illustration of the richness of mathematical science. Finally, knot theory is enormously open to manipulative presentation.

FIGURE 6. A left-handed trefoil knot.

One can ask about mirror-image knots. Is the left-hand trefoil the same as the right hand one? On some occasions we have brought along a large portable mirror to show that the one is indeed the mirror image of the other. Knots (once oriented) support a well-defined notion of (abelian) "multiplication" (having even a prime factorization theorem) that is open to engaging manipulative exploration. Here we have **symmetry** and **mathematical operations**.

4.3. Other Examples. There are many more such mathematical topics supporting rich opportunities to realize the First Four. The main point is that in really engaging in these or any other opportunities for mathematical problem-solving and communication worthy of the name, "advanced" mathematical ideas will naturally and inevitably arise and should be both expected and deepened as much as possible.

5. Parallels Between Mathematics and Literature Teaching

We have found the analogies between print literacy and mathematical literacy to be both strong and productive for generating ideas and methods for improving education in mathematics during the critical, formative elementary school years. Lacking a scapegoat and deterrent to risk-taking as formidable as The Debacle of the New Math, elementary school language arts educators have benefited from 25 years of experimentation and critical evaluation leading to teaching methodologies and classroom structures aimed less at making the student a skilled automaton with the structure of written language, and more focused on the development of the student as a literate person.

It is no less difficult to define what a literate person is than it is to describe what it means to do mathematics. For example, it is not sufficient to say that someone is literate because they know a lot of words, read fast, spell and punctuate Standard English accurately, speak several languages, or can pass tests about all of the books on a certain prescribed list. Yet so-called literate people can do many or all of these things. Likewise, in mathematics, developing a straight-forward definition of literacy is no less complicated or controversial.

Mathematics and literature have much in common. The construction, examination and communication of ideas is central to both disciplines. In each discipline these activities are carried out within forms. These forms are often misconstrued to be the discipline itself. Each discipline is so vast, with such a rich and long tradition, that no individual can claim to grasp it in its entirety, yet any aspect is accessible to the dedicated participant. In both mathematics and literature, the participants in the discipline form a community in which innovations and content are shared and examined. The most renowned and influential participants in the community achieve their position after a long period of initiation and experience, much of which, in the early years, occurs in schools.

Elementary school language arts teachers ask the following questions when they plan and evaluate their lessons [**1**]:

- How can we prepare students to become creative participants in a community where the formulation and communication of ideas is fundamental?
- How can we, with materials and tools that are on hand now, teach them to appreciate the vast and ever-changing quantity of material they will encounter in their lifetimes, to assimilate new things, and develop taste as they mature intellectually?
- What must we do so that all students acquire the complex, interrelated skills necessary to do all this?

Similar questions should be fundamental to mathematics teaching in the formative years.

The following insights borrowed from language arts teachers' examination of their goals and methodologies over the last 3 decades are most useful for considering the direction that change in mathematics education should take [**4**].

1. Children benefit from exposure to a rich variety of content without regard for hierarchical sequencing of material. Statements of "developmental appropriateness" must be taken in a large context.
2. Students are drawn forward by exposure to material that they can understand but which is beyond their capacities to produce.
3. Although skills matter, experience in the discipline cannot be secondary to mastery of them; teachers must find ways to monitor and nurture skill development within the context of meaningful and stimulating (self-selected) projects.
4. Students must be steered towards mature and independent self-selection of content materials and individual/small group projects which they undertake.
5. Peer communication about their ideas is not only critical, but inevitable. Teachers must learn to exploit, not suppress the classroom culture.
6. Students must be given large blocks of time to read, think, talk to one another, share, argue, and write down their ideas. The classroom should be a microcosm of the community into which the students are being initiated.
7. The teacher is neither spectator nor ambassador from the community into which the students are being initiated, but a participant and a practitioner.

These questions and insights are less about language teaching than about teaching in general. They are representative of a largely grass-roots movement in language teaching reform which came to be termed **Whole Language**.

The Whole Language connection [**2**] has proved to us to be an enormously useful handle in speaking to experienced elementary school teachers about mathematics education reform.[4] Many elementary school teachers have spent many years wrestling with these issues. What they typically sorely lack is any sense that mathematics *has* a literature, that it supports any kind of thinking or activity remotely resembling literacy. We can help teachers and parents appreciate and understand reform in mathematics education by appealing to their understanding and experience with print literacy.

In all kinds of contexts, the literacy connection has proved useful. Here are a few examples of what we call "standard conversations" with parents

---

[4]Current controversies regarding forms of Whole Language have done nothing except strengthen this connection, as they bring out an important and inevitable tension between skills development, and issues of motivation, participation and meaningful context. These controversies serve as a useful warning about trivializing literacy education of any kind.

and teachers, and how they can be answered by looking through the lens of literature.

"Why does my child need to know about coloring or knot theory?"

Does your child need to read *Charlotte's Web* or *Huckleberry Finn*? Does your child need to know about dinosaurs or outer space? (It is sad that mathematics is so universally associated with such a miserliness of spirit.)

"It's important to teach arithmetic. So now you are saying that it is important to teach coloring as well?"

It's important to teach spelling, but it's also important to read and enjoy books. What particular books these are is not so important, but they should be rich and interesting stories. It is much the same with mathematics.

Which topics are best for each grade level? Good mathematical topics, like good stories, are appropriate at all grade levels. A story like St. Exeupery's *The Little Prince* can be enjoyed by very young children, but is a source of profound concepts for more mature readers. Similarly, a problem like map coloring can be explored by children who are not yet able to read, yet it is a source of complex and interesting conjectures and questions for older students.

6. What Is To Be Done?

The elementary school principal who said, "These four Standards are really important — we handle them elsewhere in the curriculum!" was not (as one might first suspect) making an easy mistake. At this particular school, meaningful contexts for learning and the development of communications skills are highly valued. An intelligent and demanding (and yes, it includes phonics) Whole Language approach to print literacy is a deeply-rooted practice at this "charter" school which has served for years as an important model for curriculum innovation in British Columbia. The traditional mathematics content agenda at the elementary grade levels, however, simply does not provide adequate opportunities to realize the obviously important First Four *in mathematics*, so in order to address them, teachers must turn to opportunities elsewhere in the curriculum.

There are several things that we in the Discrete Mathematics and Computer Science communities need to do:

- We need to pay far more attention to the needs and opportunities in the early and formative years of schooling.
- We need to get the message out to the elementary schools that integrating the intellectual core of computer science (and its roots in discrete mathematics) into the curriculum is of *far* greater importance than worshiping in the expensive Cargo Cult of computers-in-the-classroom. (For further discussion of this point see [**6**].)
- We need to make the connection between mathematics education and mathematics popularization. In the areas of discrete mathematics and

computer science we have enormous resources of important, accessible mathematics for this purpose.

- We need to establish connections between mathematics education and literacy education, especially at the K-4 level. Such connections are likely to significantly strengthen *both* educational agendas. The communication of mathematical thinking and argument, and the formulation of mathematical models and conjectures constitute challenging and important kinds of writing tasks.
- We need to encourage the development of whimsical, lengthy, content-rich children's mathematical literature. We need story problems that are real stories, not "Farmer Brown wants to build a rectangular fence... ." For example, we need 30-page stories with characters, pictures, maps, and dialogue that incorporate interactive problem-solving. We need to be training mathematics/cross-disciplinary students (perhaps educational computer games designers) at the universities to *create* this kind of literature.
- We need to support teacher professionalism, and serve as (energy-efficient) catalysts for change, by organizing and involving mathematical science undergraduate and graduate students in outreach from the universities (perhaps as a component of service education programs). We need to similarly organize summer in-service institutes for teachers, and mathematical science summer camps for kids.
- We need to establish two-way communication with undergraduate departments of education. We cannot come across as the arrogant experts of the "New Math" era. We must be prepared to enlighten ourselves about the problems and goals of elementary teacher educators and the elementary school classroom itself. We must seek out and work to establish productive relationships with teacher educators and create a common ground where we can truly communicate the relevance of our discipline and our enthusiasm for it.

References

[1] Lucy McCormick Calkins, *The Art of Teaching Writing*, Heinemann, Portsmouth NH, 1994.

[2] Nancy Casey, "The Whole Language Connection," *Connections*, Carl Swenson, ed., Washington State Mathematics Council (1991), 1–13.

[3] Nancy Casey and Michael Fellows, *This is MEGA-Mathematics! Stories and Activities for Mathematical Thinking, Problem-Solving and Communication*, Los Alamos National Laboratories, 1993.

[4] Kenneth Goodman, *What's Whole in Whole Language?*, Heinemann, Portsmouth NH, 1986.

[5] Michael R. Fellows, "Computer Science in the Elementary Schools," *Mathematicians and Education Reform 1990–1991*, N. Fisher, H. Keynes and P. Wagreich, eds., Conference Board of the Mathematical Sciences, Issues in Mathematics Education 3 (1993), 143–163.

[6] Michael R. Fellows and Neal Koblitz, "Kid Krypto," *Proceedings of CRYPTO '92*, Springer-Verlag, *Lecture Notes in Computer Science*, vol. 740 (1993), 371–389.

[7] Alice Miller, *For Your Own Good: Hidden Cruelty in Child-Rearing and the Roots of Violence*, Farrar Straus, New York, 1983.
[8] National Council of Teachers of Mathematics, *Curriculum and Evaluation Standards for School Mathematics*, NCTM, Reston VA, 1989.

Department of Computer Science, University of Idaho, Moscow ID 83843
E-mail address: casey931@cs.uidaho.edu, http://www.cs.uidaho.edu/~casey931

Department of Computer Science, University of Victoria, Victoria, British Columbia, Canada V8W 3P6. Phone: 604-721-7299.
E-mail address: mfellows@csr.uvic.ca

DIMACS Series in Discrete Mathematics
and Theoretical Computer Science
Volume **36**, 1997

Discrete Mathematics: A Vehicle for Problem Solving and Excitement

Margaret B. Cozzens

Mathematics has always had the luxury and the responsibility of being recognized as a fundamental component of all school learning, one of the "three R's". However the global considerations of economic competitiveness and personal and societal decision-making place mathematics education in a premier position as we move into the 21st century. The ideal mathematics classroom of the 21st century is one where students learn to value mathematics, become confident in their own mathematical abilities, become problem solvers, and learn to communicate and reason mathematically. It is one where all students have the thrill of success through exploration and hands-on experimentation. Opportunities for learning which capitalize on curiosity, eagerness, flexibility, various levels of maturity, uniqueness of current knowledge, and comfort level in mathematics and technology must exist for all students.

It has now been seven years since the National Council of Teachers of Mathematics (NCTM) released the *Curriculum and Evaluation Standards for School Mathematics* [**1**], which describe what students should know and be able to do at various ages (or grade levels) in mathematics. The NCTM *Standards* are not prescriptions for teachers and schools to follow, but they create a coherent vision of what it means to be mathematically literate. The goals of these *Standards* are goals for students; the implementation depends on teachers, schools, and materials developers. These *Standards* are guides for the revision of school mathematics curriculum frameworks that have students explore, reflect, and discuss. These *Standards* provide guides for the development of materials that provide hands-on experiences, materials that are open-ended and flexible, but at the same time, materials that provide structure and guidance for student learning.

Where does discrete mathematics fit into the school mathematics curriculum framework and at what grade levels? NCTM Standard 12 calls for a mathematics curriculum framework in grades 9-12 that includes topics

1991 *Mathematics Subject Classification.* Primary 00A05, 00A35.

from discrete mathematics. The focus of this standard is directed at potential applications in computer technology, but the emphasis goes beyond the needs of the information industry to all areas of investigation where the domain of discourse is a finite or a countable set of objects. When viewed in this light, discrete mathematics is not independent of algebra and geometry, but it becomes a powerful representation tool that pervades all of the K-12 mathematics standards. Children of all ages are much more familiar with discrete sets than of those whose cardinality is uncountable. Children's earliest experiences with counting are little more than matching the elements of a set with a finite subset of the natural numbers, even though they know none of this terminology. Their day-to-day life is enveloped in "discrete" numbers and applications. It is through these doorways of discrete mathematics, and the opportunities that lie beyond, that they learn to be problem solvers and mathematical reasoners. It is difficult to overestimate the importance of "engagement" as a factor in learning. Activities in the area of discrete mathematics provide this "engagement" opportunity at all educational levels.

Long before coming to the Elementary, Secondary, and Informal Education Division (ESIE) at the National Science Foundation (NSF) in the area of K-12 mathematics, science, and technology education, I believed that problems in discrete mathematics were engaging for both teachers and students and accessible to a wide range of students, in particular those who may have had difficulty with algebraic manipulations. Students can not only work on problems in discrete mathematics, but they can also pose and solve their own problems that are natural extensions of ones provided by the teacher or the textbook. The nature of proofs, and the need to prove one's results, arise naturally in these problems.

Through my daily interaction with education programs as Division Director of ESIE,[1] I am now even more convinced that discrete mathematics opens the door for all students to discover the excitement and versatility of mathematics, the lure of solving problems both applied and theoretical, and the pleasure of doing things rigorously. I have seen students in classrooms throughout the country, in large urban areas, small rural schools, and high-level magnet schools, all working on the same problems, discussing them with their classmates and, in many cases, convincing their teachers that there is a "better" way to solve the problem; these students don't want to leave mathematics classes to go to other courses because they want to finish their work. In some cases, as in Philadelphia, among high school students who have been enrolled in mathematics classes with a heavy dosage of discrete mathematics, overall average student monthly attendance in school has improved by as much as 15%, and achievement scores for these students in English and Science have improved significantly, as well as in mathematics. A key component of NSF's portfolio of activities to support education reform

[1]The opinions expressed in this paper are those of the author, and not necessarily those of the National Science Foundation.

in the classroom has been the development of new mathematics instructional materials. All of the fifteen new mathematics curriculum projects funded by NSF, completed or nearing completion, have heavy doses of discrete mathematics; these span the education spectrum: elementary, middle, and high school.

For example, the problem of finding the shortest route through key cities between Boston and Miami can be posed as early as fourth grade and used as an activity to teach and reinforce scale measurement. In high school, students can develop fast algorithms to compute the shortest distance, but these algorithms still require a knowledge of the intermediate city distances. I have found that children of all ages like to compute shortest routes from real maps, rather than ones made up by the teacher or a textbook. Any road atlas provides plenty of choices. For example, using a map of the Eastern United States, ask students to identify the key cities between Boston and Miami, rather than telling them the ones to use. A discussion about which ones to choose is well worth the time. Older students can use more cities, younger students can handle only a few the first time. A simple example might include Boston, New York, Washington, Roanoke, Columbia, Talahassee, Jacksonville, and Miami. These cities are sufficient to provide choices, but are not so many as to be time consuming.

Once the students are able to work a few examples computing shortest routes, provide the map in the road atlas that gives times as well as distances and have the students compute the shortest time trip and compare the two answers. I have successfully used this example in all grades 4-12; many of the new curriculum materials have such examples.

One of the most intriguing aspects of specific discrete mathematics problems such as the shortest path problem are the many applications of the same model. For example, Mrs. Smith wants to determine the optimal time to trade in her Ford Taurus. She has consulted with the Ford dealer and determined cost and trade-in projections as indicated in the following table:

Taurus LE

year	price new	trade-in price				
		1996	1997	1998	1999	2000
1995	$22,000	$19,000	$17,500	$16,000	$14,000	$12,000
1996	$24,000		$21,000	$18,500	$16,000	$14,000
1997	$27,000			$24,000	$21,500	$19,000
1998	$29,000				$26,000	$23,500
1999	$32,000					$29,000
2000	$35,000					

TABLE 1

The dealer assumes that the average maintenance costs are $500 in the second year, $1000 in the third year, and $2000 in the fourth and fifth years of ownership. In the year 2000, Mrs. Smith assumes she will no longer need a car as she will be reassigned in Europe. The cost of keeping the car can then be computed by the following formula:

cost of keeping car between two years = cost new − trade-in price at time of sale + maintenance costs

Mrs. Smith's problem can be solved by translating the data into a graph whose vertices are the years from 1995 to 1999 and the edges are directed from lower to higher year, weighted with the cost of keeping the car during that year. The shortest route between 1995 and 2000 in the graph corresponds to the most economical times to trade in cars for total least cost; the solution is to sell the car in 1998, buy a new one, and sell that in 2000 (see Figure 1).

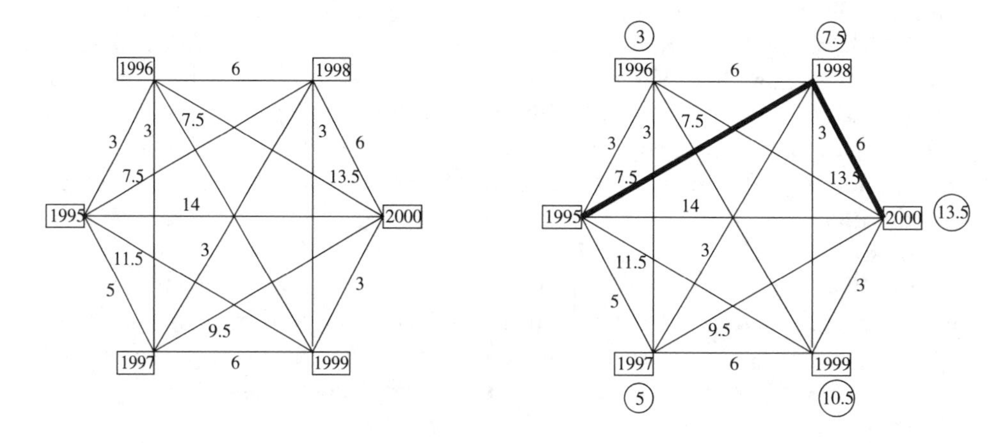

FIGURE 1.

There are a number of other problems that can be modeled using graphs and solved using shortest route techniques. The shortest route problem is an example of a problem that, until ten years ago, appeared in undergraduate or graduate courses in mathematics, operations research, or computer science, but now appears in middle school and high school mathematics curriculum materials. Students across the country are inventing new applications every day, including scheduling the activities in their own classrooms.

Verifying that a proposed solution to a problem is indeed a solution is an activity that challenges mathematicians on a daily basis. It is only recently that this activity has been taught in elementary and secondary classrooms. We traditionally have given the answer to a question in the very first paragraph of a unit by stating a theorem or giving a formula and then merely letting the student fill in numbers or apply the result to other situations, usually contrived to work easily. The correctness of Dijkstra's algorithm, the algorithm used to find shortest distances in a graph, is proved using mathematical induction, a technique accessible to high school students. For small examples, students can verify the correctness of Dijkstra's algorithm by enumerating all possible solutions, but enumerating all solutions for a few problems convinces students quickly that they don't want to do that for all problems.

Consider another example of a problem in discrete mathematics that can be posed at various levels, can be enjoyed by students of all ages, and has extensions that pose interesting mathematical problems for students in grade school through graduate school. This example is reprinted from the new *fourth grade* mathematics curriculum materials developed by TERC, and published by Dale Seymour Publications [**2**].

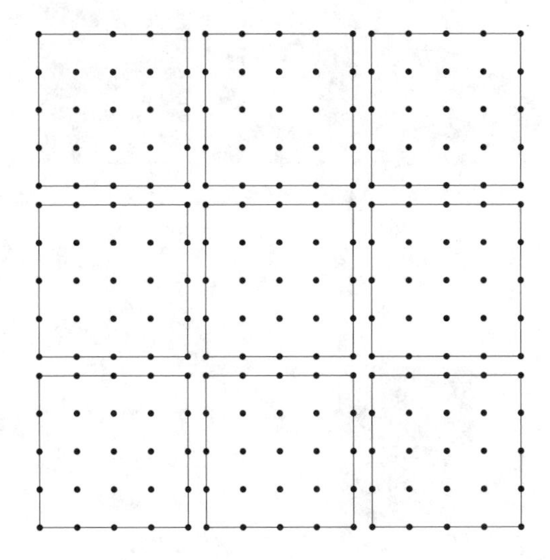

Divide each small square into fourths in a different way. Use your favorite fourths, or make up new ways of dividing into fourths. Color each square's fourths right after you divide it into four parts. Use the same four colors.

FIGURE 2. **Squares for a Quilt of Fourths**

Figure 3 gives an example of a fourths quilt made by one fourth grade student. Students may take the quilts home to finish them. One class actually took the best fourths patch from each student's quilt and sewed an actual quilt with 25 squares which was hung at the entrance of the school. Boys enjoy this exercise as much as girls.

Students are asked to prove that the squares are actually divided into fourths and, during the course of the year, learn how to verify that indeed the four pieces of the subdivision are "equal" to one another. Samples of student verifications appear in Figures 4 and 5.

Even as early as fourth grade, students learn the meaning of proof. Their level of sophistication about the nature of proof changes over time, but the notion of proof is demystified when students have the opportunity to understand methods of proof, from the early grades through high school and beyond, in the context of already familiar discrete topics.

FIGURE 3.

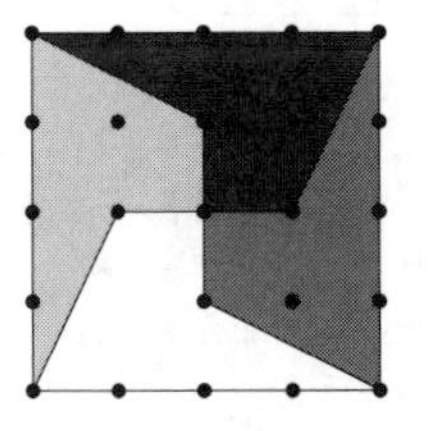

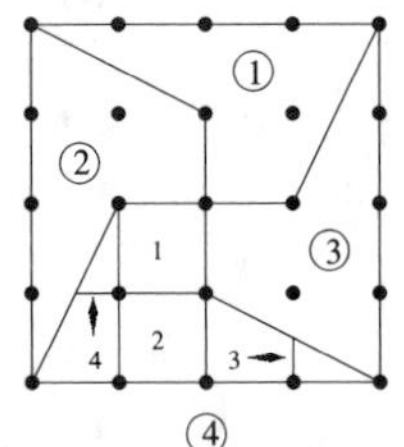

FIGURE 4. An example of a fourth grader's proof that one of his squares is divided into fourths: *This works because they are all the same, so each occupies the same amount of space and there are four parts.*

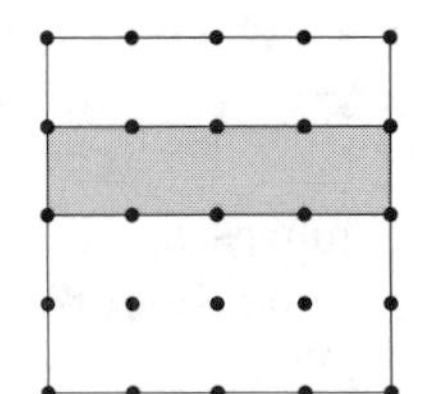

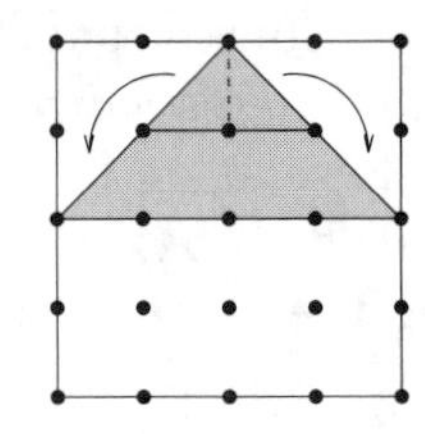

FIGURE 5. An example of a fourth grader's proof that the shapes in A and B are the same: *You can see these shapes are the same [in A and B]. You just cut off the little triangles on top of the traingle [B] and put them to fix the squares in the line.*

The activity of designing a "fourths quilt" is appealing and instructional for students at all levels. Even though the activity of designing a quilt may not be a traditional application of discrete mathematics, it has proved very engaging for students, male and female. In classrooms from fourth grade through high school, students enjoy constructing and coloring their examples. I even tried the quilt exercise at a Christmas party for the ESIE Division support and program staff. Everyone enjoyed the fun and vied for prizes, and many were so captivated with the potential combinatorial problems that they came up with interesting extensions of the original problem in the days and weeks that followed. For example, the staff came up with the following: (Some are still trying to prove their answers to some of them.)

1. How many distinctly different fourths squares, up to recoloring, are possible?
   (a) — if all edges must consist of straight line segments (SLS)?
   (b) — if all edges must be SLS and pass through points on the grid?
   (c) — if all edges must be SLS and all resulting shapes convex?
   (d) — if all edges must be SLS and all resulting shapes congruent?
   (e) — if all edges must be SLS and the resulting quilt 2-colorable?
   (f) — if all edges must be SLS and all pieces quadrilateral?
2. An ant is placed at each corner of one square. Each ant walks towards its neighbor at the same rate. The resulting paths divide the square into four congruent areas. Find the boundaries of the regions.
3. What is the minimum number of sides of the regular polygon that fits in the square and has an area of three-fourths of the square?
4. What is the radius of each of the three concentric circles that divide the square into four equal parts?

A third example of discrete mathematics in the classroom is an applied combinatorial situation from an eighth grade mathematics curriculum (*Connected Mathematics Project* [**3**]) developed at Michigan State University. Figure 6 gives a diagram for the floor plan of the Fail-Safe Warehouse where Willie has his storage locker for the *Radical Sound* shop. The warehouse has two major sections, with larger lockers in the section on the right of the floor plan and smaller lockers on the left of the floor plan. A security guard patrols the warehouse at night. Each patrol starts at checkpoint A, follows one aisle of the warehouse to checkpoint B, and another aisle to station C. One path is shown by the dashed line. To be sure that the entire Fail-Safe Warehouse is checked, the guard takes different routes on each trip. To be sure that burglars can't predict times when the guard will come by a particular point, the guard tries to take different sets of routes each night.

A sample of questions for the students to answer are the following:

1. How many different paths are there from A to C via B?
2. How many different paths are there from C to A via B?
3. How many different round-trip paths are there from A to C and back to A? If it takes the guard 2.5 minutes to walk down one aisle of

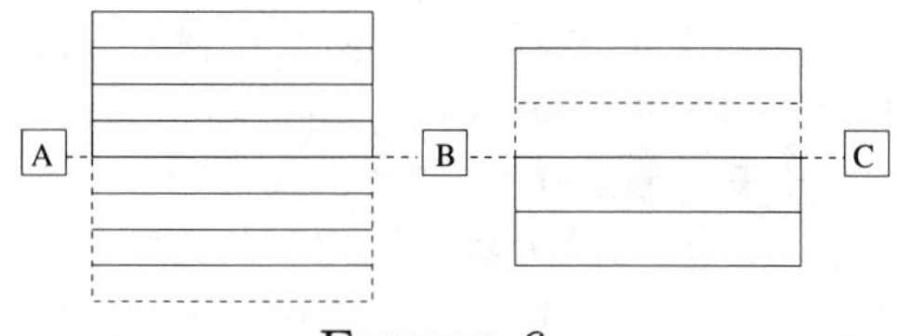

FIGURE 6.

the warehouse, how long would it take to walk all possible round-trip paths?

4. Develop a similar example for the policeman who patrols your neighborhood at night. Draw the picture first and then count the paths.

As these activities indicate, discrete mathematics in the schools is a vehicle to get students to think mathematically, become problem solvers, and become interested in mathematics. At the same time, students can work on problems that make sense to them, either because of the obvious practical application and/or because they are just plain fun and challenging.

References

[1] National Council of Teachers of Mathematics, *Curriculum and Evaluation Standards for School Mathematics*, NCTM, Reston VA, 1989.

[2] "Different Shapes, Equal Areas," *Investigations in Numbers, Data, and Space, Grade 4*, Dale Seymour Pub., Palo Alto CA, 1995.

[3] "Clever Counting," *Connected Mathematics*, Dale Seymour Pub., Palo Alto CA, 1995.

DIVISION OF ELEMENTARY, SECONDARY, AND INFORMAL EDUCATION (ESIE), NATIONAL SCIENCE FOUNDATION, 4201 WILSON BOULEVARD – ROOM 885, ARLINGTON, VA 22230

E-mail address: mcozzens@nsf.gov

DIMACS Series in Discrete Mathematics
and Theoretical Computer Science
Volume **36**, 1997

Logic and Discrete Mathematics in the Schools

Susanna S. Epp

Albert Einstein once said [**3**], "the whole of science is nothing more than a refinement of every day thinking." This quotation aptly summarizes the essential interdependence between the concrete and commonsensical and the abstract and theoretical. Developing students' abilities to shift smoothly and flexibly between these two levels, while operating effectively on each one, is arguably the central task of mathematics and science instruction.

In the language of the NCTM *Standards* [**6**], a primary goal of mathematics instruction should be to develop students' "mathematical power," which is the ability "to explore, conjecture, and reason logically, as well as ... to use a variety of mathematical methods to solve nonroutine problems." As Uri Treisman and Dick Stanley have put it [**9**], mathematics instruction should "concentrate less on the low-level use of high-level ideas and more on the high-level use of low-level ideas."

Those involved in the mathematics education reform movement have identified various specific elements that contribute to developing students' higher-level reasoning skills, such as experience working with open-ended and slightly ill-formed problems, opportunities for learning to perceive mathematical issues in a broad variety of different contexts, practice in recognizing the need for and in providing justification for mathematical assertions, increased use of cooperative learning, and employing calculators and computers to provide answers to routine parts of problems.

But success in these reformed mathematical environments requires that students—whether they know they are doing so or not—correctly apply the laws of classical logic in a variety of different settings. Specifically, in order to be able to reason effectively, students need to know

- that just because a statement of the form "if p then q" is true, one cannot conclude that "if q then p" is also true (or "if not p then not q", for that matter);
- that another way to phrase a statement of the form "if p then q" is "if not q then not p";

1991 *Mathematics Subject Classification.* Primary 00A05, 00A35, 03B65.

- what it means for statements of the following form to be false: "p and q," "p or q" and "if p then q";
- that if a property fails to hold in just one instance, then it does not hold universally;
- that to show a property holds universally, one shows that it holds in a particular but *generic* instance;
- that certain forms of argument are inherently erroneous (invalid) whereas other forms can be trusted to produce true conclusions if given true premises.

The Problem

Unfortunately, research by cognitive psychologists strongly suggests that the vast majority of students do not develop the reasoning skills described above during their high school years. Moreover, although a small proportion of the population (approximately 4%) appears seemingly spontaneously to develop a good capability for formal reasoning, most of the population does not [**1**]. Thus most students, both in high school and college, need assistance in order to improve their ability to think logically.

Perhaps in an ideal world the fundamentals of logical reasoning would be adequately conveyed to students in the context of studying other topics by teachers who know how to seize the "teachable moment" and who recognize the importance of instilling general principles of reasoning in students' minds. But the world is not ideal. For one thing, most of us are less adept at catching that elusive moment than we would wish. But, more importantly, if logical reasoning is always presented as a subtext, in an implicit rather than explicit way, how are we to convey the expertise required to teach it from one generation to the next?

One problem is that some mathematics teachers are not completely secure in their own reasoning abilities while others take correct reasoning so much for granted that they are not able to communicate effectively with students who do not think as they do. Another problem is that when instruction in logical reasoning is not made an explicit priority, it is usually subordinated to other considerations.

An ironic consequence of the attention given to mathematics instruction over the past thirty years is that, especially in algebra, clever teachers and textbook authors have devised numerous ways to help students obtain correct answers to problems by following certain mechanical procedures rather than by reasoning them through. For instance, it used to be that students were taught to solve the problem of finding all real numbers x such that $(x + 1)(x - 2) > 0$ by applying the basic principle that a product of two real numbers is positive if and only if both numbers have the same sign. Use of this approach reinforced the notion that success in mathematics results from the intelligent application of a small number of basic principles, and it taught several important methods of logical reasoning (for instance, argument by division into cases and the logic of *and* and *or*). Nowadays, however, a popular way to teach students to solve such inequalities asks them to learn that the solution consists of certain intervals and that if substitution of a value from one of these intervals makes the inequality true, then the interval is part of the solution. For most students, this method, while effective, serves no larger educational purpose than obtaining a correct answer to a particular problem.

Similarly, in an ideal situation, discussion of the "vertical line test" should help students deepen their understanding of the relationship between the analytic and geometric versions of the definition of a function. Instead the rule is often presented in such a way that students learn to get the right answer to the question "does this graph represent a function?" without making any real progress toward understanding what a function is. In practice, use of the vertical line test enables students to avoid dealing with the linguistic complexity of the function definition and thus fails to advance their ability to understand similarly complex statements in the future. In much the same way, students who are taught to find the inverse of a function f by solving $f(x) = y$ for x and then interchanging x and y are deprived of the opportunity to deepen their understanding both of functions and of the logic of quantified statements. Students not taught this short-cut are forced of necessity to use the definition of inverse function, learning to ask and answer the question, "given any y in the co-domain of f can I find an x in the domain so that $f(x) = y$?"

The Development of our Course

In 1978 at DePaul University we began developing a course to help students make the transition from traditional computationally-oriented mathematics to more abstract mathematical thinking. At the outset we thought that if we just gave students an opportunity to learn subject matter — such as set theory, relations, and function properties — that forms the basis of upper-level work in mathematics and computer science, they would be successful. What we discovered was that students had much more difficulty learning the material than we anticipated, and that to a great extent this difficulty resulted from a general lack of reasoning skills.

For instance, many applications involving one-to-one functions use one form of the definition:

for all x_1 and x_2 in the domain of f, if $f(x_1) = f(x_2)$ then $x_1 = x_2$,

whereas other applications use the alternate form

for all x_1 and x_2 in the domain of f, if $x_1 \neq x_2$ then $f(x_1) \neq f(x_2)$.

When we first started teaching the course, we merely commented on the equivalence of the definitions in passing, using whichever was most convenient in any particular situation. But we soon realized that what was obvious to us (the logical equivalence of the definitions) was a major stumbling block for many of our students. Similarly, a large number of students had difficulty determining whether or not particular functions were one-to-one, not because they didn't understand the definitions of the given functions, but because they didn't understand what it means for statements of the form displayed above to be false. That is, they did not understand (even on an intuitive level) that the negation of a universal statement is existential and that the negation of "if p then q" is "p and not q."

After several years of experimentation, we eventually settled upon the method we still use today. Cognitive psychologists have demonstrated fairly conclusively that instruction in the abstract principles of formal logic alone does not guarantee an increase in students' reasoning abilities [**7**]. Our experience also showed that in order to significantly affect cognitive processes as fundamental and broadly applicable as the correct use of the rules of formal logic, a one-shot approach is not sufficient. Just as a person's personality does not change overnight, even after the revelation and acceptance by the person of some profound personal psychological truth, neither do a person's cognitive processes undergo an instantaneous transformation even though the person may have understood and accepted (at some level) the truth of certain logical principles.

In our course, therefore, and in the book that has developed out of it [**4**], we use several methods to tie formal principles of logic to their use in actual reasoning situations. (Substantial portions of [**8**] reflect a similar approach at the high school level.) First, when we introduce the principles, we include a very large number of natural-language examples. Thus, for instance, before using truth tables to derive the law asserting that the negation of a statement of the form "p and q" is "not p or not q," we give examples of very simple *and* statements and have students think about and discuss what the negations of these statements should be. Then, after the law has been derived formally, the bulk of the exercises ask students to apply it in natural-language situations. Later on in the course, when the law is actually used as an important step in a reasoning process, we point out its occurrence to the students. And when students occasionally use the law incorrectly in their written work, mentioning their error by name helps them better understand what they did wrong.

Student difficulties dealing with negations of universal and existential statements are handled similarly. For instance, when we introduce students to proof by contradiction (a difficult topic for most of them), we might ask students to prove that the double of any irrational number is irrational. Here is a version of a common response:

> *Theorem:* The double of any irrational number is irrational.
> *Proof (by contradiction):* Suppose it is not. That is, suppose the double of any irrational number is rational. But we previously proved that $\sqrt{2}$ is irrational and also that $2\sqrt{2}$ is irrational. These results contradict our supposition. Hence the theorem is true.

When the class has not previously discussed how to negate quantified statements, a teacher has great difficulty helping students understand the error in the above "proof." In a class where there has been prior experience working with such negations, the error is less common. And when it does occur, the teacher can recall that the class previously agreed that the existence of just one college student aged 30 or over was exactly what was needed to falsify the general statement "All college students are under 30." Then the teacher can draw the parallel, pointing out how, in the same way, the existence of just one irrational number whose double is rational is exactly what one needs to negate the statement "the double of any irrational number is irrational." So that is the supposition from which a contradiction must be deduced in order to prove the theorem by contradiction.

Contrast this approach with an approach in which general logical principles have never been explicitly discussed. In such a case, the teacher is in the awkward position of having to point out an error and at the same time convince the student that it really *is* an error. We find that a student who has already thought about the particular logical principle in question and at least partially accepted its validity is much more ready to integrate an appreciation for the new instance of it than a student who has never thought about the issue before. Thus by identifying a few logical principles and giving them names early in the course, we create a basis for developing a fuller understanding of them and a means by which to communicate with students about them throughout the remainder of the course.

This approach is similar to that used in both English and foreign language courses. English teachers agree that the most important part of teaching writing is having students spend time doing it. But interspersed with actual writing practice is a certain amount of explicit instruction in the rules of grammar and organization, and an important component of writing exercises is the process of correction and revision. Similarly for foreign language instruction. Before the age of about eleven, children can learn language purely by osmosis. But after the age of eleven people seem to benefit from some formal instruction in the rules of a new language as well as from immersion in it.

Outcomes

On the whole, we and our students have been very pleased with the results of our approach. We do expect at times to have to listen to and read student explanations that are quite garbled (in the early stages of discussing set theory proofs, for instance). Deeply ingrained mental habits take time to change. But what we do see is significant growth in most students as the course progresses.

For instance, we wait to discuss equivalence relations until late in the second quarter, having interspersed the more theoretical course topics with more straightforward topics and applications earlier on. The advantage is that by the time we reach this topic, the large majority of students really understand what it means for a binary relation to be or not to be reflexive, symmetric, and transitive (which requires a well-developed sense of the logic of quantified statements, *if-then, and,* and *or*). The observation that a certain relation is, say, transitive "by default" is typically made with relish by several students simultaneously. And when we discuss the proof that an equivalence relation defined on a set partitions the set into a union of disjoint subsets, virtually the whole class participates in its development.

Similarly by the time we discuss the fact that any tree with n vertices has $n-1$ edges (see Figure 1), we find that the majority of our students have sufficient familiarity with the logic of *if-then* and quantified statements to comprehend the subtlety of the proof by mathematical induction. The difficulty in the proof comes in understanding why the proof of the inductive step proceeds as it does. In our course, the structure of the proof is seen as a natural consequence of the general logical principle that to prove a statement of the form

for all elements in a set, if (*hypothesis*) then (*conclusion*),

one assumes that one has a (particular but arbitrarily chosen) element of the set which makes the hypothesis true, and one shows that this element makes the conclusion true also. That is why in the proof of the inductive step one assumes that k is any positive integer for which property $P(k)$ holds (that is, one assumes that any tree with k vertices has $k-1$ edges), and then one shows that $P(k+1)$ must also hold (that is, one shows that any tree with $k+1$ vertices has k edges). Moreover, to show that any tree with $k+1$ vertices has k edges, application of the same logical principle leads one first to suppose that T is any (particular but arbitrarily chosen) tree with $k+1$ vertices and then to show that (this particular) T has k edges.

Even after so many years of intimate connection with this course, I am still amazed that students who are clearly bright by many measures and have done extremely well in preceding parts of the course nonetheless need to take their time and feel their way with each new topic. Given encouragement, however, and the opportunity to explore, discuss, and make mistakes, such students not only succeed but they also thoroughly enjoy their success. The point is that the ability to reason with mathematics, to deduce, to justify, and to switch back and forth between abstract definitions and theorems and concrete and applied situations, is not something that students either do or do not possess. Nor is it necessarily or primarily innate. Rather it is a conglomerate of knowledge, attitudes, and tendencies whose cultivation is the greatest challenge that mathematics educators can address.

Connection with Discrete Mathematics

The primary reason for the current interest in discrete mathematics is that it provides the theoretical foundation for the technology of the information age. The ability to reason logically in abstract settings is essential for success in computer science courses at all levels of the undergraduate curriculum. Moreover, knowledge of particular topics in formal logic is indispensible for understanding the design of digital circuits and automata,

Lemma: Any tree with more than one vertex has a vertex of degree 1.
Proof: Let T be any tree with more than one vertex. Pick a vertex v at random and search outward from v on a path along edges from one vertex to another looking for a vertex of degree 1. As each new vertex is reached, check whether it has degree 1. If so, a vertex of degree 1 has been found. If not, it is possible to exit from the new vertex along a different edge from that used to reach the vertex. Because T is a tree, it is circuit-free, and so the path never returns to a previously used vertex. Since the number of vertices of T is finite, the process of building a path must eventually terminate. When that happens, the final vertex of the path must have degree 1.

Theorem: For any positive integer n, any tree with n vertices has $n-1$ edges.
Proof: Let $P(n)$ be the property

$$\text{any tree with } n \text{ vertices has } n-1 \text{ edges}$$

We use mathematical induction to show that this property holds for all integers $n \geq 1$.

Basis Step: Let T be any tree with one vertex. Then T has zero edges (because it contains no loops). Since 0=1-1, the property holds for $n = 1$.

Inductive Step: We must show that for any positive integer k, if the property holds for k then it holds for $k+1$. Let k be a positive integer and suppose the inductive hypothesis: that any tree with k vertices has $k-1$ edges. We must show that any tree with $k+1$ vertices has k edges. Let T be any tree with $k+1$ vertices. Since k is a positive integer, $k+1 \geq 2$, and so T has more than one vertex. Hence by the lemma, T has a vertex v of degree 1. Also since T has more than one vertex, there is at least one other vertex in T besides v. Thus there is an edge e connecting v to the rest of T. Let T' be the subgraph of T consisting of all the vertices of T except v and all the edges of T except e.

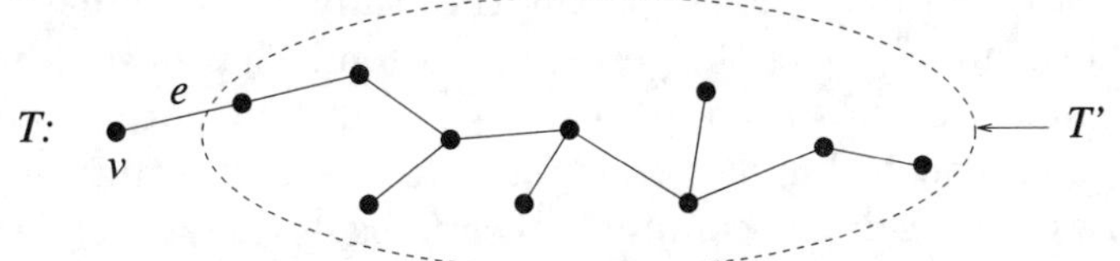

Then T' has k vertices, and T' is circuit-free (since T is circuit-free and removing an edge and a vertex cannot create a circuit) and T' is connected (since T is connected and removing a vertex of degree 1 and its adjacent edge from a graph does not disconnect the graph). Hence T' is a tree with k vertices, and so T' has $k-1$ edges by inductive hypothesis. But then, since T has one more edge than T', T has k edges.

FIGURE 1.

relational database theory, programming languages, and knowledge-based systems. Because of its importance as a topic in computer science as well as its central role in the kind of critical thinking in which computer scientists must routinely engage, logic is now a standard topic of introductory discrete mathematics courses at the college level.

Implications for the K-12 Curriculum

The majority of the reasoning skills emphasized in courses such as ours should not have to be taught for the first time at the college level. By the time students reach us, we have to expend as much effort helping them unlearn the incorrect modes of thought to which they have become accustomed as we do teaching them the correct thought processes on which mathematics is based. To achieve the lofty goals of the NCTM *Standards*, instruction in a few basic logical principles should be woven throughout the K-12 curriculum. Kindergarten is not too early for teachers to begin exploring the precise use of language with children. Indeed, even very young children can become sensitive to and enjoy making subtle linguistic distinctions. For example, starting in the primary grades, the Russian mathematics curriculum translated as part of the University of Chicago School Mathematics Project includes exercises specifically designed to develop children's logical sense [**5**]. In France, excellent materials have been developed for grades 6-10 for helping students make a transition to abstract mathematical thinking. (See, for instance, [**2**].)

In grades K-12 in the United States, however, explicit attention to the development of logical reasoning skills has been minimal or nonexistent. Our experience as described above has shown that logic can be taught explicitly and successfully within discrete mathematics. Including logic as an official topic of discrete mathematics throughout the K-12 years will not only provide a basis for more advanced study at the college level, but will help insure that the principles of formal reasoning are no longer overlooked. While there is a danger that logic will be taught in isolation, this can be avoided by well-constructed curricular materials.

The historical rationale for requiring the study of mathematics was that it sharpened the mind. Over the years this rationale has been deemphasized and greater attention has been given to the goal of acquiring specific computational skills and techniques thought to be needed in future courses or in the "real world." But the computer technology of today renders many of these computational skills less important. Using calculators and computers effectively requires general mental powers, flexibility of mind, and an understanding of concepts. Our primary goal as teachers should be to develop these abilities in our students.

References

[1] Anderson, J. R., *Cognitive Psychology and Its Implications*, 3d ed., W. H. Freeman, New York, 1990.

[2] Arsac, G., G. Chapiron et al, *Initiation au Raisonnement Déductif au Collège*, Presses Universitaires, Lyon, France, 1992.

[3] Einstein, A., "Physics and Reality." *Out of my Later Years*, Revised Reprint ed., Bonanza Books, New York, 1956, p. 59.

[4] Epp, S., *Discrete Mathematics with Applications*, Wadsworth, Belmont CA, 1990.

[5] Moro, Bantova et al., *Russian Grade 1 Mathematics, Russian Grade 2 Mathematics, Russian Grade 3 Mathematics*, University of Chicago School Mathematics Project Translation, UCSMP, Chicago, 1992.

[6] National Council of Teachers of Mathematics, *Curriculum and Evaluation Standards for School Mathematics*, NCTM, Reston VA, 1989.

[7] Nisbett, R. E., G. T. Fong, D. R. Lehman, and P. W. Cheng, "Teaching Reasoning," *Science* (238), pp. 625-631, 1987.

[8] Peressini, A., S. Epp et al., *Precalculus and Discrete Mathematics,* University of Chicago School Mathematics Project, Scott Foresman Publishing Company, Glencoe IL, 1992.

[9] "The Democratization of Undergraduate Mathematics Education." CMS-MAA Invited Address, Joint Mathematics Meetings, Vancouver, B. C., Canada, August 16, 1993.

DEPARTMENT OF MATHEMATICAL SCIENCES, DEPAUL UNIVERSITY, CHICAGO, IL 60614

E-mail address: `sepp@condor.depaul.edu`

DIMACS Series in Discrete Mathematics
and Theoretical Computer Science
Volume **36**, 1997

Writing Discrete(ly)

Rochelle Leibowitz

Discrete mathematics serves as an excellent vehicle for teaching mathematical writing. First, very little content is needed as a prerequisite, so discrete math problems can be introduced at any age. For example, the 'lockers problem', stated below[1], was given to fourth graders, high school students, and college math majors.

> *There are 1000 lockers, numbered from 1 to 1000, lining the hallways of a school. Early one Monday morning, the first student to arrive at school opens all the lockers. The second student to arrive at school that morning closes every other locker (that is, lockers numbered 2, 4, 6, ...). The third student to arrive approaches every third locker (lockers numbered 3, 6, 9, 12, ...) and closes the locker if it was open or opens the locker if it was closed. The fourth student approaches every fourth locker (lockers numbered 4, 8, 12, 16, ...) and closes the locker if it was open or opens the locker if it was closed. This process continues through the 1000th student who arrives that morning. After the 1000th student is done with the lockers, which lockers are open?*

Second, discrete mathematics offers many open-ended, real-world modeling problems, in which there is more than one possible approach. One such problem, discussed later, is the problem of scheduling final exams at college. Third, there is not yet a rigid vocabulary and symbolism in place for solving discrete math problems, so writing in English, accompanied by good pictures, is often required to communicate the solution. Fourth, when introducing students to proof techniques, which is one type of mathematical writing, discussing discrete math theorems that are intuitively obvious, for example, even integer + odd integer = odd integer, helps us focus on

1991 *Mathematics Subject Classification.* Primary 00A05, 00A35.

[1]Editors' note: See also the equivalent problem in Peter B. Henderson's article in this volume

the proof technique used rather than the material needed to understand the theorem.

One course that ties the content of discrete mathematics with the process of mathematical writing is Discrete Mathematics at Wheaton College, a sophomore-level course for mathematics and computer science majors and minors. This course's purpose is to serve as a bridge between computational mathematics and computer science (Calculus and C++) on the one side and theoretical mathematics and computer science (Linear Algebra and Data Structures) on the other side. Consequently, the emphasis is on writing algorithms and mathematical proofs. We spend the first five weeks of the semester covering algorithms, logic, and proof techniques, including mathematical induction. The remainder of the semester is spent covering other discrete mathematics topics, always keeping in mind the primary goal of improving mathematical writing skills.

The formal (graded) writing for the semester-long course consists of five problem sets and a take-home final exam. Each problem set consists of six or seven in-depth mathematical questions, which sometimes present new material. The problem sets and final exam are open text, open notes, open library books. I provide individualized responses to students' writing by making comments, corrections, and suggestions on their writing style as well as on the mathematical content of their answers.

Informal (ungraded) writing consists of homework problems put on the blackboard by students each day at the beginning of class. Students are required to put a certain number of problems on the board for the semester. The students are encouraged to work together on the homework but only one student gets credit for putting the problem on the board. They are also encouraged to put not only correct solutions, but also partial and/or incorrect attempts on the board. Lively class discussions arise from the students' board work, setting the tone for the rest of the class period.

Another example of an informal writing exercise underscores the importance of precision in technical writing. On one of the last days of the semester, I divide the class in two; half go into another room to fill out the course evaluation forms and the students remaining in the classroom become the Writers. Each Writer is given a Lego model (all models are identical) and asked to write down instructions, no pictures or diagrams allowed, on how one would build that model from a set of disassembled pieces. After 25 minutes, the students switch places. The Writers go into the other room to fill out the course evaluation forms and the students from the other room come into the classroom and become the Doers. Each Doer is given instructions written by a Writer and a set of disassembled pieces. The Doers have 25 minutes to complete their models. Try it! It's harder than you think and loads of fun. I strongly recommend it both for yourself and your class. Students and teachers learn that technical writing is not easy and that the technical writer needs to be very precise, to define all terms, and not to assume that what's intuitive to him or her is intuitive to the audience.

One advantage of teaching a writing intensive version of Discrete Mathematics is the unexpected turns that come with allowing students the freedom to write and explore mathematics. We navigate two such examples of classes following uncharted but rewarding paths.

For the first example, we look at the following homework problem. "Develop an algorithm in English to measure exactly 4 liters of water using a 3-liter container and a 5-liter container, and an unlimited supply of water. The containers have no markings on them." *Before reading on, try the problem for yourself!* One possible algorithm is:

1. Fill the 5-liter container with water.
2. Fill the 3-liter container with water from the 5-liter container. (2 liters remain in the 5-liter container.)
3. Pour out the water from the 3-liter container, leaving it empty.
4. Pour the 2 liters from the 5-liter container into the 3-liter container. (Now the 3-liter container has 2 liters and the 5-liter container is empty.)
5. Fill the 5-liter container.
6. Pour some of the water from the 5-liter container into the 3-liter container to fill up the 3-liter container. (The 3-liter container needed only one more liter to fill it up, so the 5-liter container ends up with $5 - 1 = 4$ liters.)

Did you get the same answer? Did you get a similar answer, that is, an algorithm that adds and subtracts amounts of water between the containers until the 5-liter container holds exactly 4 liters? This past year, a student gave a third type of answer; this answer adds the operation of tilting the containers.

1. Fill the 5-liter container and the 3-liter container.
2. Slowly pour out some of the water from the 5-liter container until the water surface, when the container is tilted, is tangent to both the top rim and the bottom rim of the container. (Assuming that the container is cylindrical, the 5-liter container now holds exactly $2\frac{1}{2}$ liters.)
3. Perform step 2 with the 3-liter container. (Assuming that the 3-liter container is cylindrical, it now holds exactly $1\frac{1}{2}$ liters.)
4. Pour the water from the 3-liter container into the 5-liter container. Now the 5-liter container holds $2\frac{1}{2} + 1\frac{1}{2} = 4$ liters.)

Although unplanned, the presentation of these two different algorithms provided me with the perfect opportunity to discuss versatility of algorithms. We looked at each algorithm separately and derived the generalized problem that each one solved. For the general problem, suppose container A can hold x liters of water and container B can hold y liters of water with $x \geq y$. With the restriction that $x \leq 2y$, the first algorithm will terminate with container A holding $2(x - y)$ liters of water. With the restriction that both containers are cylindrical, the second algorithm will terminate with container A holding

$\frac{1}{2}(x+y)$ liters of water. (The original homework problem has $x = 5$ and $y = 3$, therefore the coincidence $2(x-y) = \frac{1}{2}(x+y) = 4$.) We continued to discuss the versatility of these two algorithms. We also talked briefly about the efficiency of algorithms. The class went so well that I will incorporate both "water" algorithms in the future.

The second example involves graph theory. I often ask students as part of their daily homework assignment to pose a question about the material we just covered in class. The day after I presented graph coloring and applications to scheduling, a student asked about scheduling final exams which are assigned 3-hour time slots. That lead to a discussion of interval graphs and a writing assignment on unit interval graphs. Because the students were unhappy with their final exam schedule, we discussed not only the mathematics of unit interval graphs, but the problems of modeling this real world situation. After the discussion, the students were not any happier about their final exam schedule but at least they understood the difficulty in trying to please everyone and the necessity of good written communication between mathematician (in this case, the Registrar) and clients (students and faculty).

By reinforcing writing throughout the semester, students learn that writing and doing mathematics are one and the same. They come to appreciate that writing mathematics is an essential survival skill for any mathematician. Currently, for most students that I teach, this course is a first step in learning this skill. However, the connection between doing and writing mathematics can and should be emphasized much earlier. It is not only the mathematician who needs to communicate mathematically, many people need this skill. A lawyer needs to discuss probability when writing a brief concerning DNA analysis, a building contractor needs to understand management science techniques when writing a schedule of task assignments with time constraints, a conservationist needs to explain functions and graphs when writing a report on long-term effects of a hunting ban on deer population on an island, a homeowner needs to write travel directions to his/her house. All these and much more involve writing mathematically. Teaching this important skill should not be left solely in the hands of college instructors. Elementary school, middle school, and high school teachers can and should devote time to teaching this skill. They, as well as college teachers, can use discrete mathematics for this purpose. In the process of learning to write mathematically, students strengthen their ability to write clearly and logically, attributes desired of all writing. Discrete mathematics thus serves as an ideal tool for teaching mathematical writing for the lifelong writer.

Department of Mathematics, Wheaton College, Norton MA
E-mail address: `Rochelle_Leibowitz@wheatonma.edu`

DIMACS Series in Discrete Mathematics
and Theoretical Computer Science
Volume 36, 1997

Discrete Mathematics and Public Perceptions of Mathematics

Joseph Malkevitch

1. Introduction

A few years ago the Mattel Corporation marketed a talking Barbie doll, one of whose messages was "Math class is tough." Although this message from a talking doll was correctly greeted with great outrage by various sectors of the mathematics community because it conveyed a sexist message, perhaps most members of the general public would probably have agreed with Barbie. For these people, not only was math class tough, but mathematics itself was tough. Many people perceive mathematics to be tough because of what is commonly taught as mathematics in high school. Here is a list of the kinds of problems typically taught and tested for on standardized tests that attempt to measure success with high school mathematics:

Problem Set 1.

1. Factor:

$$x^3 + 5x^2 + 6x$$

2. Simplify:

$$(-2xy^2z^3)^3$$

3. Solve for x:

$$3(x-4) + 2(x-3) = x + 2$$

4. Add:

$$\frac{x+2}{x-6} + \frac{x-3}{x-4}$$

5. Find the value of the expression when $x = 3$ and $y = -2$:

$$(x^3)^2y - (xy)^2$$

1991 *Mathematics Subject Classification.* Primary 00A05, 00A35.

6. Prove that the lines through the vertices of a triangle that bisect its perimeter pass through a single point. Is the same statement true for the area bisectors that pass through the vertices of the triangle?

Compare this problem set with the following list of problems:

Problem Set 2.

A: Design an efficient route for a pot-hole inspection truck, which must inspect every stretch of street in the street network in Figure 1 at least once, and which starts and ends its tour at the location marked A. (You may assume that the streets are two-way.)

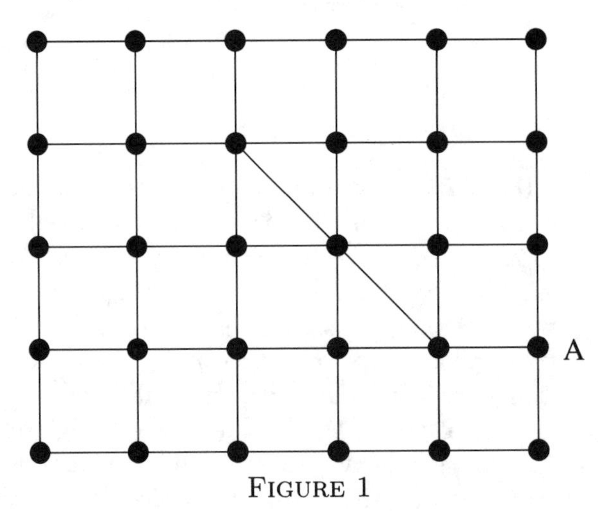

FIGURE 1

B: A small airport has three airlines that share the use of the runway at the airport. It has been decided that another runway must be constructed. What would be a fair system of allocating the cost of the new construction? (You may assume that it will be possible to obtain information such as number of flights per week, number of passengers served per week by these flights, as well as other passenger service and economic information concerning the three airlines.)

C: What would be a fair way for a divorcing couple to agree who should be given a book collection, a summer home, and some jewelry, other than selling the items and dividing the money equally?

D: A company wishes to create a decimal digit coding system for the products which it sells via a mail order catalogue. The code for each item is to consist of 9 information digits and a check digit. What are some of the considerations which might go into the design of the system?

E: The 55 ballots in Figure 2 have been collected for ranking 5 plays for a drama critics award. In this "preference schedule", the "18" at the bottom of the left column signifies that on 18 ballots the ranking of the five plays, from best to worst, was 1,4,5,3,2. Which play should be designated play of the year?

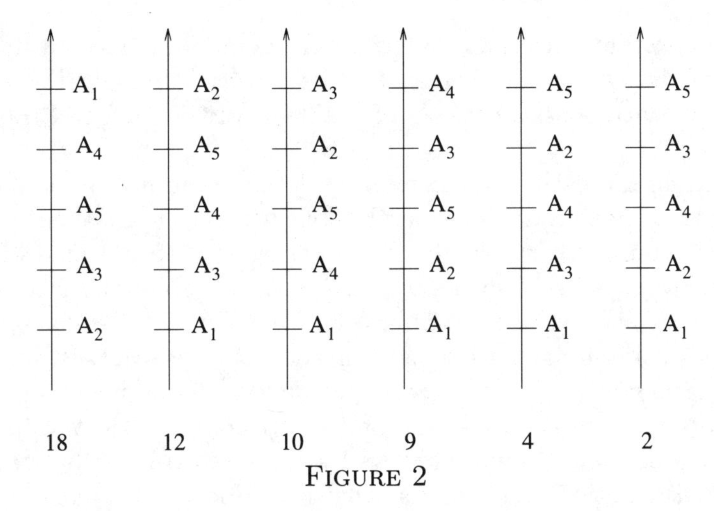

FIGURE 2

F: Figure 3 below shows the 12 courses that are being run by a small college during its summer session. An $\times$ means that the classes in that row and column have some students in common. If final examinations can be arranged in 4 time slots per day, is it possible to schedule all the final examinations in one day so that there is no conflict for any students among the times scheduled for the examinations for their courses?

	1	**2**	**3**	**4**	**5**	**6**	**7**	**8**	**9**	**10**	**11**	**12**
1	—	×		×		×		×		×	×	×
2	×	—		×			×		×	×		
3			—		×	×	×				×	
4	×	×		—		×		×				×
5			×		—		×		×	×	×	
6	×		×	×		—		×				×
7		×	×		×		—		×	×		
8	×			×		×		—			×	
9		×			×		×		—	×		
10	×	×			×		×		×	—		
11	×		×		×			×			—	×
12	×			×		×					×	—

FIGURE 3

These two problem lists are worlds apart. The first list requires successful solvers of the problems to be comfortable with the manipulation of symbols, and each of the problems (other than the geometry problem) has (essentially) one correct answer. This list also gives no hint within the problems themselves of the ways in which mathematics influences daily life. By contrast, the second list (certainly in the statement of the problems) downplays the direct role symbols play, and the problems themselves point to areas of applicability. Perhaps the general public is justified in not seeing the total picture about mathematics, when overwhelmingly the exposure they have to mathematics consists of problems of the kind in Problem Set 1.

The current high school curriculum is the cause, in my opinion, of much of the negative image that the general public attaches to mathematics. The current curriculum is surprisingly often concerned with the kind of mathematics displayed in Problem Set 1. This is true despite the fact that this type of mathematics strays far from achieving many of the goals that the mathematics community and society in general hope can be achieved by teaching mathematics. These goals include the development of thinking skills, understanding of spatial concepts, and training for the workplace (see below for a larger list). In what follows I will try to explain what features of "discrete mathematics", and the way that it can be taught, make it a useful tool for changing the widespread negative perceptions about mathematics and for achieving society's goals for teaching mathematics.

2. What is Discrete Mathematics?

In this essay I am using the phrase discrete mathematics in a special way. Here, discrete mathematics will mean that collection of non-continuous mathematical ideas that have exploded in interest and study since World War II. In many cases these mathematical ideas had roots in much earlier times (e.g., graph theory was invented by Euler in 1736), but the invention of the digital computer served as a catalyst for the flowering of these ideas. Examples of mathematical tools falling within the rubric of discrete mathematics are: matrices, graphs and digraphs, difference equations, codes, and counting techniques. Areas of mathematics which fall primarily within the domain of discrete mathematics are ranking systems and social choice, graph theory, Markov chains, discrete optimization, combinatorics, and (discrete) probability. Just as for continuous mathematics, the study of discrete mathematics can be pursued for its own intellectual content or for specific applications. However, as we shall see, discrete mathematics lends itself to achieving some of the goals for mathematics education more effectively than what is currently taught.

3. Why study mathematics?

Mathematics differs from other areas of knowledge in that society has a vested interest in having the public have a breadth of mathematical skills. More than history or anthropology, for example, mathematics fulfills special needs of large sectors of American businesses as a knowledge base for their employees. Obviously, society has many interests to be served in promoting the teaching of mathematics. Here is a list of some of the many reasons offered for the importance and value of mathematics (in no particular order):

1. Promoting skills for enlightened citizens in a democracy.

2. Providing skills for workers in an increasingly technological society.
3. Providing understanding of the physical space in which we live.
4. Teaching logical thinking and analysis.
5. Serving as the language of science and engineering.
6. Encouraging flexible thinking when exposed to new situations.

What mathematics do we teach in high schools that is designed to convey to American these important aspects of mathematics? Currently, the content of high school mathematics can loosely be described as follows:

Grade 9: Algebra
Grade 10: Geometry
Grade 11: Algebra and Trigonometry
Grade 12: Precalculus; Calculus

In the context of this over-simplified account, you may wish to take a second look at the problems in set 1 above. The reason for this content in grades 9-12, while in many ways promoting the goals mentioned above, lies greatly in society's desire to allow students who are interested in pursuing careers in mathematics, computer science, science, and engineering to have the proper skills to begin college level work in these subjects. The entry course in college for the technologically-based professions, mathematics, and science is Calculus. Success in Calculus is tied to knowledge of algebra, trigonometry, and a subtle array of skills with functions and geometry. This fact, coupled with a tradition of teaching deductive geometry (transferred to America from England) and tradition in general, has given rise to the current curriculum. However, a few moments' thought, and a look at data concerning the portion of college graduates who pursue careers in science and mathematics, show that a high price is being paid for the current curriculum. Although the current curriculum is generally successful in locating the scientifically inclined, it results in vast numbers of other students who are "at sea" with the mathematics they are exposed to.

The bottom line for many students is that despite being exposed to mathematics continuously from Kindergarten through 10th or 11th grade, the typical high school graduate can not connect the value of the study of mathematics with what mathematicians really do. Put differently, students have learned when to "call" or hire a doctor, electrician, geologist, or plumber, but not when to "call" or hire a mathematician. For example, how many high school graduates know that mathematicians study optimization problems (i.e. finding the best or most efficient way of doing something) and fairness questions?

Another major failing of the current curriculum, from society's point of view, is that it does not show the dramatic way that mathematics has been involved in the development of new technologies. It is fair to say that without 20th century mathematics it would have been impossible to accomplish the following dramatic achievements of science and engineering:

1. Landing a man on the moon.

2. Developing supersonic planes.
3. Developing more fuel-efficient cars.
4. Making CAT, PET, and MRI scans commonplace (i.e., breakthroughs in medical imaging).
5. Creating greater efficiency in American business operations (e.g., through the use of linear and integer programming models).

Although many people can in a general way see the connection between more fuel-efficient aircraft and mathematics, it would not be possible for these people to write down the mathematics involved, even in simplified terms. The reason for this is that many of the applications that people point to for demonstrating the importance of mathematics for technology involves the solution of differential and partial differential equations. This mathematics is not reasonably accessible for a high school graduate or even a college graduate (in areas outside of those with a scientific/mathematical focus). This contrasts sharply with the situation for discrete mathematics. Research problems in discrete mathematics are not likely to be resolved by typical high school students. However, for discrete mathematical problems, seeing the germ of the technical ideas of the mathematics and taking a few primitive steps with the mathematical ideas is possible with a much smaller knowledge base than would be the case for continuous mathematics. (For example, with no knowledge of algebra whatsoever one can go a long way in exploring graph theory and its applications.) Thus, altering the current curriculum to give a special role for ideas in discrete mathematics has much to recommend it.

Furthermore, many of the areas in which discrete mathematics is being applied, such as operations research, economics, and biology, are areas where average students have a richer background knowledge than for the fields where continuous mathematics is finding applications (i.e., physics, engineering, and chemistry).

4. Discrete Mathematics in our Schools

Mathematics should play an important role in our schools. Increasingly, knowledge of the role of mathematics in our technological society will be premium knowledge. This raises the issue about what concepts and ideas should be pursued as important ones in grades K-12 before differentiation of training occurs as part of career goals. My answer to this question is that we should make students aware that mathematics is involved with the following key areas and issues:

Optimization: What is the cheapest, fastest, best way of achieving a goal?

- What is the optimum blend of meats for a maker of cold cuts (i.e., salami or bologna) to put into the product, based on the costs of acquiring the meats that make up the mixture?

- What mixture of blends of gasoline should a company manufacture to optimize its profit?
- After a large storm, which forces the cancellation of many flights in a certain region of the country, what reshuffling of passengers, planes, and plane crews will restore the system to normalcy quickly and cheaply?

Fairness:

- How can one fairly divide an estate?
- How can one fairly divide property between a divorcing couple?
- What would constitute a fair way to fund schools?
- How can American election procedures be made more democratic?
- Is weighted voting a fair way to represent communities in a county legislature?
- What makes a game fair?
- What three-dimensional shapes are suitable for fair dice?
- How can two communities fairly divide the cost of constructing a water treatment plant that will benefit both communities?

Information:

- What codes would make it easy for businesses to transact their financial dealings cheaply, safely, and securely?
- How can errors in data transmission from outer space be corrected so that accurate images of planets and stellar objects are possible, even though the images are being sent with low power or unreliable transmitters?
- How can companies minimize the storage space they require for their records?
- Is it feasible to send high-definition television pictures along existing telephone wires?
- Can bar-code systems be designed that would speed the tracking of people or objects in a transportation system?

Risk:

- Is it safer to eat a vitamin that uses a non-natural color, to take a car ride, or to take an airplane ride?
- How likely is it that I will win a prize in a state lottery?
- How risky is it to gamble?
- What is the risk of using milk from cows that were fed genetically-engineered feeds?
- How dangerous are old nuclear power plants?

Growth and change:

- If current fishing patterns are continued, will the stock of a certain fish in the ocean be exhausted?
- What will the population of the world be in 50 years if current trends continue?

- How does an epidemic spread through a population?
- What pattern of market penetration should a company introducing a new product expect?
- How should a forest which contains trees that grow at different rates be managed?

Unintuitive behavior of complex systems:

- If weights in a voting game are proportional to population, is the power of the legislators proportional to the populations they represent?
- Can adding more processors to the scheduling of a collection of tasks increase the time to get the job done?
- If an additional road is built to relieve congestion, might congestion grow worse?
- Can one batter do better than another in each half of a baseball season, but do worse for the season overall?

Since discrete mathematics is a very broad area within mathematics, many more areas and application examples could be listed.

5. A Future Direction for Mathematics in our Schools

In light of the very negative view that people generally (and Barbie in particular) have of mathematics, it is highly desirable that actions be taken that would change these perceptions while at the same time providing students who are mathematically inclined with the stimulation that will allow that inclination to continue and flower. Discrete mathematics is a very fertile field to conduct experiments concerned with achieving this goal. Already at the college level, the so-called liberal arts mathematics course, historically taught with little regard to applications, has undergone a renaissance with the introduction of a new style of course based on an applied discrete mathematics curriculum (see [**1**]). There is thus reason to believe that an emphasis on discrete mathematics, delivered with teaching methods that keep the NCTM standards squarely in view, can transform the perception that mathematics students get in primary and secondary schools, while at the same time providing a steady stream of students to pursue careers in mathematics and science.

Acknowledgment

Many useful suggestions from the reviewers are gratefully appreciated.

References

[1] COMAP, *For All Practical Purposes: Introduction to Contemporary Mathematics*, 3rd ed., W. H. Freeman, New York, 1994.

[2] Hirsch, Christian R., and Margaret J. Kenney, eds. *Discrete Mathematics Across the Curriculum, K-12,* Yearbook of the National Council of Teachers of Mathematics, Reston VA, 1991.

[3] Malkevitch, J., "Mathematics' Image Problem", 1989 (Preprint available from the author).

[4] ——— (ed.), *Geometry's Future*, COMAP, Lexington MA, 1991.
[5] National Council of Teachers of Mathematics, *Curriculum and Evaluation Standards for School Mathematics*, Reston VA, 1989.

MATHEMATICS/COMPUTER SCIENCE DEPARTMENT, YORK COLLEGE (CUNY), JAMAICA, NEW YORK 11451

E-mail address: joeyc@cunyvm.cuny.edu

DIMACS Series in Discrete Mathematics
and Theoretical Computer Science
Volume **36**, 1997

Mathematical Modeling and Discrete Mathematics

Henry O. Pollak

1. What is Mathematical Modeling?

When people talk about the connection of mathematics with the rest of the world, they use a number of phrases such as "applied mathematics", "problem solving", "word problems", and "mathematical modeling", to name just a few. In order to define these more precisely, and to differentiate among them, I should like to begin by describing the series of activities which seem to take place when we try to use mathematics to examine something in the rest of the world. Some situations involving discrete mathematics to which this analysis applies will be given later.

(1) The process begins with something outside of mathematics which you would like to know or to do or to understand.
   - The result is a question in the real world, well-defined enough that you can recognize when you have made progress on it.

(2) You next select some important objects in this situation outside of mathematics, and relationships among them.
   - The result is the identification of some key concepts in the situation you want to study.

(3) You decide what to keep and what to ignore in your knowledge of the objects and their interrelationships.
   - The result is an idealized version of the question.

(4) You translate the idealized version of the question into mathematical terms.
   - The result is a mathematical version of the idealized question.

(5) You identify the field of mathematics you think you're in.
   - You bring into the forefront of your consciousness your instincts and knowledge about this field.

(6) You do mathematics.
   - The result is solutions, theorems, special cases, algorithms, estimates, open problems.

1991 *Mathematics Subject Classification.* Primary 00A71, 00A05, 00A35.

(7) You now translate back into the setting of the original problem.
  - You now have a theory of the idealized version of the question which you found in (3) above.

(8) You confront reality in the form of the original situation as represented by (1). Do you believe what is being said in (7)? In other words, do your results, when translated back to the original situation, fit the real world?
  - If yes, you have succeeded. You tell your friends, write it up, publish some papers, get a raise, get promoted, or whatever.
  - If no, go back to the beginning. Did you pick the right objects and relationships among them? Do your choices of what to keep and what to ignore need to be revisited? The way in which your theory of the idealized problem fails to satisfy you should provide some hints of where there are difficulties.

An example in which this process can be followed in detail would take us too far afield in the present context. The author's forthcoming paper [**7**] contains a detailed history of such a problem, the modeling steps, and the repeated modeling cycle.

2. Applied Mathematics, Word Problems, and Modeling

What have I just given is a brief outline of "mathematical modeling". When I use that term henceforth, this is what I mean. Now what is "applied mathematics"? The way the term is usually used, it begins with some idealized version of reality, translates it into mathematics, does a lot of mathematics, and, at its best, translates back; in other words, (4)–(7). Courses in "methods of applied mathematics" concentrate on the mathematical methods that tend to come up in (6) when you start with questions in physics.

What is a "word problem"? Typically, a word problem begins with a few words from outside mathematics to provide a semblance of (4), occurs in the textbook in a place where (5) is obvious, and concentrates on (6).

There is very little agreement on the meaning of "problem solving". It can be taken to mean doing a word problem, or applied mathematics, or modeling from beginning to end. Sometimes, "problem solving" refers to the process of solving mathematical problems with no reference to an external situation at all. When problem solving refers to word problems or to applied mathematics, i.e., beginning with (4), the earlier stages (1)–(3) are sometimes referred to as "problem finding", or "problem formulation".

Word problems have a history of being unrealistic, and the persistence of particular types lends itself to easy caricature. We have pipes of various capacities which can fill and empty bathtubs and John's age when Susie was twice as old as Sally. We have learned to say "center of mass", "moment of inertia" and "pendulum" with a straight face, but we go directly to the formulas without any thought in between. We make no attempt to see if our answers make any sense in the original situation because we had no original situation to begin with! That's very typical of many word problems, I'm afraid. On the other hand, what modeling requires is understanding of the original situation, an argument that the idealization makes sense, and the check that the results of the mathematical work carry meaning outside of mathematics.

3. Discrete Mathematics and the Teaching of Modeling

I believe that relating mathematics to the rest of the world is an essential part of mathematics education. We have not done our job if this aspect is not included. We ought to have word problems, traditional applied mathematics, and mathematical modeling—all three. Why? If modeling is what actually happens when you apply mathematics in the real world, why don't you just teach that? There are three main difficulties that I will discuss: mathematical modeling takes a lot of time, it requires a lot of knowledge on the teacher's part, and there is a lack of certainty in the results which, in the eyes of the public, is quite uncharacteristic of mathematics.

It is without doubt true that modeling is time consuming. So let us agree that not every problem with an applied flavor will go through the full (1)–(8) above. But in terms of a typical word problem, how do you tell a good problem from a bad one?

My answer depends on whether the problem *could* be the middle portion of a genuine model. What do I mean? Here is a sample word problem: "An electric fan is advertised as moving 3375 cubic feet of air per minute. How long will it take the fan to change the air in a room 27 ft. by 25 ft. by 10 ft.?" Now you all know what you are supposed to do: multiply 27 by 25 by 10 and divide the result into 3375. But the assumption behind this is that the room is hermetically sealed and that the fan evacuates all the air before any new air comes in! Absurd! This is not off by a little bit, it's off by maybe an order of magnitude. You *could* do a sensible discrete approximation to this by evacuating 10% of the air, replacing it with replacing it with fresh air and thereby diluting the old air, and repeating this process until the old air is no longer noticeable. That's a model that would make more sense. You obtain a linear recursion for the amount of "old" air that is left after k evacuations, and you ask how long it will be until the old air can no longer be perceived. This is a reasonable mathematical model; by my definition, the original word problem was not a good one.

A word of caution: there are word problems which were never meant to be taken seriously. The context is deliberately whimsical, and is intended to add lightness and humor to a heavy lesson. For example, Kolmogorov in 1966 gave the problem of a bee and a lump of sugar at two distinct points inside a triangle. The bee wishes to fly a minimum length path to the lump of sugar, under the condition that she must touch all three sides of the triangle along the way. I have no objection to such a problem—in fact, it's lovely! But nobody pretends it's about actual bees! What I object to are problems that pretend to be real but couldn't be.

Our second objection is that real modeling requires a lot of knowledge on the teacher's part, knowledge of a lot of fields outside of mathematics! That's true, but needs to be examined very carefully. Mathematics gets applied in all aspects of everyday life, intelligent citizenship, and other disciplines and occupations. Furthermore, most branches of mathematics, certainly all at the school and undergraduate level, have significant practical applications. In fact, there are unexpected and rather interesting connections between these two observations. When we worry that teachers, and students, may not know certain fields to which mathematics is applied, we often have in mind the field of physics. What mathematics is most applied to physics? Classical, continuous, analysis. Discrete mathematics is just as important for applications as continuous mathematics, and there tend to be many more applications to everyday life, operations analysis, and the social sciences, where the natural experiences of both teachers and students can give a great deal of guidance and insight. Thus discrete mathematics is an arena where we can bridge the gap between mathematical modeling in the classroom and mathematical modeling in the rest of the world with unusual effectiveness. What we are saying is that mathematical modeling can be particularly accessible when the resulting mathematical field at the heart of the development is in the area of discrete mathematics. Voting and fair division and the cleaning of streets are just as interesting mathematically as moments of inertia, and they use a lot of available intuition and experience.

Here are partial descriptions of some of my favorite modeling situations which lead to discrete mathematics and can be made accessible to high school students.

(a) Traditional private line pricing in the telephone business leads to minimal spanning trees, Cayley's theorem as well as Prim's and Kruskal's algorithms, Shamos' shortcuts, the Steiner network problem, and NP-completeness. The key modeling question: what is meant by "fair" pricing? This question drove much of the historical development. A discussion of the private-line pricing problem is given in [**7**].

(b) Building a counting circuit in a computer leads to the problem of enumerating Hamiltonian cycles for the graph which is the vertex and edge structure of an n-dimensional cube. It is easy to give an example of a single such Hamiltonian cycle, but how many different cycles are there? The graph theory soon becomes mixed with group theory. The key modeling question: when are two cycles "different"? It turns out that for engineering purposes—and this is where modeling is especially important—you want two Hamiltonian cycles to be *not* different (i.e., equivalent) if one can be obtained from the other by a symmetry of the n-dimensional cube. How many equivalence classes of Hamiltonian cycles are possible on an n-dimensional cube? The answer appears to be unknown for dimensions $n \geq 6$.

The mathematical formulation of this problem, and the complete discussion for four dimensions (i.e., counting from 0 to 15) may be found in [**3**]. This paper also relates the counting problem to the earlier Gray Code work during World War II, which was essentially a problem of analog-to-digital conversion. Martin Gardner refers to the answer for $n = 5$ in [**2**].

(c) In baseball, some of the modeling has been done for us, as in the definition of batting averages. If an additional hit takes a player's average from .299 to .306, how many at-bats and how many hits has that player had? This turns into a wonderful number theory problem, and involves Farey Series and continued fractions if we so choose. It is mathematical detective work: how do you turn a decimal into a fraction? We traditionally teach this for terminating decimals and repeating decimals, but not for arbitrary decimals known to a certain number of places—like batting averages.

The baseball example as such has not appeared in print; it is part of the author's lecture "Some Mathematics of Baseball" [**6**], which is one of the American Mathematical Society's videotaped "Selected Lectures in Mathematics". The same problem arises with free-throw percentages in basketball, and may be found in [**5**].

(d) Ed Gilbert at AT&T Bell Labs, who was involved in the research of (a) and (b), is the originator of the following problem: how do you build a perfect box? If you have six rectangular pieces of wood, what patterns of one piece covering another at an edge and at a corner are possible? There is some simple topology in this, and the Euler characteristic gives a lot of insight. Can you build a perfect box from six identical pieces of wood? The answer is "not in general", although it is possible if the dimensions of the blocks of wood satisfy certain conditions. Gilbert's article on this subject is [**4**].

(e) There are many well-known and more traditional problem areas that meet our requirements. I shall mention just one, that of coding theory. Noiseless coding, such as Huffman Codes, and group codes for the binary symmetric noisy channel, are two very accessible subjects. The combination of geometry, beginning group theory, and linear algebra at the beginning of group codes is especially appealing.

A nice exposition of the basics of group codes from just the point of view recommended in the previous paragraph may be found in [**9**]. Huffman codes at a level appropriate for high-school students may be found in [**8**]; the proofs related to Huffman codes specifically but not to noiseless coding more generally, may be found in the appendices. A more nearly complete exposition of noiseless coding appears, for example, as chapter 2 of [**1**].

Let us close with the third objection to mathematical modeling, namely the loss of certainty. There is personal judgment in the problem formulation parts (1)–(3), which is especially noticeable when, in (8), the results don't fit reality. Worse than that, there are honest differences of opinion; for example, if a problem concerns fair division, or an optimum location, what to one person looks fair may not seem fair to another. Or, to give another example, when competing criteria in an optimization problem are naturally measured in different units, such as lives and dollars, then there is no obvious way to equate them, and disagreement is inevitable. This contradicts the myth, held by many students and, alas, some teachers, that mathematics is a field of single right methods, single right answers, and unambiguous truth. This is actually not true of pure mathematics either, but it isn't even close when you apply mathematics to the rest of the world. We have to admit that this observation may be especially distressing to those who like mathematics primarily because it is a way of making a reasonable living and at the same time minimizing any danger of involvement with the real world. For such people, word problems are survivable, because of their degree of unreality, but mathematical modeling may cause great unhappiness. Their response may be to deny that modeling has a place in the mathematics curriculum. Now discrete mathematics is especially useful in applying mathematics in relatively controversial areas. Is this one of the reasons why its place in the curriculum has been hard to secure?

References

[1] Ash, R., *Information Theory*, Dover Publications, 1990.

[2] Gardner, M., *Knotted Doughnuts and other Mathematical Entertainments*, W. H. Freeman and Co., New York, 1986, chapter 2.

[3] Gilbert, E. N., "Gray Codes and Paths on the n-Cube", *Bell System Technical Journal*, v. 37, May 1958, pp. 815 - 826.

[4] ______ "The Ways to Build a Box", *Mathematics Teacher*, v. 64, Dec 1971, pp. 689-695.

[5] North Carolina School of Science and Mathematics (G. Barrett et al.), *Contemporary Precalculus through applications*, Janson Publications, Providence RI, 1991, pp. 170-172.

[6] Pollak, H. O., "Some Mathematics of Baseball", videotaped "Selected Lectures in Mathematics", American Mathematical Society.

[7] ______ "Some Thoughts on Real-World Problem Solving", *Quantitative Literacy*, The College Board, New York, 1997.

[8] Sacco, Copes, Sloyer, and Stark, *Information Theory, Saving Bits*, Janson Publications, Providence RI, 1988.

[9] Slepian, D., "Coding Theory", *Nuovo Cimento*, N. 2 del supplemento al v. 13, serie **X**, 1959, pp. 373 - 378.

TEACHERS COLLEGE, COLUMBIA UNIVERSITY, NEW YORK, NY 10027
E-mail address: 6182700@mcimail.com

DIMACS Series in Discrete Mathematics
and Theoretical Computer Science
Volume **36**, 1997

The Role of Applications in Teaching Discrete Mathematics

Fred S. Roberts

1. Using Applications Effectively in the Classroom

One of the major reasons for the great increase in interest in discrete mathematics is its importance in solving practical problems. Conversely, practical problems have stimulated the development of discrete mathematics. Applications — discrete or not — should play a major role in the mathematics classroom. They make the subject relevant. They underscore a reason for studying it. They are interesting.

With regard to the role of applications in teaching discrete mathematics, I have developed some rules of thumb over the years, based on my experience with what students respond to and on the philosophy I have developed about the role of applications in mathematics. In my opinion, these rules of thumb are appropriate at all grade levels, though most of my experience with them has been at the college level.

Rules of Thumb

1. The **Relevance Rule**: Choose applications that are relevant. There are plenty of them.
2. The **Two Are Better Than One Rule**: Never settle for one application when two are available.
3. The **Why Do Things Twice Rule**: Stress the fact that abstract methods developed for dealing with one application are often useful for another.
4. The **Get Real Rule**: Mention real uses of mathematics whenever possible.
5. The **Frontiers Rule**: Show the frontiers of the subject.
6. The **Math Is Alive Rule**: Use applications to show that mathematics is a live subject, done by real people.

1991 *Mathematics Subject Classification.* Primary 00A05, 00A35.

7. The **Motivate Rule**: Let applications motivate theory. Then apply theory to applied problems.
8. The **Don't Be Scared Off Rule**: Don't hesitate to talk about an application because you don't have a background in the subject. Most applications can be explained from general knowledge.
9. The **Modeling Rule**: Choose applications that involve model building. Illustrate the simplifying assumptions in the model and iterate to more complicated (and more realistic) models.

In this paper, I will illustrate these rules of thumb with three examples. In each case, I take one simple mathematical concept and give lots of applications of it. I have used these and similar examples in my college-level courses, but have also used them at all grade levels, including primary grades. The three examples I shall discuss are:

a: The traveling salesman problem.
b: Graph coloring.
c: Eulerian chains and paths.

Almost all of the applications I mention here are discussed in more detail in my book, Roberts [**31**]. For some of them, I will provide additional references, though many of these references are to articles that are more technical in nature.

2. The Traveling Salesman Problem

The *traveling salesman problem (TSP)*, in its traditional formulation,[1] is the following: There are n locations. A salesperson must visit all of them, in some order. There is a cost of traveling from location i to location j. What is the cheapest route? Most of those who have been exposed to discrete mathematics have seen this problem. They know it is difficult: No one has found a *good TSP algorithm*, that is, a computer algorithm for solving the TSP which is practical for very large n, and there is strong evidence that there is none. (The problem belongs to the class of problems that theoretical computer scientists call NP-complete.) Most people who teach discrete mathematics mention the TSP. But you can use it much more effectively by going to the next step: Show how this problem arises in practice in many other forms.

Let me mention some of these other forms.

> **The Automated Teller Machine Problem.** Your bank has many ATM machines. Each day, a courier goes from machine to machine to make collections, gather computer information, and so on. In what order should the machines be visited? This problem arises in practice at many banks. One of the earliest banks to use a TSP algorithm to solve it, in the early days of ATM's, was Shawmut Bank in Boston.

[1]Note that the TSP is nowadays frequently referred to as the "traveling salesperson problem". I have chosen to use the historical name.

(This example is from Margaret Cozzens (personal communication), who first developed it as an assignment for her undergraduate operations research class at Northeastern University, and assigned students to study the Shawmut Bank ATM problem, with considerable success.)

The Phone Booth Problem. Once a week, each phone booth in a region must be visited, and the coins collected. In what order should that be done?

The Problem of Robots in an Automated Warehouse. The warehouse of the future will have orders filled by a robot. Imagine a pharmaceutical warehouse with stacks of goods arranged in rows and columns. An order comes in for ten cases of Tylenol, six cases of shampoo, eight cases of bandaids, etc. Each is located by row, column, and height. In what order should the robot fill the order? The robot needs to be programmed to solve a TSP. In our programs in discrete mathematics for high school and middle school teachers and for high school students at DIMACS (the Center for Discrete Mathematics and Theoretical Computer Science), we sometimes take the students to see a Rutgers University Industrial Engineering robot, which can be used to do exactly this. See [**8**, **9**].

A Problem of X-Ray Crystallography. In x-ray crystallography, we must move a diffractometer through a sequence of prescribed angles. There is a cost in terms of time and setup for doing one move after another. How do we minimize this cost? See [**4**].

Manufacturing. In many factories, there are a number of jobs that must be performed or processes that must be run. After running process i, a certain setup cost is inferred before we can run process j, a cost in terms of time or money or labor of preparing the machinery for the next process. Sometimes this cost is minimal, for example simply amounting to making minor adjustments, and sometimes it is major, for example requiring complete cleaning of equipment or installation of new equipment. In what order should the processes be run?

These applications illustrate some of my rules of thumb. They all illustrate the **Relevance Rule** (#1) and the **Two Are Better Than One Rule** (#2). They also illustrate the **Don't Be Scared Off Rule** (#8). You don't have to know anything about x-ray crystallography to talk about that application. Yet, I know teachers who are embarrassed to bring in applications like this because they don't know what some words mean or can't pronounce the words! What is a diffractometer? One of your students might know, or be willing to find out. The entire paper will illustrate these three rules of thumb, so I will usually not explicitly mention those again.

These examples also illustrate the **Get Real Rule** (#4) – it is especially nice to be able to mention real companies (such as Shawmut Bank) that use mathematical methods. I should also note that all of these problems are, in the abstract, the identical problem we have formulated for the TSP. Once we have developed mathematical tools for dealing with the TSP, these same tools can be applied to all of these other practical problems. This illustrates the **Why Do Things Twice Rule** (#3). There are two ways I illustrate this rule. Sometimes, I formulate one version of a problem, translate it into mathematical language (with the students' help), and then develop mathematical methods needed for dealing with the problem. I then formulate another practical problem, show how, in the abstract, it is the same as the first, and then point out that little extra mathematical analysis is needed. At other times, I will formulate a large number of practical problems first, and let the students observe how they are related by formulating them all in the same abstract language, or by guessing why or how they are related.

I should point out that many of these problems in their current formulation involve simplifying assumptions. For example, in the phone booth problem, some telephone booths need to be visited more often than others, since they fill up faster; and in the manufacturing problem, some processes cannot be run before others are completed. In the first round of modeling, these complications are ignored. The next round of modeling should try to handle them. This is an illustration of the **Modeling Rule** (#9). By discussing simplifying assumptions, we teach our students to question assumptions and hypotheses, train them to be more skeptical about technical presentations, and ultimately prepare them to be better decision makers. I always try to involve my students in pinpointing oversimplifications in an initial model for a problem. I also involve them in suggesting how to modify an abstract model to take account of possible complications.

Recently, a group of researchers at four institutions, Rutgers University, AT&T Bell Labs, Bellcore, and Rice University, solved the largest TSP ever solved (up to that time). It had 3038 cities and arose from a practical problem involving the most efficient order in which to drill 3038 holes to make a circuit board (another TSP application). (For information about this, see [**1, 41**].) I like to mention this achievement, and tell my students how a real problem was solved by real people who are at the same institution as I am. This illustrates the **Get Real Rule** (#4), the **Frontiers Rule** (#5), and the **Math Is Alive Rule** (#6): It involves a real application, it is right at the frontiers of modern research, and it was done by real people. Students much prefer to see a real-world application to a make-believe one using "widgets." They get turned on by realizing that they can get to the frontiers of knowledge. They also pay more attention to things that are done by real people. I know one teacher who believes so strongly in the **Math Is Alive Rule** (#6) that he brings in slides showing pictures of mathematicians whose results he is talking about. Once you have seen a picture of a person, you somehow pay more attention to that person's results, and remember them better by associating them with the picture.

I often use the TSP to introduce the idea of complexity of computation and to motivate an interest in counting and combinatorics. It is a good example to illustrate why one needs to count the number of steps in a computation before implementing it. (Consider the brute force approach of trying all possible orders of the cities in a TSP with, say, 26 cities. Even on a computer that could check one billion orders per second, it would take us almost half a billion years to look at all possible orders.) Once I've introduced the idea of counting the number of steps in a computation, I find it much easier to interest students in methods of counting and combinatorics, which I then relate back to complexity of computation. All of this illustrates the **Motivate Rule** (#7). Students are much more interested in the rules of counting if they see a real application that requires them to be able to count.

3. Graph Coloring

A *graph* consists of a set of points or *vertices*, some of which are joined by lines or *edges*. A very old idea is to *color* the vertices of a graph so that if two vertices are joined by an edge, they get different colors. A large number of those who teach discrete mathematics talk about graph coloring. Some mention one application of graph coloring, the historically important application of **map coloring**, where the goal is to color the map with as few colors as possible, so long as countries sharing a border have different colors. We model the countries of a map by vertices of a graph and join two vertices by an edge if their countries have a common boundary. The problem of coloring a map so that countries with a common boundary must get different colors is the same as the problem of coloring the corresponding graph. This is a very important historical example. I like bringing history into my classes, especially when there is a very interesting history of over 100 years that also involves important contributions by non-mathematicians and a historically important use of computers – the first solution to the map-coloring problem used 1200 hours of computer time! That is why I like to use the map coloring example. (For more on its history, see [**2**].) However, there is much more to be said here, because graph coloring has many modern applications which students find both interesting and exciting. I start with some of these examples before going back and giving the historical example. I find that students perk up and take notice from modern and relevant examples. Here are some applications, almost all of which are expanded upon in my papers [**33, 34**].

Scheduling Meetings of Committees in a State Legislature. The problem is to assign meeting times so that if two committees have a member in common, they get different meeting times. The solution is to color an appropriate graph. To define a graph, we must say what its vertices and edges are. In this case, the vertices are the committees and there is an edge between two vertices if their corresponding committees have a common member. Then the colors are the meeting times. It should be noted that this problem arises in many places. One particular place of note is the New York State Assembly. (See [**5**] and [**31**] for more details.) This illustrates the **Get Real Rule** (#4).

Similar scheduling problems involve assigning final exam times — classes with a common student must get different exam times. Similarly, in an idealized school, students first sign up for classes and then classes are assigned meeting times so that classes with a common student get different meeting times. (This actually happens in some universities, at least for the scheduling of graduate courses in small departments.) Both of these problems are, in the abstract, the identical problem that we have just formulated for the state legislative committees. As with the TSP, once we have formulated the first scheduling problem as an abstract mathematical problem and developed tools for dealing with that problem, we can now "reduce" these new scheduling problems to the old one, in the sense that in the abstract version, they are the same problem and so are amenable to solution using the same tools. This again illustrates the **Why Do Things Twice Rule** (#3).

I usually give simple scheduling problems as examples, have the students translate them into graph problems, and have them try to find graph colorings. We usually end up using a *greedy algorithm* for doing this — color the vertices one at a time, using a new color only if no previously used color can be used. We then ask whether or not we have found a coloring with the fewest number of colors. It is not hard to give examples where such a greedy approach does not work. I point out that graph coloring is again known to be a difficult problem—as with the TSP, there is no known "good" algorithm for finding a graph coloring with the smallest number of colors, and it is unlikely that there will ever be one. This is a place where one can introduce different graph coloring algorithms and use them on practical problems. The abstract methods developed for graph coloring problems that arise from one problem are useful for others. The **Why Do Things Twice Rule** (#3) has again been illustrated. It is no surprise that software designed to solve scheduling problems is sometimes based on graph coloring. It might make a good exercise to have your students explore the software that is used in your school, or to try to write their own programs.

Practical scheduling problems involve many further complications, such as individuals' preferences for when they are to be scheduled, or certain committees being required to meet after certain others. Also, there has been little mention so far of what makes one schedule (one graph coloring) better than another. This needs to be discussed as well. Is the goal only to use the smallest number of colors? Or is it sometimes good to have a reasonable distribution of colors, i.e., to use each color approximately the same number of times? All of this illustrates the **Modeling Rule** (#9). There is a large literature on scheduling theory: several good references on the subject are the books [**3, 27, 36**].

> **The Channel Assignment Problem.** The problem is to assign channels to radio and television transmitters; transmitters that interfere must get different channels. The solution is to color an appropriate graph. The graph can be defined by letting the vertices be the transmitters and letting an edge correspond to interference. Then, the colors are the channels. Graph coloring methods for solving the channel assignment problem are widely used at such agencies as the Federal Communications Commission, the National Telecommunications and Information Administration, and NATO (the **Get Real Rule** (#4)). See [**12, 6, 34**]

I usually formulate one or two practical problems as graph coloring problems — explaining what to use for vertices and edges and what corresponds to colors. After an example or two, however, I ask the students to help, and they willingly chime in. After hearing about scheduling problems, they can readily translate the channel assignment problem into a graph coloring problem. Indeed, they are eager to think of other problems familiar to them that can be formulated as graph coloring problems.

It is worth mentioning that practical channel assignment problems as well as other applied problems have given rise to a variety of interesting variations of the ordinary concepts of graph coloring. So far, we have not considered what makes one channel assignment better than another. It is not necessarily just that one uses fewer channels than the other; it might be that one has a smaller separation between largest and smallest channel used and thus uses less of the available "spectrum." We have not considered the fact that we might have further restrictions on channels that are closer together than on channels that are further apart but still interfere, or more generally, that we might have different levels of interference. We have not considered the possibility that transmitters might be assigned more than one possible channel over which to transmit, as is the case for mobile radio telephones in cars. The removal of each of these simplifying assumptions leads to an interesting generalization of graph coloring. Some of them are called T-colorings, n-tuple colorings, and interval colorings [**33**]. Such generalizations are not difficult to explain to students, and many of them are at the forefront of modern research in graph theory. This again illustrates the **Modeling Rule** (#9) and the **Frontiers Rule** (#5).

There is another important point. Real-world channel assignment problems use graphs with thousands of vertices. It is very hard to find the "best" solution under any of a number of definitions of best. Sometimes, we should settle for a solution that can be found in a reasonable amount of time, even if it is not the best. This is a good place to bring in the idea of approximation, and perhaps to mention "heuristic" algorithms that have been developed by real people at real places such as at NATO (the **Math Is Alive Rule** (#6)).

> **Garbage Collection Problem.** Garbage trucks follow certain routes in collecting garbage. The problem is to assign each garbage truck route to a day of the week so that if two routes visit a common site, they are scheduled for different days. The solution is to color an appropriate graph. The vertices of that graph are the routes in question and an edge between two routes means that they visit a common site. The colors are the days. This particular problem arose from a more complicated garbage truck routing problem posed by the New York City Department of Sanitation. That problem involves choices of routes as well. This illustrates the **Get Real Rule** (#4) and the **Modeling Rule** (#9). See [**28, 29, 37**].
>
> **Traffic Light Phasing Problem.** We are putting in a new traffic light at a traffic intersection. We need to assign a green light time to each stream of traffic through the intersection so that two streams of traffic that interfere get different green light times. The solution is to color an appropriate graph. The vertices of that graph are the traffic streams, an edge means inteference, and the colors are the green light times. The idea of using graph coloring for phasing new traffic lights was first proposed in an article in a transportation journal, *Transportation Science* [**35**]. (See also [**28**].)

In dealing with the Traffic Light Phasing Problem, we have omitted any discussion of what makes one green light assignment better than another. Also, we are not paying attention to the duration of the green light times, and the fact that one traffic stream might require a longer green light time than another. These complications lead to generalizations of ordinary graph coloring, and in particular the generalization known as interval coloring, which is of current research interest. So, we have again illustrated the **Modeling Rule** (#9) and the **Frontiers Rule** (#5). Stoffers' algorithm for traffic light phasing, and later ones by Opsut and Roberts [**20, 21**] and Raychaudhuri [**24, 25**], are based on linear programming methods to find the best (interval) graph colorings with durations. I like to teach these algorithms to my students, then let them find some local traffic intersections to apply the algorithms to. Margaret Cozzens (personal communication) reports that when her students applied the algorithms to intersections near the campus of Northeastern University, they found much better traffic light phasings than those actually in use. To complete this really practical experience, they went and convinced the Boston department of transportation to implement their solutions! This is a wonderful example of rule of thumb #4, the **Get Real Rule**.

It should be noted that the same problem arises in scheduling other facilities, such as a classroom, computer, etc. There are different users and some of them interfere. We wish to assign green light times (permission-to-use times) so that interfering users get different green light times. In the abstract, this is the identical problem that we have already analyzed, an illustration of the **Why Do Things Twice Rule** (#3). In addition, the same problem arises in task assignment problems in the workplace. Different tasks need to be assigned times, but some of them interfere because they use the same workers or tools or resources, another illustration of the **Why Do Things Twice Rule** (#3).

> **Fleet Maintenance Problem.** Vehicles (cars, planes, ships) are coming into a facility for regular maintenance according to a fixed schedule. We wish to assign a space to each vehicle. If two vehicles are there at the same time, they must get different spaces. The solution to this problem is to color an appropriate graph. Its vertices are the vehicles and there is an edge between two vehicles if they are in the facility at the same time. The colors are the spaces. (This is the first example I have given where the colors are not times or days or something like that. Students usually see this fairly quickly.) It should be remarked that this problem was first worked on at IBM for ship maintenance (the **Get Real Rule** (#4)). See [**11, 19, 30**].

The Fleet Maintenance Problem again has its complications: What makes one assignment better than another? What if one vehicle requires more space than another? Physically, do the spaces correspond to points or to rectangles or to circles? Each complication leads to a new variation of graph coloring, much as in the channel assignment problem. Here again we illustrate the **Modeling Rule** (#9) and the **Frontiers Rule** (#5).

I have often given a talk to high school audiences that describes the many applications I have given in this section. (It certainly is an illustration of the **Two Are Better Than One Rule** (#2)!) One of these talks led to a very exciting question by one of the students: "Are there careers in graph coloring?" I think I made my point that day! (No, there are not careers in graph coloring. Yes, there are careers in applying mathematics.)

4. Eulerian Chains

Given a graph, a *chain* (or walk or path, depending on what terminology you use) arises if we follow the edges from vertex to vertex; an *eulerian chain* is a chain that uses every edge exactly once. An *eulerian closed chain* is an eulerian chain that begins and ends in the same place. Many of those who teach discrete mathematics mention the problems of finding eulerian chains and eulerian closed chains. Some people talk about their history, by describing the famous problem of the **Königsberg bridges**, which was solved by the mathematician Leonhard Euler in 1736 and gave rise to the subject of graph theory. (See [**2**], and see the article by Newman [**18**] in *Scientific American*, and the accompanying translation of the original memoir by Euler [**10**].) Some people even go beyond this, to describe the following problem (though not always connecting it to eulerian chains).

> **The "Chinese Postman Problem".** A mail carrier walking a route must hit every street in the neighborhood and use the smallest amount of time. What route should the carrier take? This problem was first analyzed using graph theoretical methods by a real postman in China, Guan Meigu (the **Get Real Rule** (#4) and the **Math Is Alive Rule** (#6).) It is not exactly the same problem as that of finding an eulerian closed chain, since the mail carrier can walk down a street a second time. However, the eulerian chain problem enters in a critical way into the solution: If there is an eulerian closed chain, this gives the solution. If not, we simply have to find the smallest number of edges to copy so that in the resulting graph there is an eulerian closed chain. See [**15, 17, 31**].

While some people teaching discrete mathematics go as far as mentioning the Chinese Postman Problem, it is so much better to go further, for instance by noting that the exact same problem arises in *street sweeping* and in *snow removal*. Certain streets in a city have to be swept or cleared, and we wish to do this in the least amount of time [**38, 16, 29, 31**]. Again, we have an illustration of the **Why Do Things Twice Rule** (#3). As it turns out, these problems have interesting complications: Only some streets need to be swept every day; there are one-way streets; it takes much longer to go down a street while sweeping it than it does to go down it when one is just passing through. These complications can be handled, and they lead to interesting variations of the Chinese Postman Problem and wonderful exercises for students [**38**]. Here again, we have illustrated the **Modeling Rule** (#9).

Here is another problem that is really the same:

> **Automated Graph Plotting by Computer.** We wish to draw a graph (with pre-specified vertex locations) by computer. When we repeat an edge, we need to pause the computer and raise the plotter pen off the paper. We draw lots of copies of the same graph and so would like to design a way of drawing it which uses as little time as possible. This is again

the Chinese Postman Problem. It has modern practical applications in chip design at IBM, drawing circuit diagrams, electrical and water networks for cities (it has been widely used in Bonn, Germany, for example), control of machines for producing lithographic masks, and so on. See [**13, 26**]. Again, we have illustrated the **Why Do Things Twice Rule** (#3) and the **Get Real Rule** (#4).)

There are other, more subtle applications of eulerian chains. For instance, eulerian chains arise in a telecommunications problem which is concerned with how to tell the position of a rotating roof antenna without going to the roof. The solution involves finding so-called deBruijn diagrams, which also can be connected to the design of computing machines through the theory of shift register sequences. See [**31**] for a discussion.

How many people know that eulerian chains have played a crucial role in the history of molecular biology? They were used in early algorithms for finding an RNA chain given fragments of it that were produced from decomposition by various enzymes. The first RNA chain was determined in 1965 by R.W. Holley and his co-workers at Cornell, using a method that soon was improved using eulerian chains and paths. The specific use of eulerian chains is a bit complicated. However, I can build up to it in several class periods, which involve some rather simple but beautiful applications of the basic counting rules of combinatorics. (See [**31**], Sections 2.13 and 11.4.4.)

After I describe, or at least mention, the use of eulerian chains in molecular biology, I usually lead into a discussion of the many applications of discrete mathematics to modern molecular biology. In particular, I mention the importance of graph theory and combinatorics in the Human Genome Project, the project of mapping and sequencing the entire human genome. For more on this subject, see for example [**7, 14, 23, 32, 39, 40**]. I like to give specific problems here that have come out of recent research in computational biology. Some of these involve eulerian chains and paths (as for example in connection with the "double digest problem" in DNA physical mapping [**22**]). Others involve a variety of questions in graph theory and combinatorics. This illustrates, once again, the **Frontiers Rule** (#5). I also mention the increasing collaboration between the biological sciences community and the mathematical sciences community and the realization on the part of biological scientists that many of their problems are basically problems that are amenable to formulation using discrete mathematics. I give examples of the many collaborations between local mathematicians/computer scientists and local biological scientists that have come about in recent years (the **Math Is Alive Rule** (#6)).

5. Concluding Remark

The main message of this paper, and the main reason that we wish to use applications in our courses, can be summed up as follows. There are so many exciting, relevant applications of discrete mathematics that if you are a good teacher, none of your students should ever again have to ask: *What is mathematics good for?*

References

[1] Applegate, D., Bixby, R., Chvatal, V., and Cook, B. "Finding Cuts in the TSP," DIMACS Technical Report 95-05, DIMACS Center, RUTGERS University, Piscataway NJ, 1995.

[2] Biggs, N.L., Lloyd, E.K., and Wilson, R.J., *Graph Theory 1736-1936*, Oxford University Press, London, 1976.

[3] Baker, K.R., *Introduction to Sequencing and Scheduling*, Wiley, New York, 1974.

[4] Bland, R.G., and Shallcross, D.F., "Large Traveling Salesman Problems Arising from Experiments in X-Ray Crystallography: A Preliminary Report on Computation," *Oper. Res. Let.*, 8 (1989), 125-128.

[5] Bodin, L.D., and Friedman, A.J., "Scheduling of Committees for the New York State Assembly," Tech. Report USE No. 71-9, Urban Science and Engineering, State University of New York, Stony Brook, 1971.

[6] Cozzens, M.B., and Roberts, F.S., "T-Colorings of Graphs and the Channel Assignment Problem," *Congr. Numer.*, 35 (1982), 191-208.

[7] DeLisi, C., "Computers in Molecular Biology: Current Applications and Emerging Trends," *Science*, 240 (1988), 47-52.

[8] Elsayed, E.A., "Algorithms for Optimal Material Handling in Automatic Warehousing Systems," *Int. J. Prod. Res.*, 19 (1981), 525-535.

[9] ———, and Stern, R.G., "Computerized Algorithms for Order Processing in Automated Warehousing Systems," *Int. J. Prod. Res.*, 21 (1983), 579-586.

[10] Euler, L., "The Königsberg Bridges," *Sci. Amer.*, 189 (1953), 66-70. (Translation from 18th century article.)

[11] Golumbic, M.C., *Algorithmic Graph Theory and Perfect Graphs*, Academic Press, New York, 1980.

[12] Hale, W.K., "Frequency Assignment: Theory and Applications," *Proc. IEEE*, 68 (1980), 1497-1514.

[13] Korte, B., "Applications of Combinatorial Optimization," in M. Iri and K. Tanabe (eds.), *Mathematical Programming: Recent Developments and Applications*, KTK Scientific Publishing, Tokyo, and Kluwer Academic Publishers, Dordrecht, 1989, pp. 1-55.

[14] Lander, E.S., and Waterman, M.S. (eds.), *Calculating the Secrets of Life: Applications of the Mathematical Sciences in Molecular Biology*, National Academy Press, Washington, DC, 1995.

[15] Lawler, E.L., *Combinatorial Optimization: Networks and Matroids*, Holt, Rinehart and Winston, New York, 1976.

[16] Liebling, T.M., *Graphentheorie in Planungs-und Tourenproblemen*, Lecture Notes in Operations Research and Mathematical Systems No. 21, Springer-Verlag, New York, 1970.

[17] Minieka, E., *Optimization Algorithms for Networks and Graphs*, Dekker, New York, 1978.

[18] Newman, J.R., "Leonhard Euler and the Königsberg Bridges," *Sci. Amer.*, 189 (1953), 66.

[19] Opsut, R.J., and Roberts, F.S., "On the Fleet Maintenance, Mobile Radio Frequency, Task Assignment, and Traffic Phasing Problems," in G. Chartrand, et al. (eds.), *The Theory and Applications of Graphs*, Wiley, New York, 1981, 479-492.

[20] ———, "I-Colorings, I-Phasings, and I-Intersection Assignments for Graphs, and their Applications," *Networks*, 13 (1983), 327-345.

[21] ______, "Optimal I-Intersection Assignments for Graphs: A Linear Programming Approach," *Networks*, 13 (1983), 317-326.
[22] Pevzner, P.A., "DNA Physical Mapping and Alternating Eulerian Cycles in Colored Graphs," *Algorithmica*, 13 (1995), 77-105.
[23] Pieper, G.W., "Computer Scientists Join Biologists in Genome Project," *SIAM News*, January 1989, 18.
[24] Raychaudhuri, A., "Optimal Scheduling of Subtasks under Compatibility and Precedence Constraints," *Congr. Numer.*, 73 (1990), 223-234.
[25] ______, "Optimal Multiple Interval Assignments in Frequency Assignment and Traffic Phasing," *Discr. Appl. Math.*, 40 (1992), 319-332.
[26] Reingold, E.M., and Tarjan, R.E., "On a Greedy Heuristic for Complete Matching," *SIAM J. Comput.*, 10 (1981), 676-681.
[27] Rinnooy Kan, A.H.G., *Machine Scheduling Problems: Classification, Complexity and Computation*, Nijhof, The Hague, 1976,
[28] Roberts, F.S., *Discrete Mathematical Models, with Applications to Social, Biological, and Environmental Problems*, Prentice-Hall, Englewood Cliffs, NJ, 1976.
[29] ______, *Graph Theory and its Applications to Problems of Society*, NSF-CBMS Monograph No. 29, Society for Industrial and Applied Mathematics, Philadelphia, 1978.
[30] ______, "On the Mobile Radio Frequency Assignment Problem and the Traffic Light Phasing Problem," *Annals NY Acad. Sci*, 319 (1979), 466-483.
[31] ______, *Applied Combinatorics*, Prentice-Hall, Englewood Cliffs, NJ, 1984.
[32] ______ (ed.), *Applications of Combinatorics and Graph Theory to the Biological and Social Sciences*, IMA Volumes in Mathematics and its Applications, Vol. 17, Springer-Verlag, New York, 1989.
[33] ______, "From Garbage to Rainbows: Generalizations of Graph Coloring and their Applications," in Y. Alavi, G. Chartrand, O.R. Oellermann, and A.J. Schwenk (eds.), *Graph Theory, Combinatorics, and Applications*, Vol. 2, Wiley, New York, 1991, pp. 1031-1052.
[34] ______, "T-Colorings of Graphs: Recent Results and Open Problems," *Discr. Math.*, 93 (1991), 229-245.
[35] Stoffers, K.E., "Scheduling of Traffic Lights – A New Approach," *Transportation Res.*, 2 (1968), 199-234.
[36] Slowinski, R., and Weglarz, J. (eds.), *Advances in Project Scheduling*, Elsevier, Amsterdam, 1989.
[37] Tucker, A.C., "Perfect Graphs and an Application to Optimizing Municipal Services," *SIAM Rev.*, 15 (1973), 585-590.
[38] ______, and Bodin, L., "A Model for Municipal Street-Sweeping Operations," in W.F. Lucas, F.S. Roberts, and R.M. Thrall (eds.), *Discrete and System Models*, Vol. 3 of *Modules in Applied Mathematics*, Springer-Verlag, New York, 1983, pp. 76-111.
[39] Waterman, M.S. (ed.), *Mathematical Methods for DNA Sequences*, CRC Press, Boca Raton, FL, 1989.
[40] Waterman, M.S., *Introduction to Computational Biology: Maps, Sequences, and Genomes*, Chapman and Hall, 1995.
[41] Zimmer, C., "And One for the Road," *Discover*, January 1993, 91-92.

DEPARTMENT OF MATHEMATICS, CENTER FOR OPERATIONS RESEARCH (RUTCOR), AND CENTER FOR DISCRETE MATHEMATICS AND THEORETICAL COMPUTER SCIENCE (DIMACS), RUTGERS UNIVERSITY, NEW BRUNSWICK, NJ 08903

E-mail address: froberts@dimacs.rutgers.edu

Section 3

What Is Discrete Mathematics: Two Perspectives

What Is Discrete Mathematics? The Many Answers
STEPHEN B. MAURER
Page 121

A Comprehensive View of Discrete Mathematics:
Chapter 14 of the New Jersey Mathematics Curriculum Framework
JOSEPH G. ROSENSTEIN
Page 133

DIMACS Series in Discrete Mathematics
and Theoretical Computer Science
Volume **36**, 1997

What is Discrete Mathematics? The Many Answers

Stephen B. Maurer

1. Introduction

We advocates of discrete mathematics have a problem: there is no agreed-on definition of our field! We are even worse off than Supreme Court justices debating pornography: we don't even agree when we see it! (Are fractals discrete mathematics? Matrices? Statistics? Number theory? Proofs? Real-world applications of high school algebra? Patterns and tiling? Construction algorithms in Euclidean geometry?) We are like the blind men feeling the elephant; each describes his own beast.

This situation is not necessarily bad. When a field is not well defined, it can blossom in many directions. But this lack of definition is different from the usual situation in most areas of mathematics. Most mathematicians have a pretty clear idea what algebra is, or calculus. (Well, they used to have a clear idea about calculus!) And though one would be hard put to define mathematics generally, there isn't too much doubt when one sees it.

However, there is a big difference between defining discrete mathematics as a field and as a course. When a mathematician invents some new mathematics, it is almost irrelevant what rubric we use to classify it. The issue is whether it is interesting and useful. But courses require decisions — is a particular topic going to be included or not? At the college level, discrete mathematics courses have been around for 20 years now, and syllabi have tended to settle into a few patterns. At the school level, discrete mathematics is still quite new and there is little agreement on content.

Thus, to help think about the K–12 curriculum, it would be useful to have a single definition or description of discrete mathematics. Alas, I can't provide one. Instead, in the first part of this paper I offer several proposed definitions and descriptions, and show shortcomings for each one.

Having several definitions can even be helpful, for the following reason: when listening to other advocates of discrete mathematics, it is important

1991 *Mathematics Subject Classification.* Primary 00A35.

to catch on quickly to what version of discrete mathematics they are talking about and to what ends they are promoting it. Similarly, when we advocate discrete mathematics to others, it is important to make clear quickly what version of discrete mathematics *we* are talking about and to what ends *we* are promoting it. This paper can help us identify the different versions and goals, and give us terminology to talk about them.

In recognizing our differences, we may recognize what is common as well. So, in the final part of the paper, I make some suggestions as to what I think we might agree should be part of discrete mathematics in the schools, and what we might agree to exclude.

2. Defining Discrete Mathematics

There are two standard approaches to defining a branch of mathematics: specifying properties of the branch and giving a list of topics. (Mathematicians usually start with the former approach but often end up with the latter.) Let's explore both approaches.

Attempts to define discrete mathematics by specifying properties. Here are several "definitions," each followed by one or two difficulties.

> **Definition 1:** Discrete mathematics is finite mathematics, that is, the mathematics of situations that can be described by finite sets.

This definition excludes all sorts of important discrete topics that require at least the set of all natural numbers: induction, difference equations, infinite graphs, and formal languages.

> **Definition 2:** Discrete mathematics is the mathematics of discrete sets, that is sets which have holes between any two elements, as do the natural numbers and the rational numbers.

Many discrete topics regularly use real numbers, such as sequences, linear programming, weighted graphs, or game theory. While many of these areas could be carried out over the rational numbers Q (for instance, the theory of linear programming is unchanged over Q), none of them *are* carried out over Q, and some of them cannot be carried out as well (e.g., linear difference equations wouldn't always have closed-form solutions).

> **Definition 3:** Discrete mathematics is any mathematics that doesn't involve limits.

This claim certainly has some merit, because continuous mathematics certainly does involve limits. But as a definition this claim is both too exclusive and too inclusive. Too exclusive: do we refuse to discuss limits of sequences in a discrete mathematics course? Do we expunge fractals? Do we refuse to mention that the (discrete) Poisson distribution in probability is the limit of binomial distributions? Too inclusive: do we claim that all of abstract algebra is part of discrete mathematics, all of logic, or for that matter, almost all of school mathematics, since limits don't appear until at least pre-calculus?

Definition 4: Discrete mathematics is whatever mathematics can be done in a finite number of steps.

I must confess that I used to offer this definition, because it emphasized that discrete mathematics is about algorithms. But life is finite, and so all mathematics is done in a finite number of steps. In short, this is the definition a discrete mathematician should use who wants to be an intellectual imperialist and take over everyone else's field!

Definition 4 can be improved by saying that discrete mathematics is mathematics where the object of study, rather than the process of studying it, is an algorithm that takes a finite number of steps. But even if we could make this distinction precise (between objects and the study of objects), Definition 4 would not be good enough. For example, bisection algorithms, in principle, might run forever. Should they be excluded from discrete math? Typically they are not.

Attempts to to define discrete mathematics by lists of topics. There are about as many proposed defining lists as there are discrete mathematics textbooks. Table 1 gives five lists. List A is typical for a discrete structures course aimed at computer science majors in college. List C is for a finite mathematics course aimed at college students interested in social science and business. Courses corresponding to these lists have been around for 20 years. List B is for an algorithms-oriented college course of more recent vintage. Lists D and E are from books for high school courses [**2, 11**]. Lists A and B are likely to be for one-year courses, lists C, D and E for a semester of material. Additional books often used for discrete mathematics in schools are listed in the references [**4, 5, 3, 9, 10**].

These lists are not all-encompassing. Topics that are on some other lists include fractals, number systems and number theory, theory of computation, simulation, block designs, and Polya counting theory. A number of terms on many lists are subtopics of ones already listed. For example, vectors are subsumed under linear algebra. Similarly, trees, networks and network algorithms come under graph theory; semigroups come under abstract algebra; coding theory comes under combinatorics and/or abstract algebra.

What positive conclusions can be drawn from these definitions and lists? Anything involving finite sets or about finite algorithms applied to discrete sets, and not traditionally covered in the curriculum, is probably discrete mathematics. Anything about graph theory, counting, recurrences or elementary logic is probably discrete mathematics also.

3. Distinguishing Approaches to Discrete Mathematics

If there are so many variants of discrete mathematics, can we at least group the variants in useful ways? As we will now show, one way is by emphases, another is by goals.

Grouping discrete mathematics approaches by emphases. To make what we mean clearer, we group emphases in contrasting pairs.

Discrete structures vs. problem-solving methodologies. A structures course emphasizes theorems about properties of various constructs. For instance, a structures course might emphasize that all Eulerian graphs are connected and have all vertices of even degree. A problem-solving course emphasizes how discrete mathematics gives concepts and techniques to solve problems. For instance, such a course might emphasize how to tell if a problem should be modeled by a graph, how to tell if that problem is solved if the graph is Eulerian, and finally, how to test if a given graph is Eulerian.

This distinction is similar to the one between real analysis and calculus. In the former, you emphasize the structure of, say, the set of differentiable functions (e.g., it is closed under addition), whereas in calculus you study the derivative and how it can help you solve problems.

Narrow clientele vs. broad clientele. If a course is offered as a service for a particular group, for example, if most of the students are planning to major in computer science, then typically the course will emphasize applications of interest to that group. On the other hand, if the students have a variety of interests, the course should offer a variety of applications. One can argue

TABLE 1. Five lists of topics for discrete mathematics courses

List A	**List B**
Logic and circuits	Algorithms and algorithmic language
Sets, relations, functions	Induction, iteration, recursion
Induction	Graph theory
Counting (combinatorics recurrences generating functions)	Difference equations
Graph theory	Probability
Boolean algebra	Logic
Automata	Linear algebra
Abstract algebra (intro)	Analysis and verification of algorithms
Partially ordered sets	Sequences and limits
	Numerical Analysis

List C	**List D**	**List E**
Logic	Election theory	Logic
Counting (elementary)	Fair division	Integers and polynomials
Finite probability	Matrices	Recursion and induction
Linear programming and games	Graphs	Combinatorics
Statistics	Counting	Graphs and circuits
Social science and business applications	Probability	Vectors
Modeling	Recursion	

that even if the clientele is narrow, a broad course should be given; the needs of a client group today may not be their needs tomorrow. In any event, both types of discrete mathematics courses exist, as indicated by the topic lists in Table 1.

Structural vs. algorithmic. The first discrete mathematics courses were about structure. For instance, planar graphs were characterized as those that contain no "homeomorph" of either K_5, the complete graph on 5 vertices, or $K_{3,3}$, the "utility graph." There was little discussion of whether there are efficient ways to check if a graph meets such a characterization, that is, whether there are good verification algorithms. Indeed, algorithms were simply not an object of study in the course. This is a bit odd, since these early courses were given mostly for computer science students, for whom algorithms are *the* object of study. Perhaps the feeling was that computer science students got enough study of algorithms in their other courses, and that mathematics courses for computer science should meet the approval of mathematicians by sticking to what was perceived as "real mathematics," that is, structure.

Pure vs. applied. Discrete mathematics has many applications. Yet, just as in other branches of mathematics, one can give a course, even a very interesting course, on purely mathematical aspects of the topic. There are discrete mathematics textbooks that do this, and others that generate everything out of applications.

Grouping discrete mathematics approaches by goals. Truth be told, most of us promoting discrete mathematics have some general goal in mind that goes beyond the particular mathematics. Sometimes the real agenda is as broad as revamping what school is like or what education is all about. In short, sometimes discrete mathematics is the means rather than the end. This is not bad, but it should be acknowledged. Below we state goals that have been advocated as reasons for teaching more discrete mathematics. We start with mathematical goals and move to more general educational goals.

To introduce proofs and abstraction. Most college students seem to have poor proof and abstraction skills. Where can they learn these skills well — that is, what topics in mathematics will convince students of the need for proofs and yet have proofs that are not too hard for beginners? Many people feel that parts of discrete mathematics fill the bill. In particular, elementary number theory and mathematical induction are mentioned.

To introduce algorithms and recursion. Certain mathematical concepts and paradigms are given short shrift in traditional studies, for instance, algorithms and recursion. Both of these concepts were around long before discrete mathematics. As for algorithms, students have always had to use them (but rarely think about them). As for recursion, who hasn't heard of "reduce to the previous case"; recursion is a careful formulation of this idea. Discrete mathematics brings algorithms and recursion to the fore, by making them objects of study and providing precise ways to discuss them.

To emphasize applications. Few students are turned on by pure mathematics. We may regret this — most of us *were* turned on by pure mathematics — but we can't deny it. To take a more positive attitude, mathematics has a double appeal: it is simultaneously beautiful *and* useful. In any event, for most students to begin to appreciate mathematics, they have to see that it is useful. Teaching discrete mathematics can show them this, because so many real applications are accessible at an elementary level.

To introduce modeling. Traditionally, in both school and college, doing mathematics was a process that began with a mathematically formulated problem and ended with the mathematical solution of that problem – even if the problem was applied. Modeling emphasizes that this traditional view is but one step (often the easiest) of several:

1. A problem is given in amorphous real-world terms
2. The problem is idealized into a mathematical form, the initial model
3. That model is solved (this is the traditional activity)
4. The results are interpreted in the original context
5. The cycle is repeated until the solution is deemed helpful.

Having mathematics presented in this broader way makes many more students value it. Many mathematicians feel discrete problems are the best for introducing modeling; modeling is by nature complicated and discrete models provide some of the simpler instances.

To introduce operations research. There are many sorts of optimization that cannot be touched by calculus, for instance, maximizing flow in a network, minimizing the number of colors needed for a map, and maximizing a linear function when the domain is restricted by inequalities. Many such optimization problems are intimately tied to the modeling approach and are highly relevant to business and management. Yet until recently most students, even at the college level, never heard that mathematics has anything to say about optimization except for the very specialized sort in calculus. Discrete mathematics is where they can learn the good news.

To entice more students into a mathematical sciences major. Many students enter college with their minds almost made up about a major. Fields not seen before college attract few students. Therefore, not enough students will choose mathematical science majors with a discrete flavor unless they see some discrete mathematics in school.

To introduce computers into school mathematics. There are many contexts for introducing computers into mathematics class, for instance, graphing functions, doing algebra manipulations, and drill. However, discrete mathematics is probably the most natural context for introducing computers, since discrete mathematics *is* the mathematics of computation.

To give students something fresh and relevant to them. Too many students have been turned off to mathematics as they keep seeing more of the same, where "the same" is usually senseless manipulation (or so it seems to them). For instance, much of high school mathematics seems to be repeated algebraic computation concerning rates, time, distance, area, volume, etc. If a student is successful and gets to calculus, he or she does the same calculations over again, only more of them, to handle the optimization aspect as well.

Many educators feel students are turned off *because* they see so much repetition, and that they won't be turned off if instead they see something completely different, especially if it is obviously relevant. Discrete mathematics can certainly be completely different and relevant. For instance, the mathematics of fair division, apportionment and election methods is an eye-opener. Here is something important in the struggles over equity in today's world, and most students would never have thought mathematics has something to say about social equity.

To give students a chance to be creative and do research. We are told that, in the future, most employment will require creative approaches to open-ended problems. Therefore, education should involve such creative work. In mathematics and science, this means research. Some educators go further and suggest that becoming active junior researchers is the primary thing kids should do in school. In most sciences it is possible to show students what research is like, and perhaps get them actively involved in their own research, early on. Using discrete topics, the same can be done in mathematics. This is because there are parts of discrete mathematics where it is easy to state problems that are beyond what the students have learned how to solve (or in some cases, beyond what anybody has solved).

To introduce important, active areas of mathematics. Whether or not the goal should be to make kids researchers, certainly they deserve to be shown what is going on at the frontiers. Discrete mathematics is one area of mathematics where this is possible.

To promote experimental mathematics. Part of the reason students can do research in science much earlier than in traditional mathematics is because you can make progress in science by experiments, even if you have not developed a theory, whereas in traditional mathematics the only way to make progress was to conceive and prove theorems. But now, with computers, there is opportunity for experimental mathematics, especially within discrete mathematics. The intellectual effort that goes into creating programs to generate mathematical data is substantial, and from this data students can make conjectures that they would not arrive at otherwise. Professional mathematicians are making much more use of experiments. So should students.

To promote cooperative learning and other new classroom approaches. Traditionally, schoolwork is done alone. In mathematics, there has been a very competitive aspect to this approach, as indicated by the interest in mathematics competitions and the emphasis on individual scores. Just as there is now more group work at the professional level in mathematics, so can there be more group work in school. Group work is most appropriate for larger, open-ended problems. Discrete mathematics is an excellent source at the school level for such problems.

There are many other ways in which some classrooms today are very different from traditional classrooms. Take assessment for example. In some places tests have largely been replaced by broader methods such as portfolios. For a portfolio to assess more than a test, the portfolio must involve items more open-ended than traditional test problems. Thus once again the opportunity discrete mathematics provides for open-ended problems makes it a good context in which to introduce the new approach.

To teach students to think. Traditional school mathematics emphasizes technique, technique that can be mastered in a mechanical way. Thus many students have coped with mathematics by learning how to "turn the crank" instead of learning how to think. Mechanical strategies are not so successful in discrete mathematics, for there are many fewer parts of discrete mathematics that can be routinized. For instance, there are endless varieties of counting problems. Also, the first time a student sees a graph theory problem, no previously learned solution method will help directly. This lack of standard techniques can have a downside: students may get frustrated and give up. But as long as the difficulty level of material is carefully monitored, the lack of standard techniques can make students think.

Some words of caution: Some of the goals just listed are contradictory. For instance, topics that are relevant to school students or which provide accessible unsolved problems are often not particularly deep or active mathematics. Example: many discrete courses at the school level include substantial material on the theory of elections. However, this is not a large, very active, or central area of discrete mathematics; it does not have many connections to other parts of discrete mathematics and the solution methods do not generalize to other areas. Theory of elections is rarely included in college discrete mathematics courses, and it would not be included in school courses in order to meet the previously described goals, for instance, of introducing proofs and abstraction, introducing algorithms and recursion, or introducing active areas of mathematics.

Also, to achieve many of these goals it is not necessary to use discrete mathematics. We should promote any goal we feel is important, but we should not equate discrete mathematics with those goals. To do so only obscures the issues and hinders the effort.

4. Suggestions for Common Ground

Let me make some proposals.[1] I said at the start that one purpose of describing the many meanings of discrete mathematics is so that each of us can make our positions clear to others. So let me identify my views. I'm primarily interested in discrete mathematics because of its content, not as

[1]These proposals were well received at the conference. Of course, it's easy to be well received if you are sufficiently vague!

a means towards pedagogical goals. The content that interests me (at least for introductory courses) is not the formal structure but rather the concepts, the problem-solving paradigms (like recursion), and the role of algorithms (see [**6, 7**]). I have a broad audience in mind, not just mathematics and computer science majors.

Also, I am uncomfortable with the tone of much pedagogical discussion in the mathematics community today. I don't think that the sole role of mathematics education is to get students to think, or that experimentation is central (e.g., students should discover all key ideas for themselves through experimentation, and any topic for which experimental confirmation can be obtained is appropriate to study), or that every topic taught should be one linked to real-world applications. I feel there has been too much bashing of traditional methods; there is much good in them, at least in the hands of good teachers. I *do* feel there is need for pedagogical change. It is incumbent on us all to verse ourselves in new methods and give them a fair try; but it remains to be seen what the right mix of old and new will be.

Because I take this view, I limit my proposals for common ground to content.

Principles for selecting discrete mathematics topics for schools.
Some discrete mathematics is appropriate in schools for each student. This is in fact a principle of the NCTM Standards [**8**], where examples are given of discrete mathematics topics appropriate for students at various grade levels.

Relevance to calculus should not be the main criterion for selecting mathematics to teach in school. The traditional means for deciding what to put in the school curriculum was "is it good background for calculus?" To maintain this point of view severely crimps any change, and besides, most students are (or will be) as likely to take some sort of discrete mathematics in college as continuous mathematics.

Concepts from some non-traditional areas of mathematics belong in the student repertoire, beginning early, sometimes in elementary school. Some discrete topcs, not traditionally taught in schools, should be taught there, sometimes beginning in elementary school. Table 2 gives my proposals.

TABLE 2. Discrete Math Topics for Schools

"Definite"	"Maybes"
algorithmic language	computer programming
graph theory	counting/combinatorics
probability and statistics	logic and proof concepts
vectors and matrices	modeling
	recursion and iteration

For instance, graph theory is a "Definite" because so many situations can be pictured with graphs — any binary relation (e.g., adjacency of countries), any network problem, any problem about transition between configurations (e.g., almost any puzzle where you have to move pieces). Thus, students ought to become familiar with the concept of a graph itself and with various properties a graph can have, such as being connected.

The importance of probability/statistics and matrix algebra is by now well known (see the Standards [**8**]), but algorithmic language may need some explanation. This refers to the sort of language needed to discuss algorithms in precise ways. Computer scientists refer to such language as *pseudocode.* Students need to be familiar with concepts like loop (i.e., for-next) and if-then statements, and they need to have terminology to use such concepts carefully.

Why then is computer programming a "Maybe"? Certainly students should use computers — there is much good mathematical software — but writing programs in a computer language is another matter, even if these programs are merely translations of ideas the students have already expressed in algorithmic language. In programming there are always so many technical details one can get hung up on. Such implementation of algorithms might best be left optional, or left for a computer science course.

As for counting, of course students should do some, and will do some as part of probability, but a detailed study of formal counting methods is what I classify as "Maybe". Some students love to count, others regard it as boring abstraction. For most students, this is perhaps best left to college discrete mathematics.

As for logic and proof, these ideas must appear at least informally; the experimental approach should not push them out entirely. But to present them explicitly and at length, as in traditional Euclidean geometry courses, may have the same stultifying effect on many students as that course has had. [2]

As for modeling, again my concern is with a full head-on approach. For instance, every time one turns an applied problem into a graph problem one is doing modeling, and I am all for this. But it is probably too much to elaborate on the explicit stages of modeling (as described under goal 4), or to deal with all the length, detail, special cases, and partial results of real modeling of real problems.

Recursion is my own hobbyhorse, but you can do quite well without it until the point where you are serious about "algorithmics" – not just using algorithmic constructs but actually creating, verifying and analyzing the efficiency of algorithms — and these activities may not be appropriate until late high school or college. To the extent that one is doing recursion when one devises a recurrence relation for a sequence (say, the formula $f_n = f_{n-1} + f_{n-2}$ for the Fibonacci numbers), then of course one should introduce recursion early. But as for recursive algorithms, or for proofs

[2]In discussion, some conferees were inclined to move logic and proof to the "Definites", and to move recursion also. In school there are too few places to practice proofs, especially with the decline of Euclidean geometry.

obtained by first restating a problem in a recursive formulation, this might be postponed.

One might list several more "Maybes", but please note the sort of things from Table 1 that I have left out. First, I have left out topics of interest to special groups only, e.g., automata (computer scientists), circuits (electrical and computer engineers), business applications. School students haven't or shouldn't define themselves so narrowly, and school courses should not cater to narrow interests. Second, I have left out abstract topics, e.g., sets and relations, abstract algebra. Third, I have left out minor areas of discrete mathematics that have few ties to major areas, e.g., election theory, though I realize there are fervent champions of such areas because of their relevance to real life (see goal 8 earlier).

What general impressions should be fostered. Returning to principles, my last concern is with what people retain after their mathematics education has ended. Someone has said that learning is what remains after detailed techniques have been forgotten and fundamental concepts have gotten rusty. Let us refer to this remainder as *general impressions.* We should strive hard to instill our students with certain correct general impressions, especially the majority of our students who will not use mathematics directly in their work but who will need to have some sense of how others are using mathematics for them. For instance, all students should get the general impression that mathematics is very useful, though I fear that many leave school with the contrary impression that mathematics is a useless sorcery with x, y and z.

Students should leave school with several general impressions about discrete mathematics:

- Mathematical models can be continuous or discrete.
- Much optimization does not use calculus.
- Computation and the use of computers involves interesting mathematics.
- A key theme in mathematics is the method of reducing to the previous case (recursion).

What school discrete mathematics should not be. The guidelines above leave a lot of room, so let me narrow things somewhat by suggesting some restrictions. A high school discrete mathematics course should *not* be

- A computer science oriented course — there is a much broader clientele.
- Billed as an advanced placement course — there isn't such an advanced placement test and though the College Board has considered it, one isn't planned [**1**].
- Too formal (discrete structures) — this is not appropriate at the school level.

5. Concluding Remarks

What is discrete mathematics? If you wanted a 30-second definition, I have left you no better off than before you read this article. But if you sought some examples, the flavor, and the goals of discrete mathematics, and you wanted to recognize the different varieties, then I hope I have helped. If you wanted some ideas for what to include from discrete mathematics in the schools, I hope I have helped you as well.

References

[1] Bailey, Harold F., "The Status of Discrete Mathematics in the High Schools", this volume.

[2] Crisler, Nancy, Patience Fisher and Gary Froelich, *Discrete Mathematics Through Applications*, W. H. Freeman for COMAP (Consortium for Mathematics and its Applications), New York, 1994.

[3] COMAP, *For All Practical Purposes: Introduction to Contemporary Mathematics*, 3rd ed., W. H. Freeman, New York, 1994.

[4] Cozzens, Margaret B., and Richard D. Porter, *Mathematics with Calculus*, D. C. Heath, Lexington MA, 1987.

[5] Dossey, John A., Albert D. Otto, Lawrence E. Spence, and Charles Vanden Eynden, *Discrete Mathematics*, 2nd ed., Scott, Foresman, Glenview IL, 1993.

[6] Maurer, Stephen B., and Anthony Ralston, *Discrete Algorithmic Mathematics*, Addison-Wesley, Reading MA, 1991.

[7] ——— "Algorithms: You Can't Teach Discrete Mathematics without Them", *Discrete Mathematics Across the Curriculum, K–12*, 1991 NCTM Yearbook (Margaret J. Kenney and Christian R. Hirsch, eds.), NCTM, Reston VA, 1991, pp. 195–206.

[8] NCTM, *Curriculum and Evaluation Standards for School Mathematics*, NCTM, Reston VA, 1989.

[9] North Carolina School of Science and Mathematics (G. Barrett et al.), *Contemporary Precalculus through Applications*, Janson Publications, Providence RI, 1991.

[10] Sandefur, James T., *Discrete Dynamical Systems*, Oxford University Press, New York, 1990.

[11] University of Chicago School Mathematics Project (A. Peressini et al.), *Precalculus and Discrete Mathematics*, Scott-Foresman, Glenview IL, 1991.

Department of Mathematics and Statistics, Swarthmore College, Swarthmore PA 19081-1397

E-mail address: smaurer1@cc.swarthmore.edu

DIMACS Series in Discrete Mathematics
and Theoretical Computer Science
Volume **36**, 1997

A Comprehensive View of Discrete Mathematics: Chapter 14 of the New Jersey Mathematics Curriculum Framework

Joseph G. Rosenstein

Introduction

This article contains the chapter of the *New Jersey Mathematics Curriculum Framework*[1] which deals with discrete mathematics. The first three pages of the article describes what this document is and why it was written.

On May 1, 1996, the New Jersey Board of Education adopted core curriculum content standards in seven content areas, including mathematics.

These standards describe what all New Jersey students need to know and be able to do at the end of grades 4, 8, and 12. Statewide assessments reflecting these standards are being developed at these grade levels, and students will be expected to demonstrate that they meet these standards in order to graduate from high school.

The standards for mathematics includes a discrete mathematics standard; thus all New Jersey students will be expected to demonstrate understanding and proficiency in discrete mathematics.

The development and adoption of standards extended over a period of three years, and, as Director of the New Jersey Mathematics Coalition, I was very much involved at every step along the way. The mathematics standards represent what New Jersey mathematics educators believe are high achievable standards for all students in the state.

How will New Jersey teachers ensure that their students can meet these standards? During the past four years, the New Jersey Mathematics Coalition, working in collaboration with the New Jersey Department of Education and with an Eisenhower grant from the United States Department of Education, has developed a resource book, the *New Jersey Mathematics Curriculum Framework*; this 688-page document was developed to assist teachers

1991 *Mathematics Subject Classification.* Primary 00A05, 00A35.

[1]Rosenstein, Joseph G., Janet H. Caldwell, and Warren D. Crown, *New Jersey Mathematics Curriculum Framework*, New Jersey Mathematics Coalition, 1996.

and administrators in implementing the mathematics standards at both the classroom and the district level. The preliminary version was published in Spring 1995, and a revised version in December 1996. The preliminary version included the contributions of many New Jersey educators; the revised version incorporated the suggestions of many reviewers and reflected the standards adopted by the Board.

I am pleased to have spearheaded and directed this effort; we have produced a valuable guide for New Jersey teachers and, through its availability on the World Wide Web,[2] for those of other states. The *New Jersey Mathematics Curriculum Framework* is not intended to be a curriculum; rather it is intended to be a structure (i.e., a "framework") around which a district can build its own curriculum (or curricula). This particular framework, however, provides much more detail about the content of K-12 mathematics than any other state framework of which I am aware; for that reason, it should be a valuable resource to all teachers of mathematics.

What should students be expected to know and be able to do? The discrete mathematics standard, like the other mathematics standards (and those in other content areas), consists of a general statement about discrete mathematics followed by five or six statements, called "cumulative progress indicators", which describe what students should be able to do at each of the three grade levels. The discrete mathematics standard and cumulative progress indicators appear at the end of this Introduction.

How will teachers be able to reflect these indicators in their curricula? The discrete mathematics chapter of the *Framework* (like each of the other chapters) is intended to respond to this question. The chapter consists of a K-12 overview of discrete mathematics, followed by sections addressing five different grade levels; for each grade level there is a (self-contained) overview of discrete mathematics for that grade level, followed by a number of classroom activities that illustrate how each indicator could be addressed at that grade level. These materials are arranged in this article in the following sections:

1. Grades K-12 Overview
2. Grades K–2 Overview
3. Grades K–2 Indicators and Activities
4. Grades 3–4 Overview
5. Grades 3–4 Indicators and Activities
6. Grades 5–6 Overview
7. Grades 5–6 Indicators and Activities
8. Grades 7–8 Overview
9. Grades 7–8 Indicators and Activities
10. Grades 9–12 Overview
11. Grades 9–12 Indicators and Activities

[2] `http://dimacs.rutgers.edu/nj_math_coalition/framework.html/`

Note that because the materials for each grade level are self-contained, there is considerable overlap between the overviews (even numbered sections). Note also that all references for each grade level are provided at the end of the odd numbered sections.

The activities in this chapter are based on activities used by teachers in the DIMACS-sponsored and NSF-funded Leadership Program in Discrete Mathematics, which I have directed since its inception in 1989 (see article in this volume). The organization of discrete mathematics into five areas and the list of indicators, one for each area at each grade level, emerged from a series of discussions in 1993 by Rutgers University faculty associated with DIMACS. Although I have been responsible for the selection and writing of the activities, as well as the overall organization of the material, I would like to acknowledge the assistance I received from a number of people, including many participants in the Leadership Program, who reviewed and commented on drafts of this chapter. The expectation is that, through the wonders of the Web, the entire *Framework* and this chapter on discrete mathematics in particular will continue to evolve.

And now, Chapter 14 of the *New Jersey Mathematics Curriculum Framework*, which addresses the following standard and cumulative progress indicators of the *New Jersey Core Curriculum Content Standards:*

All students will apply the concepts and methods of discrete mathematics to model and explore a variety of practical situations.

Cumulative Progress Indicators

By the end of Grade 4, students:

1. Explore a variety of puzzles, games, and counting problems.
2. Use networks and tree diagrams to represent everyday situations.
3. Identify and investigate sequences and patterns found in nature, art, and music.
4. Investigate ways to represent and classify data according to attributes, such as shape or color, and relationships, and discuss the purpose and usefulness of such classification.
5. Follow, devise, and describe practical lists of instructions.

Building upon knowledge and skills gained in the preceding grades, by the end of Grade 8, students:

6. Use systematic listing, counting, and reasoning in a variety of contexts.
7. Recognize common discrete mathematical models, explore their properties, and design them for specific situations.
8. Experiment with iterative and recursive processes, with the aid of calculators and computers.
9. Explore methods for storing, processing, and communicating information.

10. Devise, describe, and test algorithms for solving optimization and search problems.

Building upon knowledge and skills gained in the preceding grades, by the end of Grade 12, students:

11. Understand the basic principlies of iteration, recursion, and mathematical induction.
12. Use basic principles to solve combinatorial and algorithmic problems.
13. Use discrete models to represent and solve problems.
14. Analyze iterative processes with the aid of calculators and computers.
15. Apply discrete methods to storing, processing, and communicating information.
16. Apply discrete methods to problems of voting, apportionment, and allocations, and use fundamental strategies of optimizaion to solve problems.

1. Grades K-12 Overview

Descriptive Statement. Discrete mathematics is the branch of mathematics that deals with arrangements of distinct objects. It includes a wide variety of topics and techniques that arise in everyday life, such as how to find the best route from one city to another, where the objects are cities arranged on a map. It also includes how to count the number of different combinations of toppings for pizzas, how best to schedule a list of tasks to be done, and how computers store and retrieve arrangements of information on a screen. Discrete mathematics is the mathematics used by decision-makers in our society, from workers in government to those in health care, transportation, and telecommunications. Its various applications help students see the relevance of mathematics in the real world.

Meaning and Importance. During the past 30 years, discrete mathematics has grown rapidly and has evolved into a significant area of mathematics. It is the language of a large body of science and provides a framework for decisions that individuals will need to make in their own lives, in their professions, and in their roles as citizens. Its many practical applications can help students see the relevance of mathematics to the real world. It does not have extensive prerequisites, yet it poses challenges to all students. It is fun to do, is often geometry based, and can stimulate an interest in mathematics on the part of students at all levels and of all abilities.

K-12 Development and Emphases. Although the term "discrete mathematics" may seem unfamiliar, many of its themes are already present in the classroom. Whenever objects are counted, ordered, or listed, whenever instructions are presented and followed, whenever games are played and analyzed, teachers are introducing themes of discrete mathematics. Through understanding these themes, teachers will be able to recognize and introduce them regularly in classroom situations. For example, when calling three students to work at the three segments of the chalkboard, the teacher might ask *In how many different orders can these three students work at the board?* Another version of the same question is *How many different ways, such as ABC, can you name a triangle whose vertices are labeled A, B, and C?* A similar, but slightly different question is *In how many different orders can three numbers be multiplied?*

Two important resources on discrete mathematics for teachers at all levels are the 1991 NCTM Yearbook *Discrete Mathematics Across the Curriculum K-12* and the 1997 DIMACS Volume *Discrete Mathematics in the Schools.* The material in this chapter is drawn from activities that have been reviewed and classroom-tested by the K-12 teachers in the Rutgers University Leadership Program in Discrete Mathematics over the past nine years; this program is funded by the National Science Foundation.

Students should learn to recognize examples of discrete mathematics in familiar settings, and explore and solve a variety of problems for which discrete techniques have proved useful. These ideas should be pursued throughout the school years. Students can start with many of the basic ideas in concrete settings, including games and general play, and progressively develop these ideas in more complicated settings and more abstract forms. Five major themes of discrete mathematics should be addressed at all K-12 grade levels — **systematic listing, counting, and reasoning; discrete mathematical modeling using graphs (networks) and trees; iterative (that is, repetitive) patterns and processes; organizing and processing information; and following and devising lists of instructions, called "algorithms," and using algorithms to find the best solution to real-world problems.** These five themes are discussed in the paragraphs below.

Students should use a variety of strategies to **systematically list and count** the number of ways there are to complete a particular task. For example, elementary school students should be able to make a list of all possible outcomes of a simple situation such as the number of outfits that can be worn using two coats and three hats. Middle school students should be able to systematically list and count the number of different four-block-high towers that can be built using blue and red blocks (see example below), or the number of possible routes from one location on a map to another, or the number of different "words" that can be made using five letters. High school students should be able to determine the number of possible orderings of an arbitrary number of objects and to describe procedures for listing and counting all such orderings. These strategies for listing and counting should be applied by both middle school and high school students to solve problems in probability.

Following is a list of all four-block-high towers that can be built using clear blocks and solid blocks. The 16 towers are presented in a systematic list — the first 8 towers have a clear block at the bottom and the second 8 towers have a solid block at the bottom; within each of these two groups, the first 4 towers have the second block clear, and the second 4 towers have the second block solid; etc.

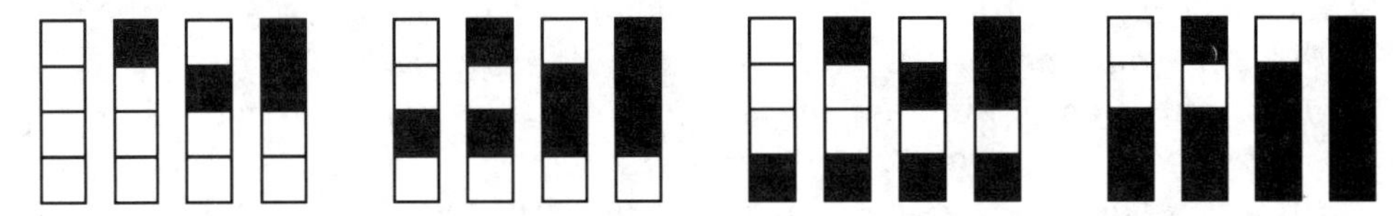

If each tower is described alphabetically as a sequence of `C`'s and `S`'s, representing "clear" and "solid" — the tower at the left, for example, would be `C-C-C-C`, and the third tower from the left would be `C-C-S-C`, reading from the bottom up — then the sixteen towers would be in alphabetical order:

```
C-C-C-C  C-S-C-C  S-C-C-C  S-S-C-C
C-C-C-S  C-S-C-S  S-C-C-S  S-S-C-S
C-C-S-C  C-S-S-C  S-C-S-C  S-S-S-C
C-C-S-S  C-S-S-S  S-C-S-S  S-S-S-S
```

There are other ways of systematically listing the 16 towers; for example, the list could contain first the one tower with no solid blocks, then the four towers with one solid block, then the six towers with two solid blocks, then the four towers with three solid blocks, and finally the one tower with four solid blocks.

Discrete mathematical models such as graphs (networks) and trees (such as those pictured below) can be used to represent and solve a variety of problems based on real-world situations.

In the left-most graph of the figures above, all seven dots are linked into a network consisting of the six line segments emerging from the center dot; these six line segments form the tree at the far right which is said to "span" the original graph since it reaches all of its points. Another example: if we think of the second graph as a street map and we make the streets one way, we can represent the situation using a directed graph where the line segments are replaced by arrows.

Elementary school students should recognize that a street map can be represented by a graph and that routes can be represented by paths in the graph; middle school students should be able to find cost-effective ways of linking sites into a network using spanning trees; and high school students should be able to use efficient methods to organize the performance of individual tasks in a larger project using directed graphs.

Iterative patterns and processes are used both for describing the world and in solving problems. An iterative pattern or process is one which involves repeating a single step or sequence of steps many times. For example, elementary school students should understand that multiplication corresponds to repeatedly adding the same number a specified number of times. They should investigate how decorative floor tilings can often be described as the repeated use of a small pattern, and how the patterns of rows in pine cones follow a simple mathematical rule. Middle school students should explore how simple repetitive rules can generate interesting patterns by using spirolaterals or Logo commands, or how they can result in extremely complex behavior by generating the beginning stages of fractal curves. They should investigate the ways that the plane can be covered by repeating patterns, called tessellations. High school students should understand how many processes describing the change of physical, biological, and economic systems over time can be modeled by simple equations applied repetitively, and use these models to predict the long-term behavior of such systems.

Students should explore different methods of **arranging, organizing, analyzing, transforming, and communicating information**, and understand how these methods are used in a variety of settings. Elementary school students should investigate ways to represent and classify data according to attributes such as color or shape, and to organize data into structures like tables or tree diagrams or Venn diagrams. Middle school students should be able to read, construct, and analyze tables, matrices, maps and other data structures. High school students should understand the application of discrete methods to problems of information processing and computing such as sorting, codes, and error correction.

Students should be able to **follow and devise lists of instructions, called "algorithms," and use them to find the best solution to real-world problems** — where "best" may be defined, for example, as most cost-effective or as most equitable. For example, elementary school students should be able to carry out instructions for getting from one location to another, should discuss different ways of dividing a pile of snacks, and should determine the shortest path from one site to another on a map laid out on the classroom floor. Middle school students should be able to plan an optimal route for a class trip (see the vignette in the Introduction to this *Framework* entitled *Short-circuiting Trenton*), write precise instructions for adding two two-digit numbers, and, pretending to be the manager of a fast-food restaurant, devise work schedules for employees which meet specified conditions yet minimize the cost. High school students should be conversant with fundamental strategies of optimization, be able to use flow charts to describe algorithms, and recognize both the power and limitations of computers in solving algorithmic problems.

IN SUMMARY, discrete mathematics is an exciting and appropriate vehicle for working toward and achieving the goal of educating informed citizens who are better able to function in our increasingly technological society; have better reasoning power and problem-solving skills; are aware of the importance of mathematics in our society; and are prepared for future careers which will require new and more sophisticated analytical and technical tools. It is an excellent tool for improving reasoning and problem-solving abilities.

Note: *Although each content standard is discussed in a separate chapter, it is not the intention that each be treated separately in the classroom. Indeed, as noted in the Introduction to this Framework, an effective curriculum is one that successfully integrates these areas to present students with rich and meaningful cross-strand experiences.*

References.

- Kenny, M. J., Ed. *Discrete Mathematics Across the Curriculum K-12.* 1991 Yearbook of the National Council of Teachers of Mathematics (NCTM). Reston, VA, 1991.
- Rosenstein, J. G., D. Franzblau, and F. Roberts, Eds. *Discrete Mathematics in the Schools.* Proceedings of a 1992 DIMACS Conference on "Discrete Mathematics in the Schools." DIMACS Series on Discrete Mathematics and Theoretical Computer Science. Providence, RI: American Mathematical Society (AMS), 1997.

On-Line Resources.

`http://dimacs.rutgers.edu/nj_math_coalition/framework.html/`

The *Framework* will be available at this site during Spring 1997. In time, we hope to post additional resources relating to this standard, such as grade-specific activities submitted by New Jersey teachers, and to provide a forum to discuss the *Mathematics Standards.*

2. Grades K-2 Overview

The five major themes of discrete mathematics, as discussed in the K-12 Overview,[3] are **systematic listing, counting, and reasoning; discrete mathematical modeling using graphs (networks) and trees; iterative (that is, repetitive) patterns and processes; organizing and processing information; and following and devising lists of instructions, called "algorithms," and using them to find the best solution to real-world problems.**

Despite their formidable titles, these five themes can be addressed with activities at the K-2 grade level which involve purposeful play and simple analysis. Indeed, teachers will discover that many activities they already are using in their classrooms reflect these themes. These five themes are discussed in the paragraphs below.

[3]Since K-2 grade level teachers may not read the K-12 Overview, and, more generally, teachers at other grade levels will begin their review of this chapter of the *Framework* by turning to the section addressing their own grade levels, the grade level overviews have significant overlap.

Activities involving **systematic listing, counting, and reasoning** can be done very concretely at the K-2 grade level. For example, dressing cardboard teddy bears with different outfits becomes a mathematical activity when the task is to make a list of all possible outfits and count them; pictured below are the six outfits that can be arranged using one of two types of shirts and one of three types of shorts. Similarly, playing any game involving choices becomes a mathematical activity when children reflect on the moves they make in the game.

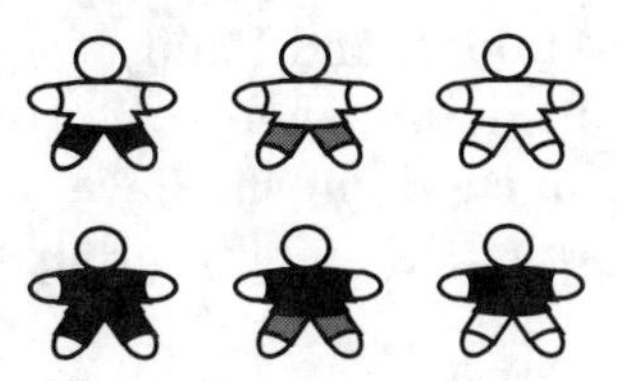

An important **discrete mathematical model** is that of a **network or graph**, which consists of dots and lines joining the dots; the dots are often called *vertices* (*vertex* is the singular) and the lines are often called *edges.* (This is different from other mathematical uses of the term "graph.") The two terms "network" and "graph" are used interchangeably for this concept. An example of a graph with seven vertices and twelve edges is given below. You can think of the vertices of this graph as islands in a river and the edges as bridges. You can also think of them as buildings and roads, or houses and telephone cables, or people and handshakes; wherever a collection of things are joined by connectors, the mathematical model used is that of a network or graph. At the K-2 level, children can recognize graphs and use life-size models of graphs in various ways. For example, a large version of this graph, or any other graph, can be "drawn" on the floor using paper plates as vertices and masking tape as edges. Children might select two "islands" and find a way to go from one island to the other island by crossing exactly four "bridges." (This can be done for any two islands in this graph, but not necessarily in another graph.)

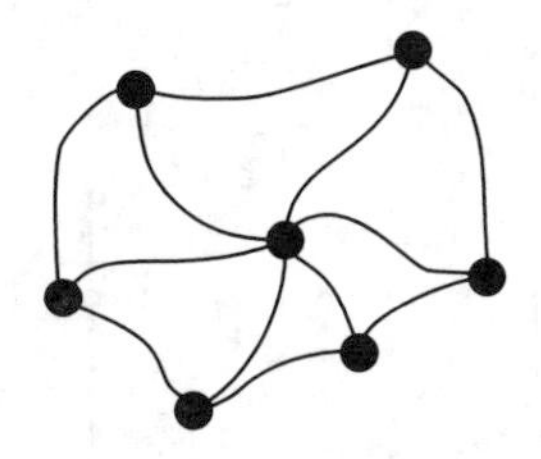

Children can recognize and work with **repetitive patterns and processes** involving numbers and shapes, using objects in the classroom and in the world around them. For example, children at the K-2 level can create (and decorate) a pattern of triangles or squares (as pictured here) that cover a section of the floor (this is called a "tessellation"), or start with a number and repeatedly add three, or use clapping and movement to simulate rhythmic patterns.

Children at the K-2 grade levels should investigate ways of **sorting items** according to attributes like color, shape, or size, and ways of **arranging data** into charts, tables, and family trees. For example, they can sort attribute blocks or stuffed animals by color or kind, as in the diagram, and can count the number of children who have birthdays in each month by organizing themselves into birthday-month groups.

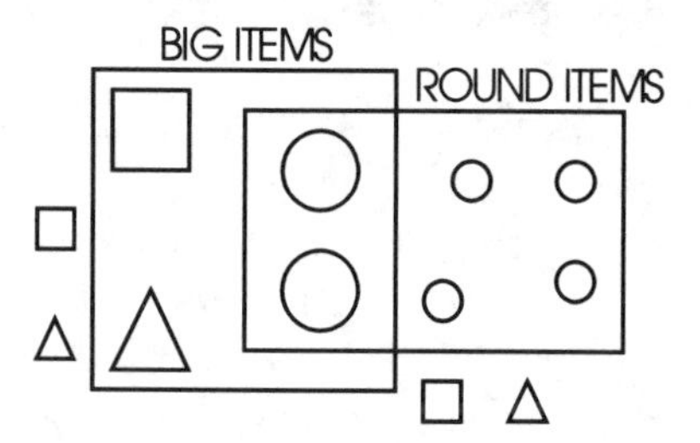

Finally, at the K-2 grade levels, children should be able to **follow and describe simple procedures** and determine and discuss **what is the best solution** to a problem. For example, they should be able to follow a prescribed route from the classroom to another room in the school (as pictured below) and to compare various alternate routes, and in the second grade should determine the shortest path from one site to another on a map laid out on the classroom floor.

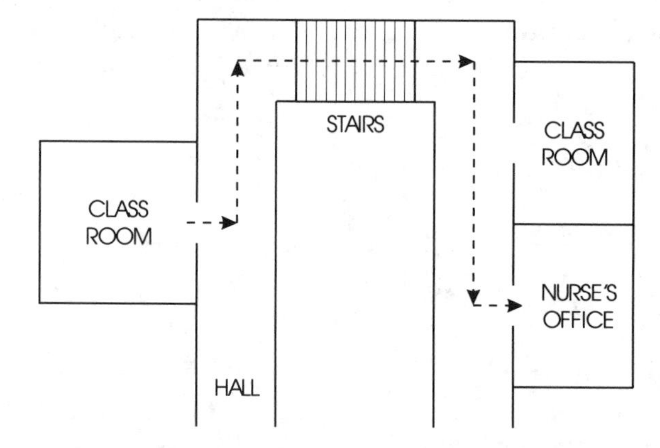

Two important resources on discrete mathematics for teachers at all levels are the 1991 NCTM Yearbook *Discrete Mathematics Across the Curriculum K-12* and the 1997 DIMACS Volume *Discrete Mathematics in the Schools.* Another important resource for K-2 teachers is *This Is MEGA-Mathematics!*

3. Grades K-2 Indicators and Activities

The cumulative progress indicators for grade 4 appear below in boldface type. Each indicator is followed by activities which illustrate how it can be addressed in the classroom in kindergarten and grades 1 and 2.

Experiences will be such that all students in grades K-2:

1. **Explore a variety of puzzles, games, and counting problems.**

- Students use teddy bear cut-outs with, for example, shirts of two colors and shorts of three colors, and decide how many different outfits can be made by making a list of all possibilities and arranging them systematically. (See illustration in K-2 Overview.)
- Students use paper faces or Mr. Potato Head type models to create a "regular face" given a nose, mouth, and a pair of eyes. Then they use another pair of eyes, then another nose, and then another mouth (or other parts) and explore and record the number of faces that can be made after each additional part has been included.
- Students read *A Three Hat Day* and then try to create as many different hats as possible with three hats, a feather, a flower, and a ribbon as decoration. Students count the different hats they've made and discuss their answers.
- Students count the number of squares of each size (1×1, 2×2, 3×3) that they can find on the square grid below. They can be challenged to find the numbers of small squares of each size on a larger square or rectangular grid.

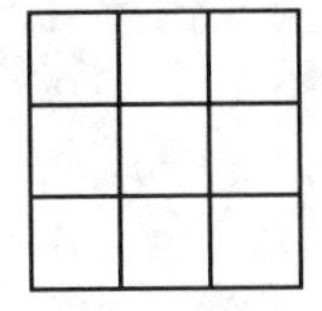

- Students work in groups to figure out the rules of addition and placement that are used to pass from one row to the next in the diagram below, and use these rules to find the numbers in the next few rows.

$$\begin{array}{c} 1 \\ 1\ 1 \\ 1\ 2\ 1 \\ 1\ 3\ 3\ 1 \\ 1\ 4\ 6\ 4\ 1 \end{array}$$

In this diagram, called Pascal's triangle, each number is the sum of the two numbers that are above it, to its left and right; the numbers on the left and right edges are all 1.

- Students cut out five "coins" labeled 1¢, 2¢, 4¢, 8¢, and 16¢. For each number in the counting sequence $1, 2, 3, 4, 5, \ldots$ (as far as is appropriate for a particular group of students), students determine how to obtain that amount of money using a combination of different coins.

- Students play simple games and discuss why they make the moves they do. For example, two students divide a six-piece domino set (with 0-0, 0-1, 0-2, 1-1, 1-2, and 2-2) and take turns placing dominoes so that dominoes which touch have the same numbers and so that all six dominoes are used in the chain.

2. **Use networks and tree diagrams to represent everyday situations.**

- Students find a way of getting from one island to another, in the graph described in the K-2 Overview laid out on the classroom floor with masking tape, by crossing exactly four bridges. They make their own graphs, naming each of the islands, and make a "from-to" list of islands for which they have found a four-bridge-route. (Note: it may not always be possible to find four-bridge-routes.)
- Students count the number of edges at each vertex (called the **degree** of the vertex) of a network and construct graphs where all vertices have the same degree, or where all the vertices have one of two specified degrees.
- On a pattern of islands and bridges laid out on the floor, students try to find a way of visiting each island exactly once; they can leave colored markers to keep track of islands already visited. Note that for some patterns this may not be possible! Students can be challenged to find a way of visiting each island exactly once which returns them to their starting point. Similar activities can be found in *Inside, Outside, Loops, and Lines* by Herbert Kohl.

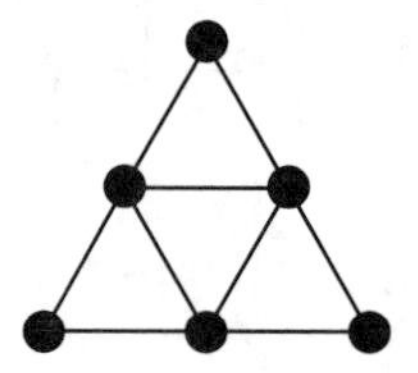

- Students create a map with make-believe countries (see example below), and color the maps so that countries which are next to each other have different colors. *How many colors were used? Could it be done with fewer colors? with four colors? with three colors? with two colors?* A number of interesting map coloring ideas can be found in *Inside, Outside, Loops and Lines* by Herbert Kohl.

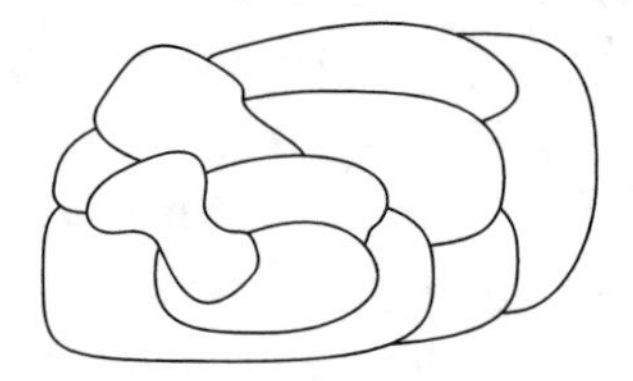

3. **Identify and investigate sequences and patterns found in nature, art, and music.**

   - Students use a calculator to create a sequence of ten numbers starting with zero, each of which is three more than the previous one; on some calculators, this can be done by pressing $0 + 3 = = = . . .$, where = is pressed ten times. As they proceed, they count one 3, two 3s, three 3s, etc.
   - Students "tessellate" the plane, by using groups of squares or triangles (for example, from sets of pattern blocks) to completely cover a sheet of paper without overlapping; they record their patterns by tracing around the blocks on a sheet of paper and coloring the shapes.

   - Students listen to or read *Grandfather Tang's Story* by Ann Tompert and then use tangrams to make the shape-changing fox fairies as the story progresses. Students are then encouraged to do a retelling of the story with tangrams or to invent their own tangram characters and stories.
   - Students read *The Cat in the Hat* or *Green Eggs and Ham* by Dr. Seuss and identify the pattern of events in the book. Students could create their own books with similar patterns.
   - Students collect leaves and note the patterns of the veins. They look at how the veins branch off on each side of the center vein and observe that their branches are smaller copies of the original vein pattern. Students collect feathers, ferns, Queen Anne's lace, broccoli, or cauliflower and note in each case how the pattern of the original is repeated in miniature in each of its branches or clusters.
   - Students listen for rhythmic patterns in musical selections and use clapping, instruments, and movement to simulate those patterns.
   - Students take a "patterns walk" through the school, searching for patterns in the bricks, the play equipment, the shapes in the classrooms, the number sequences of classrooms, the floors and ceilings, etc.; the purpose of this activity is to create an awareness of all the patterns around them.

4. **Investigate ways to represent and classify data according to attributes, such as shape or color, and relationships, and discuss the purpose and usefulness of such classification.**

- Students sort themselves by month of birth, and then within each group by height or birth date. (Other sorting activities can be found in *Mathematics Their Way*, by Mary Baratta-Lorton.)
- Each student is given a card with a different number on it. Students line up in a row and put the numbers in numerical order by exchanging cards, one at a time, with adjacent children. (After practice, this can be accomplished without talking.)
- Students draw stick figures of members of their family and arrange them in order of size.
- Students sort stuffed animals in various ways and explain why they sorted them as they did. Students can use *Tabletop, Jr.* software to sort characters according to a variety of attributes.
- Using attribute blocks, buttons, or other objects with clearly distinguishable attributes such as color, size, and shape, students develop a sequence of objects where each differs from the previous one in only one attribute. *Tabletop, Jr.* software can also be used to create such sequences of objects.
- Students use two Hula Hoops (or large circles drawn on paper so that a part of their interiors overlap) to assist in sorting attribute blocks or other objects according to two characteristics. For example, given a collection of objects of different colors and shapes, students are asked to place them so that all red items go inside hoop #1 and all others go on the outside, and so that all square items go inside hoop #2 and all others go on the outside. *What items should be placed in the overlap of the two hoops? What is inside only the first hoop? What is outside both hoops?*

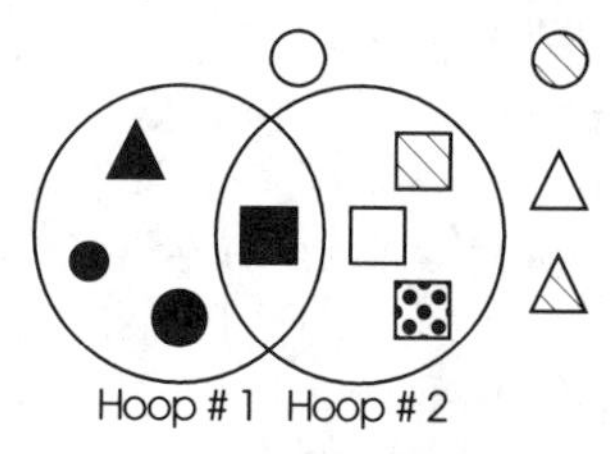

This is an example of a Venn diagram. Students can also use Venn diagrams to organize the similarities and differences between the information in two stories by placing all features of the first story in hoop #1 and all features of the second story in hoop #2, with common features in the overlap of the two hoops. A similar activity can be found in the *Shapetown* lesson that is described in the First Four Standards of this *Framework*. *Tabletop, Jr.* software allows students to arrange and sort data, and to explore these concepts easily.

5. **Follow, devise, and describe practical lists of instructions.**

- Students follow directions for a trip within the classroom — for example, students are asked where they would end up if they started at a given spot facing in a certain direction, took three steps forward, turned left, took two steps forward, turned right, and moved forward three more steps.

- Students follow oral directions for going from the classroom to the lunchroom, and represent these directions with a diagram. (See K-2 Overview for a sample diagram.)
- Students agree on a procedure for filling a box with rectangular blocks. For example, a box with dimensions 4"×4"×5" can be filled with 10 blocks of dimensions 1"×2"×4". (Linking cubes can be used to create the rectangular blocks.)
- Students explore the question of finding the shortest route from school to home on a diagram like the one pictured below, laid out on the floor using masking tape, where students place a number of counters on each line segment to represent the length of that segment. (The shortest route will depend on the placement of the counters; what appears to be the most direct route may not be the shortest.)

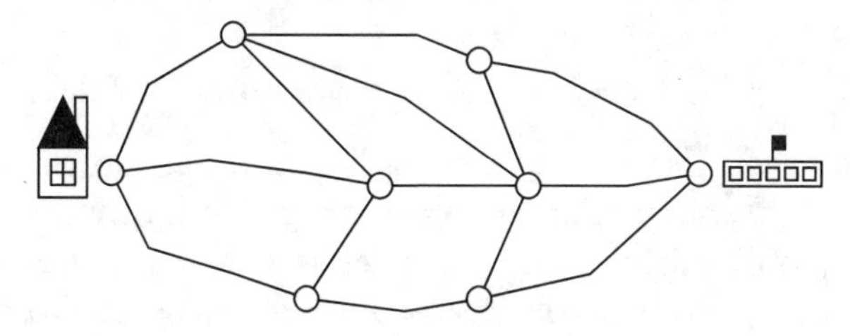

- Students find a way through a simple maze. They discuss the different paths they took and their reasons for doing so.
- Students use Logo software to give the turtle precise instructions for movement in specified directions.

References.

- Baratta-Lorton, Mary. *Mathematics Their Way.* Menlo Park, CA: Addison Wesley, 1993.
- Casey, Nancy, and Mike Fellows. *This is MEGA-Mathematics! — Stories and Activities for Mathematical Thinking, Problem-Solving, and Communication.* Los Alamos, CA: Los Alamos National Laboratories, 1993. (A version is available online at http://www.c3.lanl.gov/mega-math)
- Geringer, Laura. *A Three Hat Day.* New York: Harper Row Junior Books, 1987.
- Kenney, M. J., Ed. *Discrete Mathematics Across the Curriculum K-12.* 1991 Yearbook of the National Council of Teachers of Mathematics (NCTM). Reston, VA: 1991.
- Kohl, Herbert. *Insides, Outsides, Loops, and Lines.* New York: W. H. Freeman, 1995.

- Murphy, Pat. *By Nature's Design.* San Francisco, CA: Chronicle Books, 1993.
- Rosenstein, J. G., D. Franzblau, and F. Roberts, Eds. *Discrete Mathematics in the Schools.* Proceedings of a 1992 DIMACS Conference on "Discrete Mathematics in the Schools." DIMACS Series on Discrete Mathematics and Theoretical Computer Science. Providence, RI: American Mathematical Society (AMS), 1997.
- Seuss, Dr. *Cat in the Hat.* Boston, MA: Houghton Mifflin, 1957.
- Seuss, Dr. *Green Eggs and Ham.* Random House.
- Tompert, Ann. *Grandfather Tang's Story.* Crown Publishing, 1990.

Software.

- *Logo.* Many versions of Logo are commercially available.
- *Tabletop, Jr.* Broderbund Software. TERC.

4. Grades 3-4 Overview

The five major themes of discrete mathematics, as discussed in the K-12 Overview, are **systematic listing, counting, and reasoning; discrete mathematical modeling using graphs (networks) and trees; iterative (that is, repetitive) patterns and processes; organizing and processing information; and following and devising lists of instructions, called "algorithms," and using them to find the best solution to real-world problems.**

Despite their formidable titles, these five themes can be addressed with activities at the 3-4 grade level which involve purposeful play and simple analysis. Indeed, teachers will discover that many activities that they already are using in their classrooms reflect these themes. These five themes are discussed in the paragraphs below.

The following discussion of activities at the 3-4 grade levels in discrete mathematics presupposes that corresponding activities have taken place at the K-2 grade levels. Hence 3-4 grade teachers should review the K-2 grade level discussion of discrete mathematics and might use activities similar to those described there before introducing the activities for this grade level.

Activities involving **systematic listing, counting, and reasoning** should be done very concretely at the 3-4 grade levels, building on similar activities at the K-2 grade levels. For example, the children could systematically list and count the total number of possible combinations of dessert and beverage that can be selected from pictures of those two types of foods they have cut out of magazines or that can be selected from a restaurant menu. Similarly, playing games like Nim, dots and boxes, and dominoes becomes a mathematical activity when children systematically reflect on the moves they make in the game and use those reflections to decide on the next move.

An important **discrete mathematical model** is that of a **graph**, which is used whenever a collection of things are joined by connectors — such as buildings and roads, islands and bridges, or houses and telephone cables — or, more abstractly, whenever the objects have some defined relationship to each other; this kind of model is described in the K-2 Overview. At the 3-4 grade levels, children can recognize and use models of graphs in various ways, for example, by finding a way to get from one island to another by crossing exactly four bridges, or by finding a route for a city mail carrier which uses each street once, or by constructing a collaboration graph for the class which describes who has worked with whom during the past week. A special kind of graph is called a "tree." Three views of the same tree are pictured in the diagram below; the first suggests a family tree, the second a tree diagram, and the third a "real" tree.

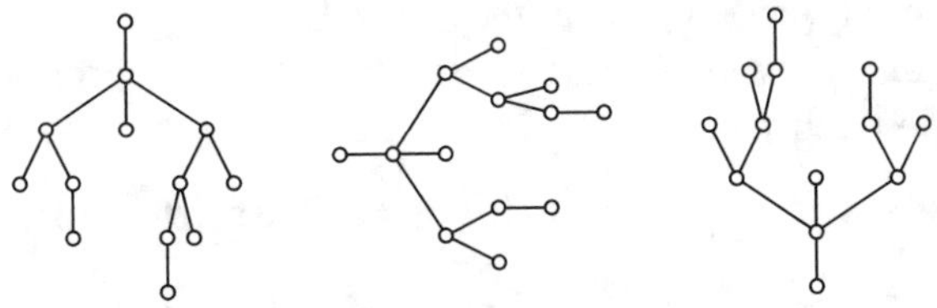

At the 3-4 grade levels, students can use a tree diagram to organize the six ways that three people can be arranged in order. (See the Grades 3-4 Indicators and Activities for an example.)

Students can recognize and work with **repetitive patterns and processes** involving numbers and shapes, with classroom objects and in the world around them. Children at the 3-4 grade levels are fascinated with the Fibonacci sequence of numbers 1, 1, 2, 3, 5, 8, 13, 21, 34, 55, 89, ... where every number is the sum of the previous two numbers. This sequence of numbers turns up in petals of flowers, in the growth of populations (see the activity involving rabbits), in pineapples and pine cones, and in lots of other places in nature. Another important sequence to introduce at this age is the doubling sequence 1, 2, 4, 8, 16, 32, ... and to discuss different situations in which it appears.

Students at the 3-4 grade levels should investigate ways of **sorting items** according to attributes like color or shape, or by quantitative information like size, **arranging data** using tree diagrams and building charts and tables, and **recovering hidden information** in games and encoded messages. For example, they can sort letters into zip code order or sort the class alphabetically, create bar charts based on information obtained experimentally (such as soda drink preferences of the class), and play games like hangman to discover hidden messages.

Students at the 3-4 grade levels should **describe and discuss simple algorithmic procedures** such as providing and following directions from one location to another, and should in simple cases determine and discuss **what is the best solution** to a problem. For example, they might follow a recipe to make a cake or to assemble a simple toy from its component parts.

Or they might find the best way of playing tic-tac-toe or the shortest route that can be used to get from one location to another.

Two important resources on discrete mathematics for teachers at all levels are the 1991 NCTM Yearbook *Discrete Mathematics Across the Curriculum K-12* and the 1997 DIMACS Volume *Discrete Mathematics in the Schools*. Another important resource for 3-4 teachers is *This Is MEGA-Mathematics!*

5. Grades 3-4 Indicators and Activities

The cumulative progress indicators for grade 4 appear below in boldface type. Each indicator is followed by activities which illustrate how it can be addressed in the classroom in grades 3 and 4.

Building upon knowledge and skills gained in the preceding grades, experiences will be such that all students in grades 3-4:

1. **Explore a variety of puzzles, games, and counting problems.**

- Students read *One Hundred Hungry Ants* by Elinor Pinczes and then illustrate and write their own story books (perhaps titled *18 Ailing Alligators* or *24 Furry Ferrets*) in a style similar to the book using as many different arrangements of the animals as possible in creating their books. They read their books to students in the lower grades.

- Students count the number of squares of each size (1×1, 2×2, 3×3, 4×4, 5×5) that they can find on a geoboard, and in larger square or rectangular grids.

- Students determine the number of possible combinations of dessert and beverage that could be selected from pictures of those two types of foods they have cut out of magazines. Subsequently, they determine the number of possible combinations of dessert and beverage that could be chosen from a restaurant menu, and how many of those combinations could be ordered if they only have $4.

- Students find the number of different ways to make a row of four flowers each of which could be red or yellow. They can model this with Unfix cubes and explain how they know that all combinations have been obtained.

- Students determine the number of different ways any three people can be arranged in order, and use a tree diagram to organize the information. The tree diagram below represents the six ways that Barbara (B), Maria (M), and Tarvanda (T), can be arranged in order. The three branches emerging from the "start" position represent the three people who could be first; each path from left to right represents the arrangement of the three people listed to the right.

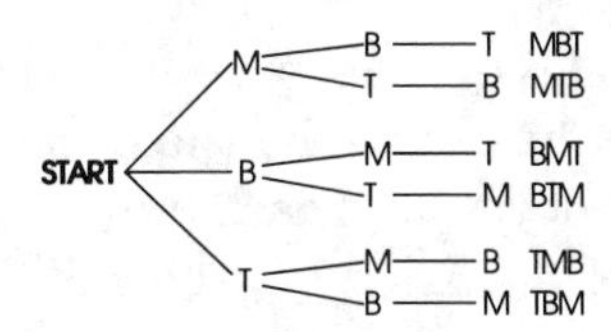

- Each student uses four squares to make designs where each square shares an entire side with at least one of the other three squares. Geoboards, attribute blocks or Linker cubes can be used. *How many different shapes can be made?* These shapes are called "tetrominoes."
- Each group of students receives a bag containing four colored beads. One group may be given 1 red, 1 black and 2 green beads; other groups may have the same four beads or different ones. Students take turns drawing a bead from the bag, recording its color, and replacing it in the bag. After 20 beads are drawn, each group makes a bar graph illustrating the number of beads drawn of each color. They make another bar graph illustrating the number of beads of each color actually in the bag, and compare the two bar graphs. As a follow-up activity, students should draw 20 or more times from a bag containing an unknown mixture of beads and try to guess, and justify, how many beads of each color are in the container.
- Students determine what amounts of postage can and cannot be made using only 3¢ and 5¢ stamps.
- Students generate additional rows of Pascal's triangle (below). They color all odd entries one color and all even entries another color. They examine the patterns that result, and try to explain what they see. They discuss whether their conclusions apply to a larger version of Pascal's triangle.

$$\begin{array}{c} 1 \\ 1\ 1 \\ 1\ 2\ 1 \\ 1\ 3\ 3\ 1 \\ 1\ 4\ 6\ 4\ 1 \end{array}$$

- Students make a table indicating which stamps of the denominations 1¢, 2¢, 4¢, 8¢, 16¢, 32¢ would be used (with no repeats) to obtain each amount of postage from 1¢ to 63¢. For the table, they list the available denominations across the top and the postage amounts from 1¢ to 63¢ at the left; they put a checkmark in the appropriate spot if they need the stamp for that amount, and leave it blank otherwise. They try to find a pattern which could be used to decide which amounts of postage could be made if additional stamps (like 64¢ and 128¢) were used.
- Students play games like Nim and reflect on the moves they make in the game. (See *Math for Girls and Other Problem Solvers*, by D. Downie et al., for other games for this grade level.) In Nim, you start

with a number of piles of objects — for example, you could start with two piles, one with five buttons, the other with seven buttons. Two students alternate moves, and each move consists of taking some or all of the buttons from a single pile; the child who takes the last button off the table wins the game. Once they master this game, students can try Nim with three piles, starting with three piles which have respectively 1, 2, and 3 buttons.

- Students play games like *dots and boxes* and systematically think about the moves they make in the game. In dots and boxes, you start with a square (or rectangular) array of dots, and two students alternate drawing a line which joins two adjacent dots. Whenever all four sides of a square have been drawn, the student puts her or his initial in the square and draws another line; the person with initials in more squares wins the game.

2. **Use networks and tree diagrams to represent everyday situations.**

- Students make a collaboration graph for the members of the class which describes who has worked with whom during the past week.
- Students draw specified patterns on the chalkboard without retracing, such as those below. Alternatively, they may trace these patterns in a small box of sand, as done historically in African cultures. (See *Ethnomathematics, Drawing Pictures With One Line*, or *Insides, Outsides, Loops, and Lines.*) Alternatively, on a pattern of islands and bridges laid out on the floor with masking tape, students might try to take a walk which involves crossing each bridge exactly once (leaving colored markers on bridges already crossed); note that for some patterns this may not be possible. The patterns given here can be used, but students can develop their own patterns and try to take such a walk for each pattern that they create.

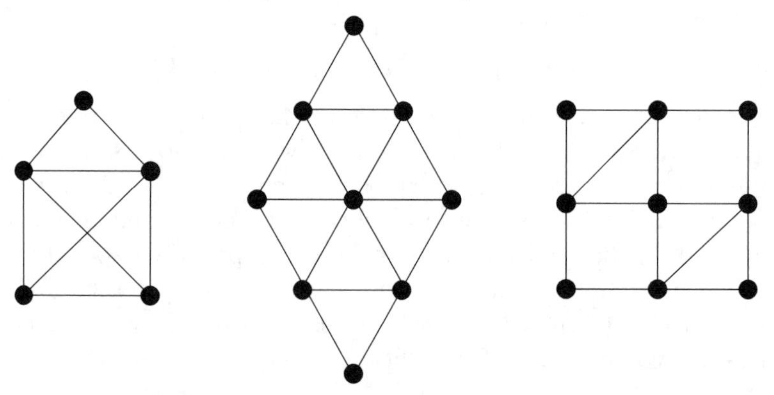

- Students create "human graphs" where they themselves are the vertices and they use pieces of yarn (several feet long) as edges; each piece of yarn is held by two students, one at each end. They might create graphs with specified properties; for example, they might create a human graph with four vertices of degree 2, or, as in the figure below, with six vertices of which four have degree 3 and two have degree 2. (The *degree* of a vertex is the total number of edges that meet at the vertex.) They might count the number of different shapes of human graphs they can form with four students (or five, or six).

- Students use a floor plan of their school to map out alternate routes from their classroom to the school's exits, and discuss whether the fire drill route is in fact the shortest route to an exit.
- Students draw graphs of their own neighborhoods, with edges representing streets and vertices representing locations where roads meet. *Can you find a route for the mail carrier in your neighborhood which enables her to walk down each street, without repeating any streets, and which ends where it begins? Can you find such a route if she needs to walk up and down each street in order to deliver mail on both sides of the street?*
- Students color maps (e.g., the 21 counties of New Jersey) so that adjacent counties (or countries) have different colors, using as few colors as possible. The class could then share a NJ cake frosted accordingly. (See *The Mathematician's Coloring Book.*)
- Students recognize and understand family trees in social and historical studies, and in stories that they read. Where appropriate, they create their own family trees.

3. **Identify and investigate sequences and patterns found in nature, art, and music.**

- Students read *A Cloak for a Dreamer* by A. Friedman, and make outlines of cloaks or coats like those worn by the sons of the tailor in the book by tracing their upper bodies on large pieces of paper. Students could use pattern blocks or pre-cut geometric shapes to cover (tessellate) the paper cloaks with patterns like those in the book or try to make their own cloth designs.

- Students read *Sam and the Blue Ribbon Quilt* by Lisa Ernst, and by rotating, flipping, or sliding cut-out squares, rectangles, triangles, etc., create their own symmetrical designs on quilt squares similar to those found in the book. The designs from all the members of the class are put together to make a patchwork class quilt or to form the frame for a math bulletin board.
- Students take a "pattern walk" through the neighborhood, searching for patterns in the trees, the houses, the buildings, the manhole covers (by the way, *why are they always round?*), the cars, etc.; the purpose of this activity is to create an awareness of the patterns around us. *By Nature's Design* is a photographic journey with an eye for many of these natural patterns.
- Students "tessellate" the plane using squares, triangles, or hexagons to completely cover a sheet of paper without overlapping. They also tessellate the plane using groups of shapes, like hexagons and triangles as in the figure below.

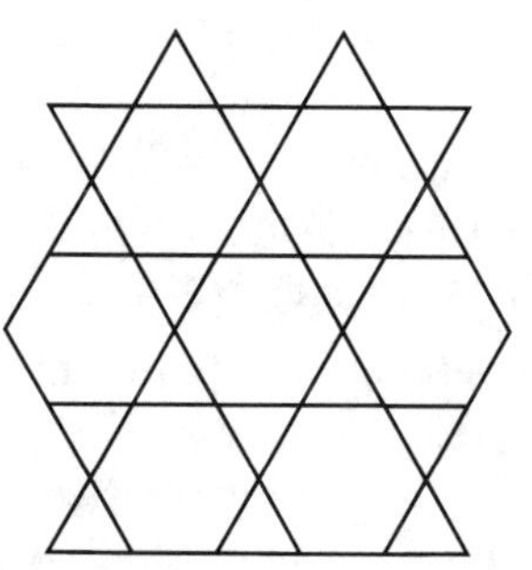

- Students might ask if their parents would be willing to give them a penny for the first time they do a particular chore, two pennies for the second time they do the chore, four pennies for the third time, eight pennies for the fourth time, and so on. Before asking, they should investigate, perhaps using towers of Unifix cubes that keep doubling in height, how long their parents could actually afford to pay them for doing the chore.
- Students cut a sheet of paper into two halves, cut the resulting two pieces into halves, cut the resulting four pieces into halves, etc. *If they do this a number of times, say 12 times, and stacked all the pieces of paper on top of each other, how high would the pile of paper be?* Students estimate the height before performing any calculations.
- Students color half a large square, then half of the remaining portion with another color, then half of the remaining portion with a third color, etc. *Will the entire area ever get colored? Why, or why not?*
- Students count the number of rows of bracts on a pineapple or pine cone, or rows of petals on an artichoke, or rows of seeds on a sunflower, and verify that these numbers all appear in the sequence 1, $1, 2, 3, 5, 8, 13, 21, 34, \ldots$ of Fibonacci numbers, where each number is

the sum of the two previous numbers on the list. Students find other pictures depicting Fibonacci numbers as they arise in nature, referring, for example, to *Fibonacci Numbers in Nature.* In *Mathematical Mystery Tour* by Mark Wahl, an elementary school teacher provides a year's worth of Fibonacci explorations and activities.

- Using a large equilateral triangle provided by the teacher, students find and connect the approximate midpoints of the three sides, and then color the triangle in the middle. (See Stage 1 picture.) They then repeat this procedure with each of the three uncolored triangles to get the Stage 2 picture, and then repeat this procedure again with each of the nine uncolored triangles to get the Stage 3 picture. These are the first three stages of the Sierpinski triangle; subsequent stages become increasingly intricate. *How many uncolored triangles are there in the Stage 3 picture? How many would there be in the Stage 4 picture if the procedure were repeated again?*

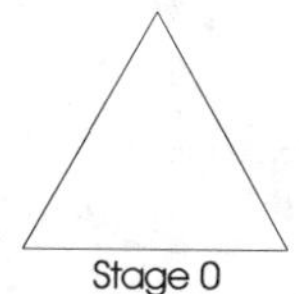
Stage 0

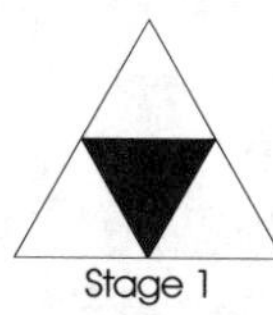
Stage 1

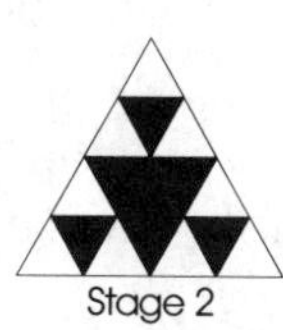
Stage 2

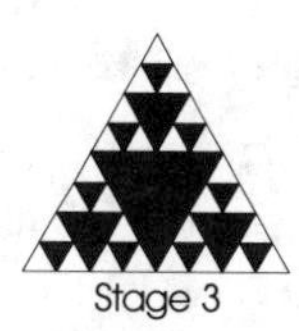
Stage 3

4. **Investigate ways to represent and classify data according to attributes, such as shape or color, and relationships, and discuss the purpose and usefulness of such classification.**

- Students are provided with a set of index cards on each of which is written a word (or a number). Working in groups, students put the cards in alphabetical (or numerical) order, explain the methods they used to do this, and then compare the various methods that were used.

- Students bring to class names of cities and their zip codes where their relatives and friends live, paste these at the appropriate locations on a map of the United States, and look for patterns which might explain how zip codes are assigned. Then they compare their conclusions with post office information to see whether they are consistent with the way that zip codes actually are assigned.

- Students send and decode messages in which each letter has been replaced by the letter which follows it in the alphabet (or occurs two letters later). Students explore other coding systems described in *Let's Investigate Codes and Sequences* by Marion Smoothey.

- Students collect information about the soft drinks they prefer and discuss various ways of presenting the resulting information, such as tables, bar graphs, and pie charts, displayed both on paper and on a computer.

- Students play the game of *Set* in which participants try to identify three cards from those on display which, for each of four attributes (number, shape, color, and shading), all share the attribute or are all different. Similar ideas can be explored using *Tabletop, Jr.* software.

5. **Follow, devise, and describe practical lists of instructions.**

- Students follow a recipe to make a cake or to assemble a simple toy from its component parts, and then write their own versions of those instructions.
- Students give written and oral directions for going from the classroom to another room in the school, and represent these directions with a diagram drawn approximately to scale.
- Students read *Anno's Mysterious Multiplying Jar* by Mitsumasa Anno. During a second reading they devise a method to record and keep track of the increasing number of items in the book and predict how that number will continue to grow. Each group explains its method to the class.
- Students write step-by-step directions for a simple task like making a peanut butter and jelly sandwich, and follow them to prove that they work.
- Students find and describe the shortest path from the computer to the door or from one location in the school building to another.
- Students find the shortest route from school to home on a map (see figure below), where each edge has a specified numerical length in meters; students modify lengths to obtain a different shortest route.

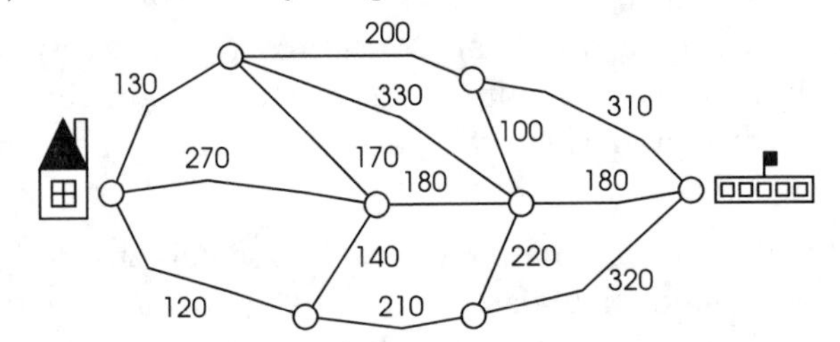

- Students write a program which will create specified pictures or patterns, such as a house or a clown face or a symmetrical design. Logo software is well-suited to this activity. In *Turtle Math,* students use Logo commands to go on a treasure hunt, and look for the shortest route to complete the search.
- Working in groups, students create and explain a fair way of sharing a bagful of similar candies or cookies. (See also the vignette entitled *Sharing A Snack* in the Introduction to this *Framework.*) For example, if the bag has 30 brownies and there are 20 children, then they might suggest that each child gets one whole brownie and that the teacher divide each of the remaining brownies in half. Or they might suggest that each pair of children figure out how to share one

brownie. *What if there were 30 hard candies instead of brownies? What if there were 25 brownies? What if there were 15 brownies and 15 chocolate chip cookies?* The purpose of this activity is for students to brainstorm possible solutions in the situations where there may be no solution that *everyone* perceives as fair.

- Students devise a strategy for never losing at tic-tac-toe.
- Students find different ways of paving just enough streets of a "muddy city" (like the street map below, perhaps laid out on the floor) so that a child can walk from any one location to any other location along paved roadways. In "muddy city" none of the roads are paved, so that whenever it rains all streets turn to mud. The mayor has asked the class to propose different ways of paving the roads so that a person can get from any one location to any other location on paved roads, but so that the fewest number of roads possible are paved.

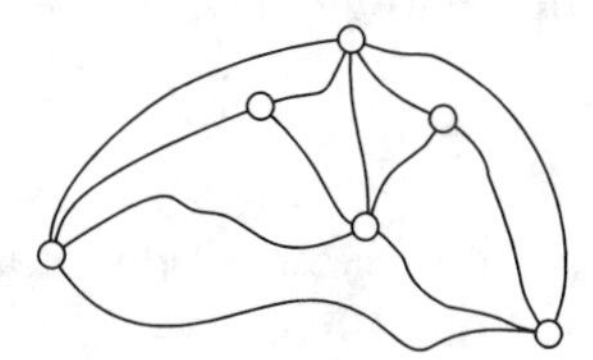

- Students divide a collection of Cuisenaire rods of different lengths into two or three groups whose total lengths are equal (or as close to equal as possible).

References.

- Anno, M. *Anno's Mysterious Multiplying Jar.* Philomel Books, 1983.
- Asher, M. *Ethnomathematics.* Brooks/Cole Publishing Company, 1991.
- Casey, Nancy, and Mike Fellows. *This is MEGA-Mathematics! - Stories and Activities for Mathematical Thinking, Problem-Solving, and Communication.* Los Alamos, CA: Los Alamos National Laboratories, 1993. (A version is available online at `http://www.c3.lanl.gov/mega-math`)
- Chavey, Darrah. *Drawing Pictures with One Line: Exploring Graph Theory.* Consortium for Mathematics and Its Applications (COMAP), Module #21, 1992.
- Downie, D., T. Slesnick, and J. Stenmark. *Math for Girls and Other Problem Solvers.* EQUALS. Lawrence Hall of Science, 1981.
- Ernst, L. *Sam Johnson and the Blue Ribbon Quilt.* Mulberry Paperback Book, 1992.
- Francis, R. *The Mathematician's Coloring Book.* Consortium for Mathematics and Its Applications (COMAP), Module #13, 1989.
- *Fibonacci Numbers in Nature.* Dale Seymour Publications.

- Friedman, A. *A Cloak for a Dreamer.* Penguin Books. Scholastic.
- Kenney, M. J., Ed. *Discrete Mathematics Across the Curriculum K-12.* 1991 Yearbook of the National Council of Teachers of Mathematics (NCTM). Reston, VA: 1991.
- Kohl, Herbert. *Insides, Outsides, Loops, and Lines.* New York: W. H. Freeman, 1995.
- Murphy, P. *By Nature's Design.* San Francisco, CA: Chronicle Books, 1993.
- Pinczes, E. J. *One Hundred Hungry Ants.* Houghton Mifflin Company, 1993.
- Rosenstein, J. G., D. Franzblau, and F. Roberts, Eds. *Discrete Mathematics in the Schools.* Proceedings of a 1992 DIMACS Conference on "Discrete Mathematics in the Schools." DIMACS Series on Discrete Mathematics and Theoretical Computer Science. Providence, RI: American Mathematical Society (AMS), 1997.
- *Set.* Set Enterprises.
- Smoothey, Marion. *Let's Investigate Codes and Sequences.* New York: Marshall Cavendish Corporation, 1995.
- Tompert, Ann. *Grandfather Tang's Story.* Crown Publishing, 1990.
- Wahl, Mark. *Mathematical Mystery Tour: Higher-Thinking Math Tasks.* Tucson, AZ: Zephyr Press, 1988.

Software.

- *Logo.* Many versions of Logo are commercially available.
- *Tabletop, Jr.* Broderbund Software. TERC.
- *Turtle Math.* LCSI.

6. Grades 5-6 Overview

The five major themes of discrete mathematics, as discussed in the K-12 Overview, are **systematic listing, counting, and reasoning; discrete mathematical modeling using graphs (networks) and trees; iterative (that is, repetitive) patterns and processes; organizing and processing information; and following and devising lists of instructions, called "algorithms," and using them to find the best solution to real-world problems.** Two important resources on discrete mathematics for teachers at all levels are the 1991 NCTM Yearbook *Discrete Mathematics Across the Curriculum K-12* and the 1997 DIMACS Volume *Discrete Mathematics in the Schools.*

Despite their formidable titles, these five themes can be addressed with activities at the 5-6 grade level which involve both the purposeful play and simple analysis suggested for elementary school students and experimentation and abstraction appropriate at the middle grades. Indeed, teachers will discover that many activities that they already are using in their classrooms reflect these themes. These five themes are discussed in the paragraphs below.

The following discussion of activities at the 5-6 grade levels in discrete mathematics presupposes that corresponding activities have taken place at the K-4 grade levels. Hence 5-6 grade teachers should review the K-2 and 3-4 grade level discussions of discrete mathematics and might use activities similar to those described there before introducing the activities for this grade level.

Activities involving **systematic listing, counting, and reasoning** at K-4 grade levels can be extended to the 5-6 grade level. For example, they might determine the number of possible license plates with two letters followed by three numbers followed by one letter, and decide whether this total number of license plates is adequate for all New Jersey drivers. They need to become familiar with the idea of permutations, that is, the different ways in which a group of items can be arranged. Thus, for example, if three children are standing by the blackboard, there are altogether six different ways, call permutations, in which this can be done; for example, if the three children are Amy (A), Bethany (B), and Coriander (C), the six different **permutations** can be described as ABC, ACB, BAC, BCA, CAB, and CBA. Similarly, the total number of different ways in which three students out of a class of thirty can be arranged at the blackboard is altogether $30 \times 29 \times 28$, or 24,360 ways, an amazing total!

An important **discrete mathematical model** is that of a **network or graph**, which consists of dots and lines joining the dots; the dots are often called *vertices* (*vertex* is the singular) and the lines are often called *edges*. (This is different from other mathematical uses of the term "graph"; the two terms "network" and "graph" are used interchangeably for this concept.) An example of a graph with 24 vertices and 38 edges is given below. Graphs can be used to represent islands and bridges, or buildings and roads, or houses and telephone cables; wherever a collection of things are joined by connectors, the mathematical model used is that of a graph. At the 5-6 level, students should be familiar with the notion of a graph and recognize situations in which graphs can be an appropriate model. For example, they should be familiar with problems involving routes for garbage pick-ups, school buses, mail deliveries, snow removal, etc.; they should be able to model such problems by using graphs, and be able to solve such problems by finding suitable paths in these graphs, such as in the town whose street map is the graph below.

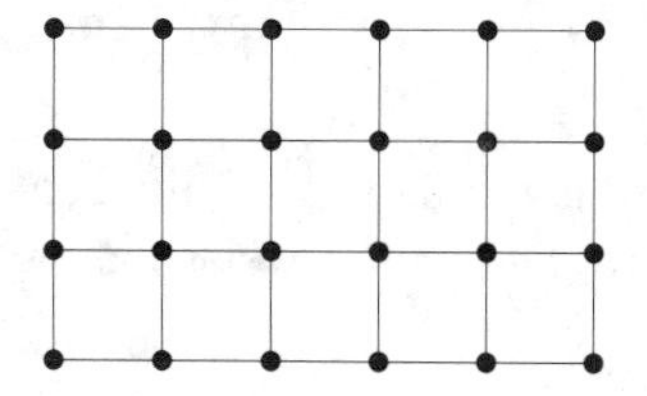

Students should recognize and work with **repetitive patterns and processes** involving numbers and shapes, with objects found in the classroom and in the world around them. Building on these explorations, fifth- and sixth-graders should also recognize and work with **iterative and recursive processes**. They explore iteration using Logo software, where they recreate a variety of interesting patterns (such as a checkerboard) by iterating the construction of a simple component of the pattern (in this case a square). As with younger students, 5th and 6th graders are fascinated with the Fibonacci sequence $1, 1, 2, 3, 5, 8, 13, 21, 34, 55, 89, \ldots$ where every number is the sum of the previous two numbers. Although the Fibonacci sequence starts with small numbers, the numbers in the sequence become large very quickly. Students can now also begin to understand the Fibonacci sequence and other sequences recursively — where each term of the sequence is described in terms of preceding terms.

Students in the 5th and 6th grade should investigate **sorting items** using Venn diagrams, and continue their explorations of **recovering hidden information** by decoding messages. They should begin to **explore how codes are used to communicate information**, by traditional methods such as Morse code or semaphore (flags used for ship-to-ship messages) and also by current methods such as zip codes, which describe a location in the United States by a five-digit (or nine-digit) number. Students should also explore modular arithmetic through applications involving clocks, calendars, and binary codes.

Finally, at grades 5-6, students should be able to **describe, devise, and test algorithms for solving a variety of problems**. These include finding the shortest route from one location to another, dividing a cake fairly, planning a tournament schedule, and planning layouts for a class newspaper.

Two important resources on discrete mathematics for teachers at all levels is the 1991 NCTM Yearbook *Discrete Mathematics Across the Curriculum K-12* and the 1997 DIMACS Volume *Discrete Mathematics in the Schools*. Another important resource for 5-6 teachers is *This Is MEGA-Mathematics!*

7. Grades 5-6 Indicators and Activities

The cumulative progress indicators for grade 8 appear below in boldface type. Each indicator is followed by activities which illustrate how it can be addressed in the classroom in grades 5 and 6.

Building upon knowledge and skills gained in the preceding grades, experiences will be such that all students in grades 5-6:

6. **Use systematic listing, counting, and reasoning in a variety of different contexts.**

   - Students determine the number of different sandwiches or hamburgers that can be created at local eateries using a combination of specific ingredients.

- Students find the number of different ways to make a row of flowers each of which is red or yellow, if the row has 1, 2, 3, 4, or 5 flowers. Modeling this with Unifix cubes, they discover that adding an additional flower to the row doubles the number of possible rows, provide explanations for this, and generalize to longer rows. Similar activities can be found in the *Pizza Possibilities* and *Two-Toned Towers* lessons that are described in the First Four Standards of this *Framework.*

- Students find the number of ways of asking three different students in the class to write three homework problems on the blackboard.

- Students understand and use the concept of permutation. They determine the number of ways any five items can be arranged in order, justify their conclusion using a tree diagram, and use factorial notation, 5!, to summarize the result.

- Students find the number of possible telephone numbers with a given area code and investigate why several years ago the telephone company introduced a new area code (908) in New Jersey, and why additional area codes are being introduced in 1997. *Is the situation the same with zip codes?*

- Students estimate and then calculate the number of possible license plates with two letters followed by three numbers followed by one letter. They investigate why the state license bureau tried to introduce license plates with seven characters and why this attempt might have been unsuccessful.

- Students explore the sequence of triangular numbers 1, 1+2, 1+2+3, 1+2+3+4, ... which represent the number of dots in the triangular arrays below, and find the location of the triangular numbers in Pascal's triangle.

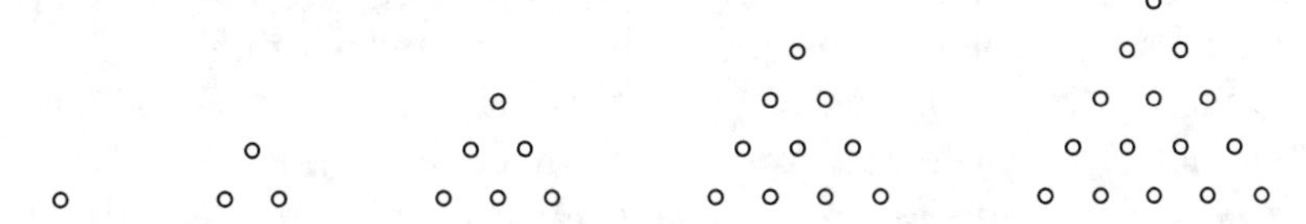

- Students look for patterns in the various diagonals of Pascal's triangle, and in the differences between consecutive terms in these diagonals. *Patterns in Pascal's Triangle Poster* is a nice resource for introducing these ideas.

- Students analyze simple games like the following: Beth wins the game whenever the two dice give an even total, and Hobart wins whenever the two dice give an odd total. They play the game a number of times, and using experimental evidence, decide whether the game is fair, and, if not, which player is more likely to win. They then try to justify their conclusions theoretically, by counting the number of combinations of dice that would result in a win for each player.

- Students create a table in the form of a grid which indicates how many of each of the coins of the fictitious country "Ternamy" — in denominations of 1, 3, 9, 27, and 81 "terns" — are needed to make up any amount from 1 to 200. They list the denominations in the columns at the top of the table and the amounts they are trying to make in the rows at the left. They write the number of each coin needed to add up to the desired amount in the appropriate squares in that row. The only "rule" to be followed is that the least number of coins must be used; for example, three 1's should always be replaced by one 3. This table can be used to introduce base 3 ("ternary") numbers, and then numbers in other bases.

7. **Recognize common discrete mathematical models, explore their properties, and design them for specific situations.**

- Students experiment with drawing make-believe maps which can be colored with two, three, and four colors (where adjacent countries must have different colors), and explain why their fictitious maps, and real maps like the map of the 50 states, cannot be colored with fewer colors. Note that it was proven in 1976 that no map can be drawn on a flat surface which requires more than four colors. *The Mathematician's Coloring Book* contains a variety of map-coloring activities, as well as historical background on the map coloring problem.

- Students play games using graphs. For example, in the strolling game, two players stroll together on a path through the graph which never repeats itself; they alternate in selecting edges for the path, and the winner is the one who selects the last edge on the path. Who wins? In the game below, Charles and Diane both start at V, Charles picks the first edge (marked 1) and they both stroll down that edge. Then Diane picks the second edge (marked 2) and the game continues. Diane has won this play of the game since the path cannot be continued after the sixth edge without repeating itself. *Does Diane have a way of always winning this game, or does Charles have a winning strategy? What if there was a different starting point? What if a different graph was used? What if the path must not cross itself (instead of requiring that it not repeat itself)?* Students should try to explain in each case why a certain player has a winning strategy.

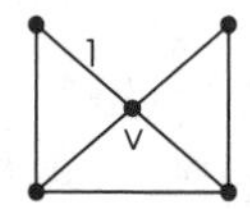

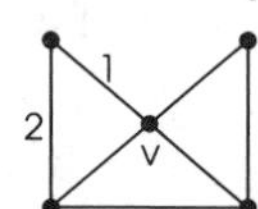

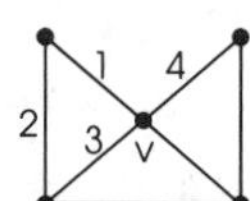

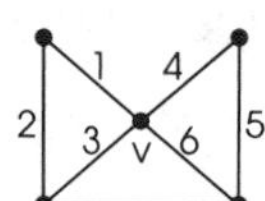

- Students find paths in graphs which utilize each edge exactly once; a path in a graph is a sequence of edges each of which begins where the previous one ends. They apply this idea by converting a street map to a graph where vertices on the graph correspond to intersections on the street map, and by using this graph to determine whether a garbage truck can complete its sector without repeating any streets. See the segment *Snowbound: Euler Circuits* on the videotape *Geometry: New Tools for New Technologies*; the module *Drawing Pictures With One Line* provides a strong background for problems of this kind.

- Students plan emergency evacuation routes at school or from home using graphs.

- All of the students together create a "human graph" where each child in the class is holding two strings, one in each hand. This can be accomplished by placing in the center of the room a number of pieces of yarn (each six feet long) equal to the number of students, and having each student take the ends of two strings. The children are asked to untangle themselves, and discuss or write about what happens.

- Students play the game of Sprouts, in which two students take turns in building a graph until one of them (the winner!) completes the graph. The rules are: start the game with two or three vertices; each person adds an edge (it can be a curved line!) joining two vertices, and then adds a new vertex at the center of that edge; no more than three edges can occur at a vertex; edges may not cross. In the sample game below, the second player (B) wins because the first player (A) cannot draw an edge connecting the only two vertices that have degree less than three without crossing an existing edge.

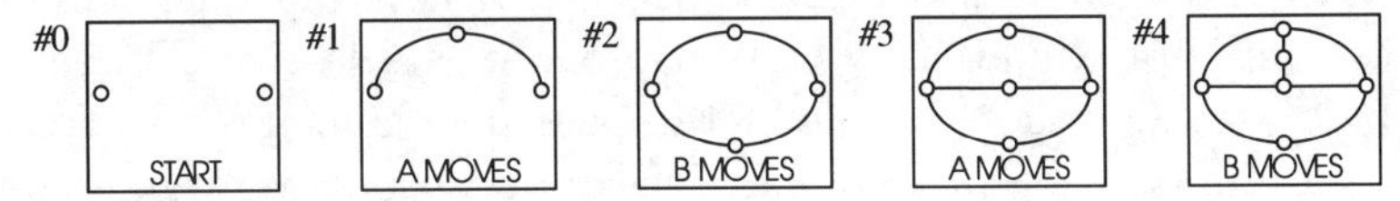

8. **Experiment with iterative and recursive processes, with the aid of calculators and computers.**

   - Students develop a method for solving the Tower of Hanoi problem: There are three pegs, on the first of which is stacked five disks, each smaller than the ones underneath it (see diagram below); the problem is to move the entire stack to the third peg, moving disks, one at a time, from any peg to either of the other two pegs, with no disk ever placed upon a smaller one. *How many moves are required to do this?*

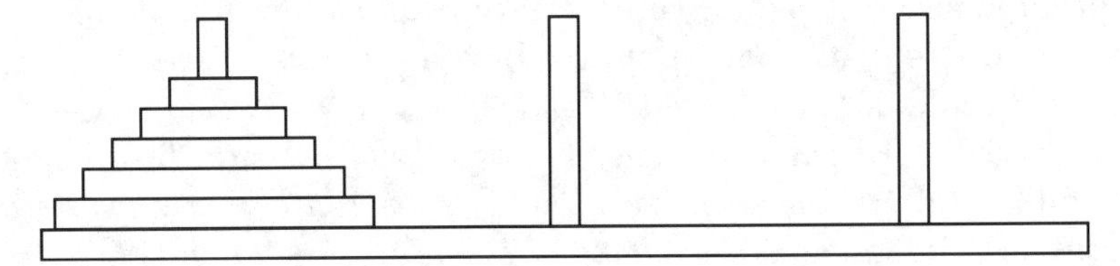

• Students use iteration in Logo software to draw checkerboards, stars, and other designs. For example, they iterate the construction of a simple component of a pattern, such as a square, to recreate an entire checkerboard design.

• Students use paper rabbits (prepared by the teacher) with which to simulate Fibonacci's 13th century investigation into the growth of rabbit populations: *If you start with one pair of baby rabbits, how many pairs of rabbits will there be a year later?* Fibonacci's assumption was that each pair of baby rabbits results in another pair of baby rabbits two months later — allowing a month for maturation and a month for gestation. Once mature, each pair has baby rabbits monthly. (Each pair of students should be provided with 18 cardboard pairs each of baby rabbits, not-yet-mature rabbits, and mature rabbits.) *The Fascinating Fibonaccis* by Trudi Garland illustrates the rabbit problem and a number of other interesting Fibonacci facts. In *Mathematics Mystery Tour* by Mark Wahl, an elementary school teacher provides a year's worth of Fibonacci explorations and activities.

• Students use calculators to compare the growth of various sequences, including counting by 4's $(4, 8, 12, 16, \ldots)$, doubling $(1, 2, 4, 8, 16, \ldots)$, squaring $(1, 4, 9, 16, 25, \ldots)$, and Fibonacci $(1, 1, 2, 3, 5, 8, 13, \ldots)$.

• Students explore their surroundings to find rectangular objects whose ratio of length to width is the "golden ratio." Since the golden ratio can be approximated by the ratio of two successive Fibonacci numbers, students should cut a rectangular peephole of dimensions 21mm x 34 mm out of a piece of cardboard, and use it to "frame" potential objects; when it "fits," the object is a golden rectangle. They describe these activities in their math journals.

• Students study the patterns of patchwork quilts, and make one of their own. They might first read *Eight Hands Round.*

• Students make equilateral triangles whose sides are 9", 3", and 1" (or other lengths in ratio 3:1), and use them to construct "Koch snowflakes of stage 2" (as shown below) by pasting the 9" triangle on a large sheet of paper, three 3" triangles at the middle of the three sides of the 9" triangle (pointing outward), and twelve 1" triangles at the middle of the exposed sides of the twelve 3" segments (pointing outward). To get Koch snowflakes of stage 3, add forty-eight 1/3" equilateral triangles. *How many 1/9" equilateral triangles would be needed for the Koch snowflake of stage 4? Fractals for the Classroom* is a valuable resource for these kinds of activities and explorations.

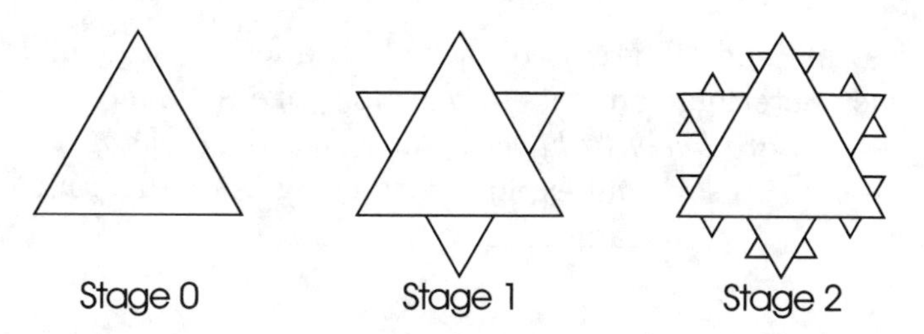

- Students mark one end of a long string and make another mark midway between the two ends. They then continue marking the string by following some simple rule such as "make a new mark midway between the last midway mark and the marked end" and then repeat this instruction. Students investigate the relationship of the lengths of the segments between marks. *How many marks are possible in this process if it is assumed that the marks take up no space on the string? What happens if the rule is changed to "make a new mark midway between the last two marks?"*

9. **Explore methods for storing, processing, and communicating information.**

- After discussing possible methods for communicating messages across a football field, teams of students devise methods for transmitting a short message (using flags, flashlights, arm signals, etc.). Each team receives a message of the same length and must transmit it to members of the team at the other end of the field as quickly and accurately as possible.

- Students devise rules so that arithmetic expressions without parentheses, such as $5 \times 8 - 2/7$, can be evaluated unambiguously. They then experiment with calculators to discover the calculators' built-in rules for evaluating these expressions.

- Students explore binary arithmetic and arithmetic for other bases through applications involving clocks (base 12), days of the week (base 7), and binary (base 2) codes.

- Students assign each letter in the alphabet a numerical value (possibly negative) and then look for words worth a specified number of points.

- Students send and decode messages in which letters of the message are systematically replaced by other letters. *The Secret Code Book* by Helen Huckle shows these coding systems as well as others.

- Students use Venn diagrams to sort and then report on their findings in a survey. For example, they can seek responses to the question, *When I grow up I want to be a) rich and famous, b) a parent, c) in a profession I love*, where respondents can choose more than one option. The results can be sorted into a Venn diagram like that below, where entries "m" and "f" are used for male and female students. The class can then determine answers to questions like *Are males or females in our class more likely to have a single focus? Tabletop, Jr.* software can be used to sort and explore data using Venn diagrams.

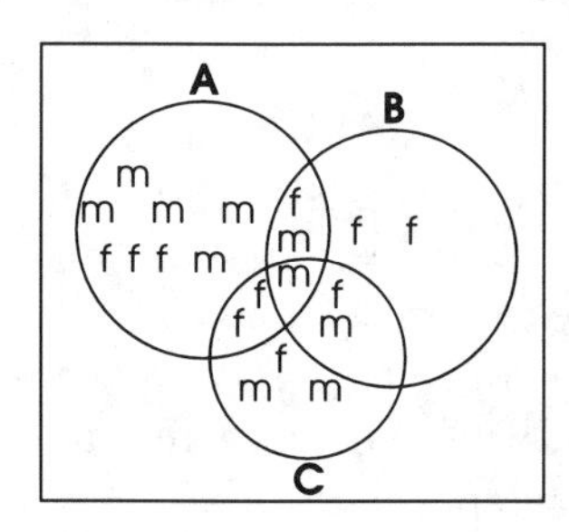

10. **Devise, describe, and test algorithms for solving optimization and search problems.**

- Students use a systematic procedure to find the total number of routes from one location in their town to another, and the shortest such route. (See *Problem Solving Using Graphs.*)

- In *Turtle Math*, students use Logo commands to go on a treasure hunt, and look for the shortest route to complete the search.

- Students discuss and write about various methods of dividing a cake fairly, such as the "divider/chooser method" for two people (one person divides, the other chooses) and the "lone chooser method" for three people (two people divide the cake using the divider/chooser method, then each cuts his/her half into thirds, and then the third person takes one piece from each of the others). *Fair Division: Getting Your Fair Share* can be used to explore methods of fairly dividing a cake or an estate.

- Students conduct a class survey for the top ten songs and discuss different ways to use the information to select the winners.

- Students devise a telephone tree for disseminating messages to all 6th grade students and their parents.

- Students schedule the matches of a volleyball tournament in which each team plays each other team once.

- Students use flowcharts to represent visually the instructions for carrying out a complex project, such as scheduling the production of the class newspaper.

- Students develop an algorithm to create an efficient layout for a class newspaper.

References.

- Bennett, S., et al. *Fair Division: Getting Your Fair Share.* Consortium for Mathematics and Its Applications (COMAP). Module #9, 1987.
- Casey, Nancy, and Mike Fellows. *This is MEGA-Mathematics! - Stories and Activities for Mathematical Thinking, Problem-Solving, and Communication.* Los Alamos, CA: Los Alamos National Laboratories, 1993. (A version is available online at `http://www.c3.lanl.gov/mega-math`)
- Chavey, D. *Drawing Pictures With One Line.* Consortium for Mathematics and Its Applications (COMAP). Module #21, 1992.
- Cozzens, M., and R. Porter. *Problem Solving Using Graphs.* Consortium for Mathematics and Its Applications (COMAP). Module #6, 1987.
- Francis, R. *The Mathematician's Coloring Book.* Consortium for Mathematics and Its Applications (COMAP). Module #13.
- Garland, Trudi. *The Fascinating Fibonaccis.* Palo Alto, CA: Dale Seymor Publications, 1987.
- Huckle, Helen. *The Secret Code Book.* Dial Books.
- Kenney, M. J., Ed. *Discrete Mathematics Across the Curriculum K-12.* 1991 Yearbook of the National Council of Teachers of Mathematics (NCTM). Reston, VA, 1991.
- Paul, A. *Eight Hands Round.* New York: Harper Collins, 1991.
- Peitgen, Heinz-Otto, et al. *Fractals for the Classroom: Strategic Activities Volume One & Two.* Reston, VA: NCTM and New York: Springer-Verlag, 1992.
- Rosenstein, J. G., D. Franzblau, and F. Roberts, Eds. *Discrete Mathematics in the Schools.* Proceedings of a 1992 DIMACS Conference on "Discrete Mathematics in the Schools." DIMACS Series on Discrete Mathematics and Theoretical Computer Science. Providence, RI: American Mathematical Society (AMS), 1997.
- Wahl, Mark. *Mathematical Mystery Tour: Higher-Thinking Math Tasks.* Tucson, AZ: Zephyr Press, 1988.

Software.

- *Logo.* Many versions of Logo are commercially available.
- *Tabletop, Jr.* Broderbund, TERC.
- *Turtle Math.* LCSI.

Video.

- *Geometry: New Tools for New Technologies*, videotape by the Consortium for Mathematics and Its Applications (COMAP). Lexington, MA, 1992.

8. Grades 7-8 Overview

The five major themes of discrete mathematics, as discussed in the K-12 Overview, are **systematic listing, counting, and reasoning; discrete mathematical modeling using graphs (networks) and trees; iterative (that is, repetitive) patterns and processes; organizing and processing information; and following and devising lists of instructions, called "algorithms," and using them to find the best solution to real-world problems.**

Despite their formidable titles, these five themes can be addressed with activities at the 7-8 grade level which involve both the purposeful play and simple analysis suggested for elementary school students and experimentation and abstraction appropriate at the middle grades. Indeed, teachers will discover that many activities that they already are using in their classrooms reflect these themes. These five themes are discussed in the paragraphs below.

The following discussion of activities at the 7-8 grade levels in discrete mathematics presupposes that corresponding activities have taken place at the K-6 grade levels. Hence 7-8 grade teachers should review the K-2, 3-4, and 5-6 grade level discussions of discrete mathematics and might use activities similar to those described there before introducing the activities for this grade level.

Students in 7th and 8th grade should be able to **use permutations and combinations and other counting strategies in a wide variety of contexts**. In addition to working with permutations, where the order of the items is important (see Grades 5-6 Overview and Activities), they should also be able to work with combinations, where the order of the items is irrelevant. For example, the number of different three digit numbers that can be made using three different digits is $10 \times 9 \times 8$ because each different ordering of the three digits results in a different number. However, the number of different pizzas that can be made using three of ten available toppings is $(10 \times 9 \times 8)/(3 \times 2 \times 1)$ because the *order* in which the toppings are added is irrelevant; the division by $3 \times 2 \times 1$ eliminates the duplication.

An important **discrete mathematical model** is that of a **network or graph**, which consists of dots and lines joining the dots; the dots are often called *vertices* (*vertex* is the singular) and the lines are often called *edges*. (This is different from other mathematical uses of the term "graph.") Graphs can be used to represent islands and bridges, or buildings and roads, or houses and telephone cables; wherever a collection of things are joined by connectors, the mathematical model used is that of a graph. Students in the 7th and 8th grades should be able to **use graphs to model situations and solve problems using the model**. For example, students should be able to use graphs to schedule a school's extracurricular activities so that, if at all possible, no one is excluded because of conflicts. This can be done by creating a graph whose vertices are the activities, with two activities joined by an edge if they have a person in common, so that the activities should be scheduled for different times. Coloring the vertices of the graph so that adjacent vertices have different colors, using a minimum number of colors, then provides an efficient solution to the scheduling problem — a separate time slot is needed for each color, and two activities are scheduled for the same time slot if they have the same color.

Students can recognize and work with **iterative and recursive processes**, extending their earlier explorations of **repetitive patterns and procedures**. In the 7th and 8th grade, they can combine their understanding of exponents and iteration to solve problems involving compound interest with a calculator or spreadsheet. Topics which before were viewed iteratively — arriving at the present situation by repeating a procedure n times — can now be viewed recursively - arriving at the present situation by modifying the previous situation. They can apply this understanding to Fibonacci numbers, to the Tower of Hanoi puzzle, to programs in Logo, to permutations and to other areas.

Students in the 7th and 8th grades should **explore how codes are used to communicate information**, by traditional methods such as Morse code or semaphore (flags used for ship-to-ship messages) and also by current methods such as zip codes. Students should investigate and report about various codes that are commonly used, such as binary codes, UPCs (universal product codes) on grocery items, and ISBN numbers on books. They should also **explore how information is processed**. A useful metaphor is how a waiting line or queue is handled (or "processed") in various situations; at a bank, for example, the queue is usually processed in first-in-first-out (FIFO) order, but in a supermarket or restaurant there is usually a pre-sorting into smaller queues done by the shoppers themselves before the FIFO process is activated.

In the 7th and 8th grade, students should be able to **use algorithms to find the best solution in a number of situations** — including the shortest route from one city to another on a map, the cheapest way of connecting sites into a network, the fastest ways of alphabetizing a list of words, the optimal route for a class trip (see the *Short-Circuiting Trenton* lesson in the Introduction to this *Framework*), or optimal work schedules for employees at a fast-food restaurant.

Two important resources on discrete mathematics for teachers at all levels are the 1991 NCTM Yearbook *Discrete Mathematics Across the Curriculum K-12* and the 1997 DIMACS Volume *Discrete Mathematics in the Schools.* Teachers of grades 7-8 would also find useful the textbook *Discrete Mathematics Through Applications.*

9. Grades 7-8 Indicators and Activities

The cumulative progresses indicators for grade 8 appear below in boldface type. Each indicator is followed by activities which illustrate how it can be addressed in the classroom in grades 7 and 8.

Building upon knowledge and skills gained in the preceding grades, experiences will be such that all students in grades 7-8:

6. **Use systematic listing, counting, and reasoning in a variety of different contexts.**

- Students determine the number of possible different sandwiches or hamburgers that can be created at local eateries using a combination of specified ingredients. They find the number of pizzas that can be made with three out of eight available toppings and relate the result to the numbers in Pascal's triangle.

- Students determine the number of dominoes in a set that goes up to 6:6 or 9:9, the number of candles used throughout Hannukah, and the number of gifts given in the song "The Twelve Days of Christmas," and connect the results through discussion of the triangular numbers. (Note that in a 6:6 set of dominoes there is exactly one domino with each combination of dots from 0 to 6.)

- Students determine the number of ways of spelling "Pascal" in the array below by following a path from top to bottom in which each letter is directly below, and just to the right or left of the previous letter.

```
     P
    A A
   S S S
  C C C C
 A A A A A
L L L L L L
```

- Students design different license plate systems for different population sizes; for example, *how large would the population be before you would run out of plates which had only three numbers, or only five numbers, or two letters followed by three numbers?*

- Students find the number of different ways of making a row of six red and yellow flowers, organize and tabulate the possibilities according to the number of flowers of the first color, and explain the connection with the numbers in the sixth row of Pascal's triangle. (See also *Visual Patterns in Pascal's Triangle.*)

- Students pose and act out problems involving the number of different ways a group of people can sit around a table, using as motivation the scene of the Mad Hatter at the tea party. (See *Mathematics, a Human Endeavor*, p. 394.)

- Students count the total number of different cubes that can be made using either red or green paper for each face. (To solve this problem, they will have to use a "break up the problem into cases" strategy.)

- Students determine the number of handshakes that take place if each person in a room shakes hands with every other person exactly once, and relate this total to the number of line segments joining the vertices in a polygon, to the number of two-flavor ice-cream cones, and to triangular numbers.

- Students count the number of triangles or rectangles in a geometric design. For example, they should be able to count systematically the number of triangles (and trapezoids) in the figure below to the left, noting that there are triangles of three sizes, and the number of rectangles in the 4×5 grid pictured below to the right, listing first all dimensions of rectangles that are present.

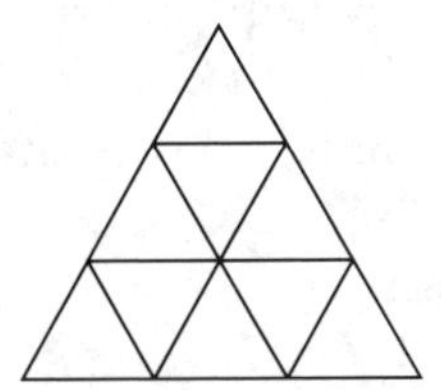
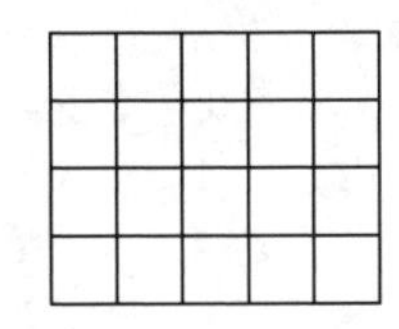

7. **Recognize common discrete mathematical models, explore their properties, and design them for specific situations.**

- Students find the minimum number of colors needed to assign colors to all vertices in a graph so that any two adjacent vertices are assigned different colors and justify their answers. For example, students can explain why one of the graphs below requires four colors while for the other, three colors are sufficient.

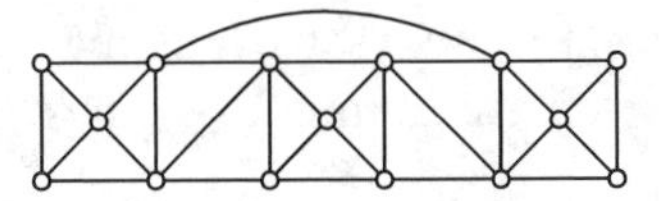
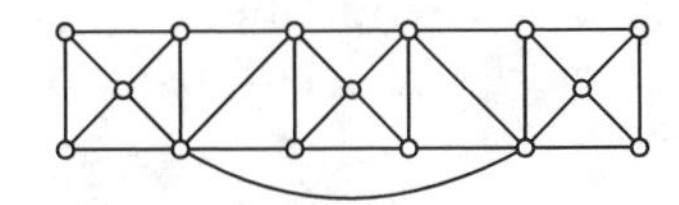

- Students use graph coloring to solve problems which involve avoiding conflicts such as: scheduling the school's extra curricular activities; scheduling referees for soccer games; determining the minimum number of aquariums needed for a specified collection of tropical fish; and assigning channels to radio stations to avoid interference. In the graph below an edge between two animals indicates that they cannot share a habitat. The videotape, *Geometry: New Tools for New Technologies* has a segment *Connecting the Dots: Vertex Coloring* which discusses the minimum number of habitats required for this situation.

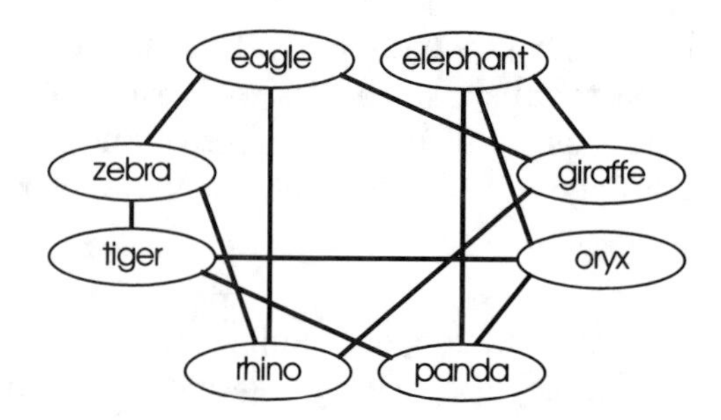

- Students use tree diagrams to represent and analyze possible outcomes in counting problems, such as tossing two dice.

- Students determine whether or not a given group of dominoes can be arranged in a line (or in a rectangle) so that the number of dots on the ends of adjacent dominoes match. For example, the dominoes (03), (05), (12), (14), (15), (23), (34) can be arranged as (12), (23), (30), (05), (51), (14), (43); and if an eighth domino (13) is added, they can be formed into a rectangle. *What if instead the eighth domino was (24) — could they then be arranged in a rectangle or in a line?*

- Students determine the minimum number of blocks that a police car has to repeat if it must try to patrol each street exactly once on a given map. *Drawing Pictures With One Line* contains similar real-world problems and a number of related game activities.

- Students find the best route for collecting recyclable paper from all classrooms in the school, and discuss different ways of deciding what is the "best." (See *Drawing Pictures With One Line.*)

- Students make models of various polyhedra with straws and string, and explore the relationship between the number of edges, faces, and vertices.

8. **Experiment with iterative and recursive processes, with the aid of calculators and computers.**

- Students develop a method for solving the Tower of Hanoi problem: There are three pegs, on the first of which five disks are stacked, each smaller than the ones underneath it (see diagram below); the problem is to move the entire stack to the third peg, moving disks, one at a time, from any peg to either of the other two pegs, with no disk ever placed upon a smaller one. *How many moves are required to do this? What if there were 6 disks? How long would it take to do this with 64 disks?* (An ancient legend predicts that when this task is completed, the world will end; should we worry?)

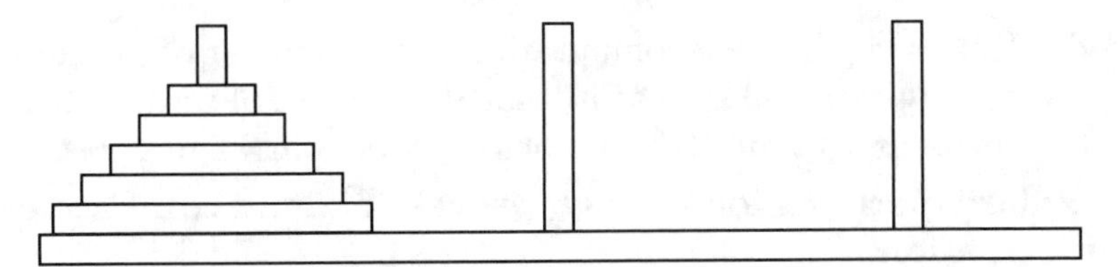

Students view recursively Tower of Hanoi puzzles with various numbers of disks so that they can express the number of moves needed to solve the puzzle with one more disk in terms of the number of moves needed for the puzzle with the current number of disks.

- Students attempt to list the different ways they could travel 10 feet in a straight line if they were a robot which moved only in one or two foot segments, and then thinking recursively determine the number of different ways this robot could travel n feet.

- Students develop arithmetic and geometric progressions on a calculator.

- Students find square roots using the following iterative procedure on a calculator. Make an estimate of the square root of a number B, divide the estimate into B, and average the result with the estimate to get a new estimate. Then repeat this procedure until an adequate estimate is obtained. *For example, if the first estimate of the square root of 10 is 3, then the second would be the average of 3 and 10/3, or 19/6 = 3.166. What is the next estimate of the square root of 10? How many repetitions are required to get the estimate to agree with the square root of 10 provided by the calculator?*

- Students develop the sequence of areas and perimeters of iterations of the constructions of the Sierpinski triangle (top figures) and the Koch snowflake (bottom figures), and discuss the outcome if the process were continued indefinitely. (These are discussed in more detail in the sections for earlier grade levels. See Unit 1 of *Fractals for the Classroom* for related activities.)

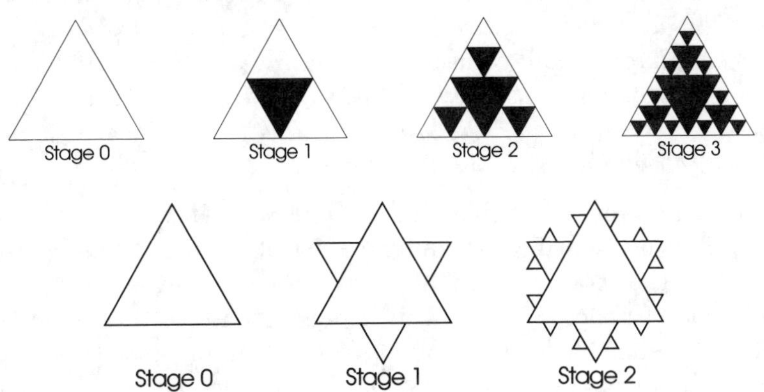

- Students recognize the computation of the number of permutations as a recursive process — that is, that the number of ways of arranging 10 students is 10 times the number of ways of arranging 9 students.

9. **Explore methods for storing, processing, and communicating information.**

- Students conjecture which of the following (and other) methods is the most efficient way of handing back corrected homework papers which are already sorted alphabetically: (1) the teacher walks around the room handing to each student individually; (2) students pass the papers around, each taking their own; (3) students line themselves up in alphabetical order. Students test their conjectures and discuss the results.

- Students investigate and report about various codes that are commonly used, such as zip codes, UPCs (universal product codes) on grocery items, and ISBN numbers on books. (A good source for information about these and other codes is *Codes Galore* by J. Malkevitch, G. Froelich, and D. Froelich.)

- Students write a Logo procedure for making a rectangle that uses variables, so that they can use their rectangle procedure to create a graphic scene which contains objects, such as buildings, of varying sizes.

- Students are challenged to guess a secret word chosen by the teacher from the dictionary, using at most 20 yes-no questions. *Is this always, or only sometimes possible?*

- Students use Venn diagrams to solve problems like the following one from the New Jersey Department of Education's *Mathematics Instruction Guide* (p. 7-13). *Suppose the school decided to add the springtime sport of lacrosse to its soccer and basketball offerings for its 120 students. A follow-up survey showed that: 35 played lacrosse, 70 played soccer, 40 played basketball, 20 played both soccer and basketball, 15 played both soccer and lacrosse, 15 played both basketball and lacrosse, and 10 played all three sports. Using this data, complete a Venn diagram and answer the following questions: How many students played none of the three sports? What percent of the students played lacrosse as their only sport? How many students played both basketball and lacrosse, but not soccer?*

- Students keep a scrapbook of different ways in which information is stored or processed. For example a list of events is usually stored by date, so the scrapbook might contain a picture of a pocket calendar; a queue of people at a bank is usually processed in first-in-first-out (FIFO) order, so the scrapbook could contain a picture of such a queue. *(How is this different from the waiting lines in a supermarket, or at a restaurant?)*

- Students determine whether it is possible to have a year in which there is no Friday the 13th, and the maximum number of Friday the 13th's that can occur in one calendar year.
- Students predict and then explore the frequency of letters in the alphabet through examination of sample texts, computer searches, and published materials.
- Students decode messages where letters are systematically replaced by other letters without knowing the system by which letters are replaced; newspapers and games magazines are good sources for "cryptograms" and students can create their own. They also explore the history of code-making and code-breaking. The videotape *Discrete Mathematics: Cracking the Code* provides a good introduction to the uses of cryptography and the mathematics behind it.

10. **Devise, describe, and test algorithms for solving optimization and search problems.**

- Students find the shortest route from one city to another on a New Jersey map, and discuss whether that is the best route. (See *Problem Solving Using Graphs.*)
- Students write and solve problems involving distances, times, and costs associated with going from towns on a map to other towns, so that different routes are "best" according to different criteria.
- Students use binary representations of numbers to find winning strategy for Nim. (See *Mathematical Investigations* for other mathematical games.)
- Students plan an optimal route for a class trip. (See the *Short-circuiting Trenton* lesson in the Introduction to this *Framework.*)
- Students devise work schedules for employees of a fast-food restaurant which meet specified conditions yet minimize the cost.
- Students compare strategies for alphabetizing a list of words, and test to see which strategies are more efficient.
- Students find a network of roads which connects a number of sites and involves the smallest cost. *In the example below, what roads should be built so as to minimize the total cost, where the number on each road reflects the cost of building that road (in hundreds of thousands of dollars)?*

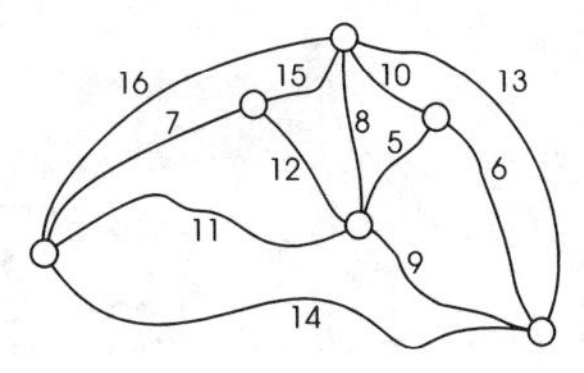

- Students develop a precise description of the standard algorithm for adding two two-digit integers.
- Students devise strategies for dividing up the work of adding a long list of numbers among the members of the team.

References.

- Chavey, D. *Drawing Pictures with One Line.* Consortium for Mathematics and Its Applications (COMAP), Module #21, 1987.
- Cozzens, M., and R. Porter. *Problem Solving Using Graphs.* Consortium for Mathematics and Its Applications (COMAP), Module #6, 1987.
- Crisler, N., P. Fisher, and G. Froelich, *Discrete Mathematics Through Applications.* W. H. Freeman and Company, 1994.
- Jacobs, H. R. *Mathematics: A Human Endeavor.* W. H. Freeman and Company, 1982.
- Kenney, M. J., Ed. *Discrete Mathematics Across the Curriculum K-12,* 1991 Yearbook of the National Council of Teachers of Mathematics (NCTM). Reston, VA: 1991.
- Malkevitch, J., G. Froelich, and D. Froelich, *Codes Galore*, Consortium for Mathematics and Its Applications (COMAP), Module #18, 1991.
- New Jersey Department of Education. *Mathematics Instruction Guide.* D. Varygiannis, Coord. January 1996.
- Peitgen, Heinz-Otto, et al. *Fractals for the Classroom: Strategic Activities Volume One & Two.* Reston, VA: NCTM and New York: Springer-Verlag, 1992.
- Rosenstein, J. G., D. Franzblau, and F. Roberts, Eds. *Discrete Mathematics in the Schools.* Proceedings of a 1992 DIMACS Conference on "Discrete Mathematics in the Schools." DIMACS Series on Discrete Mathematics and Theoretical Computer Science. Providence, RI: American Mathematical Society (AMS), 1997.
- Seymour, D. *Visual Patterns in Pascal's Triangle.* Palo Alto, CA: Dale Seymour Publications, 1986.
- Souviney, R., et al. *Mathematical Investigations.* Book One, Dale Seymour Publications, 1990.

Video.

- *Discrete Mathematics: Cracking the Code*, Consortium for Mathematics and Its Applications.
- *Geometry: New Tools for New Technologies*, videotape by the Consortium for Mathematics and Its Applications (COMAP). Lexington, MA, 1992.

10. Grades 9-12 Overview

The five major themes of discrete mathematics, as discussed in the K-12 Overview, are **systematic listing, counting, and reasoning; discrete mathematical modeling using graphs (networks) and trees; iterative (that is, repetitive) patterns and processes; organizing and processing information; and following and devising lists of instructions, called "algorithms," and using them to find the best solution to real-world problems.**

The following discussion of activities at the 9-12 grade levels in discrete mathematics presupposes that corresponding activities have taken place at the K-8 grade levels. Hence high school teachers should review the discussions of discrete mathematics at earlier grade levels and might use activities similar to those described there before introducing the activities for these grade levels.

At the high school level, students are becoming familiar with algebraic and functional notation, and their understanding of all of the themes of discrete mathematics and their ability to generalize earlier activities should be **enhanced by their algebraic skills and understandings**. Thus, for example, they should use formulas to express the results of problems involving permutations and combinations, relate Pascal's triangle to the coefficients of the binomial expansion of $(x + y)^n$, explore models of growth using various algebraic models, explore iterations of functions, and discuss methods for dividing an estate among several heirs.

At the high school level, students are particularly interested in applications; they ask *What is all of this good for?* In all five areas of discrete mathematics, students should **focus on how discrete mathematics is used to solve practical problems.** Thus, for example, they should be able to apply their understanding of counting techniques, to analyze lotteries; of graph coloring, to schedule traffic lights at a local intersection; of paths in graphs, to devise patrol routes for police cars; of iterative processes, to analyze and predict fish populations in a pond or concentration of medicine in the bloodstream; of codes, to understand how bar-code scanners detect errors and how CD's correct errors; and of optimization, to understand the 200 year old debates about apportionment and to find efficient ways of scheduling the components of a complex project.

Two important resources on discrete mathematics for teachers at all grade levels are the 1991 NCTM Yearbook, *Discrete Mathematics Across the Curriculum K-12* and the DIMACS Volume, *Discrete Mathematics in the Schools* edited by J. Rosenstein, D. Franzblau, and F. Roberts. Useful resources at the high school level are *Discrete Mathematics Through Applications* by N. Crisler, P. Fisher, and G. Froelich; *For All Practical Purposes: Introduction to Contemporary Mathematics*, by the Consortium for Mathematics and its Applications; and *Excursions in Modern Mathematics* by P. Tannenbaum and R. Arnold.

11. Grades 9-12 Indicators and Activities

The cumulative progress indicators for grade 12 appear below in boldface type. Each indicator is followed by activities which illustrate how it can be addressed in the classroom in grades 9, 10, 11, and 12.

Building upon knowledge and skills gained in the preceding grades, experiences will be such that all students in grades 9-12:

11. **Understand the basic principles of iteration, recursion, and mathematical induction.**

   • Students relate the possible outcomes of tossing five coins with the binomial expansion of $(x + y)^5$ and the fifth row of Pascal's triangle, and generalize to values of n other than 5.

   • Students develop formulas for counting paths on grids or other simple street maps.

   • Students find the number of cuts needed in order to divide a giant pizza so that each student in the school gets at least one piece.

   • Students develop a precise description, using iteration, of the standard algorithm for adding two integers.

12. **Use basic principles to solve combinatorial and algorithmic problems.**

   • Students determine the number of ways of spelling "mathematics" in the array below by following a path from top to bottom in which each letter is directly below, and just to the right or left of the previous letter.

```
          M
        A   A
      T   T   T
    H   H   H   H
  E   E   E   E   E
M   M   M   M   M   M
  A   A   A   A   A
    T   T   T   T
      I   I   I
        C   C
          S
```

   • Students determine the number of ways a committee of three members could be selected from the class, and the number of ways three people with specified roles could be selected. They generalize this activity to finding a formula for the number of ways an n person committee can be selected from a class of m people, and the number of ways n people with specified roles can be selected from a class of m people.

- Students find the number of ways of lining up thirty students in a class, and compare that to other large numbers; for example, they might compare it to the number of raindrops (volume = .1 cc) it would take to fill a sphere the size of the earth (radius = 6507 KM).

- Students determine the number of ways of dividing 52 cards among four players, as in the game of bridge, and compare the number of ways of obtaining a flush (five cards of the same suit) and a full house (three cards of one denomination and two cards of another) in the game of poker.

- Students play Nim (and similar games) and discuss winning strategies using binary representations of numbers.

13. **Use discrete models to represent and solve problems.**

- Students study the four color theorem and its history. (*The Mathematicians' Coloring Book* provides a good background for coloring problems.)

- Students using graph coloring to determine the minimum number of guards (or cameras) needed for museums of various shapes (and similarly for placement of lawn sprinklers or motion-sensor burglar alarms).

- Students use directed graphs to represent tournaments (where an arrow drawn from A to B represents "A defeats B") and food webs (where an arrow drawn from A to B represents "A eats B"), and to construct one-way orientations of streets in a given town which involve the least inconvenience to drivers. (A directed graph is simply a graph where each edge is thought of as an arrow pointing from one endpoint to the other.)

- Students use tree diagrams to analyze the play of games such as tic-tac-toe or Nim, and to represent the solutions to weighing problems. Example: Given 12 coins one of which is "bad," find the bad one, and determine whether it is heavier or lighter than the others, using three weighings.

- Students use graph coloring to schedule the school's final examinations so that no student has a conflict, if at all possible, or to schedule traffic lights at an intersection.

- Students devise graphs for which there is a path that covers each edge of the graph exactly once, and other graphs which have no such paths, based on an understanding of necessary and sufficient conditions for the existence of such paths, called "Euler paths," in a graph. *Drawing Pictures With One Line* provides background and applications for Euler path problems.

- Students make models of polyhedra with straws and string, and explore the relationship between the numbers of edges, faces, and vertices, and generalize the conclusion to planar graphs.

- Students use graphs to solve problems like the "fire-station problem": *Given a city where the streets are laid out in a grid composed of many square blocks, how many fire stations are needed to provide adequate coverage of the city if each fire station services its square block and the four square blocks adjacent to that one?* The Maryland Science Center in Baltimore has a hands-on exhibit involving a fire-station problem for 35 square blocks arranged in a six-by-six grid with one corner designated a park.

14. **Analyze iterative processes with the aid of calculators and computers.**

- Students analyze the Fibonacci sequence $1, 1, 2, 3, 5, 8, 13, 21, \ldots$ as a recurrence relation $A_{n+2} = A_n + A_{n+1}$ with connections to the golden ratio. *Fascinating Fibonaccis* illustrates a variety of connections between Fibonacci numbers and the golden ratio.

- Students solve problems involving compound interest using iteration on a calculator or on a spreadsheet.

- Students explore examples of linear growth, using the recursive model based on the formula $A_{n+1} = A_n + d$, where d is the common difference, and convert it to the explicit linear formula, $A_{n+1} = A_1 + n \cdot d$.

- Students explore examples of population growth, using the recursive model based on the formula $A_{n+1} = A_n \times r$, where r is the common multiple or growth rate, convert it to the explicit exponential formula $A_{n+1} = A_1 \times r^n$, and apply it to both economics (such as interest problems) and biology (such as concentration of medicine in blood supply).

- Students explore logistic growth models of population growth, using the recursive model based on the formula $A_{n+1} = A_n \times (1 - A_n) \times r$, where r is the growth rate and A_n is the fraction of the carrying capacity of the environment, and apply this to the population of fish in a pond. Using a spreadsheet, students experiment with various values of the initial value A_1 and of the growth rate, and describe the relationship between the values chosen and the long term behavior of the population.

- Students explore the pattern resulting from repeatedly multiplying $\begin{bmatrix} 1 & 1 \\ 1 & 0 \end{bmatrix}$ by itself.

- Students use a calculator or a computer to study simple Markov chains, such as weather prediction and population growth models. (See Chapter 7.3 of *Discrete Mathematics Through Applications.*)

- Students explore graphical iteration by choosing a function key on a calculator and pressing it repeatedly, after choosing an initial number, to get sequences of numbers like $2, 4, 8, 16, 32, \ldots$ or $2, \sqrt{2}, \sqrt{\sqrt{2}}, \sqrt{\sqrt{\sqrt{2}}}, \ldots$. They use the graphs of the functions to explain the behavior of the sequences obtained. They extend these explorations by iterating functions they program into the calculator, such as linear functions, where slope is the predictor of behavior, and quadratic functions $f(x) = ax(1 - x)$, where $0 < x < 1$ and $1 < a < 4$, which exhibit chaotic behavior.

- Students explore iteration behavior using the function defined by the two cases

$$\begin{array}{ll} f(x) = x + \frac{1}{2} & \text{for } x \text{ between } 0 \text{ and } \frac{1}{2} \\ f(x) = 2 - 2x & \text{for } x \text{ between } \frac{1}{2} \text{ and } 1 \end{array}$$

They use the initial values 1/2, 2/3, 5/9, and 7/10, and then, with a calculator or computer, the initial values .501, .667, and .701 (which differ by a small amount from the first group of "nice" initial values). They compare the behavior of the sequences generated by these values to the sequences generated by the previous initial values.

- Students play the *Chaos Game*. Each pair of students is provided with an identical transparency on which have been drawn the three vertices L, T, and R of an equilateral triangle. Each team starts by selecting any point on the triangle. They roll a die and create a new point halfway to L if they roll 1 or 2, halfway to R if they roll 3 or 4, and halfway to T if they roll 5 or 6. They repeat 20 times, each time using the new point as the starting point for the next iteration. The teacher overlays all of the transparencies and out of this chaos comes ... the familiar Sierpinski triangle. (The Sierpinski triangle is discussed in detail in the sections for earlier grade levels. Also see Unit 2 in *Fractals for the Classroom. The Chaos Game* software allows students to try variations and explore the game further.)

15. **Apply discrete methods to storing, processing, and communicating information.**

- Students discuss various algorithms used for sorting large numbers of items alphabetically or numerically, and explain why some sorting algorithms are substantially faster than others. To introduce the topic of sorting, give each group of students 100 index cards each with one word on it, and let them devise strategies for efficiently putting the cards into alphabetical order.

- Students discuss how scanners of bar codes (zip codes, UPCs, and ISBNs) are able to detect errors in reading the codes, and evaluate and compare how error-detection is accomplished in different codes. (See the COMAP Module *Codes Galore* or Chapter 9 of *For All Practical Purposes*.)

• Students investigate methods of error correction used to transmit digitized pictures from space (Voyager or Mariner probes, or the Hubble space telescope) over noisy or unreliable channels, or to ensure the fidelity of a scratched CD recording. (See Chapter 10 of For All Practical Purposes.)

• Students read about coding and code-breaking machines and their role in World War II.

• Students research topics that are currently discussed in the press, such as public-key encryption, enabling messages to be transmitted securely, and data-compression, used to save space on a computer disk.

16. **Apply discrete methods to problems of voting, apportionment, and allocations, and use fundamental strategies of optimization to solve problems.**

• Students find the best route when a number of alternate routes are possible. For example: *In which order should you pick up the six friends you are driving to the school dance? In which order should you make the eight deliveries for the drug store where you work? In which order should you visit the seven "must-see" sites on your vacation trip?* In each case, you want to find the "best route," the one which involves the least total distance, or least total time, or least total expense. Students create their own problems, using actual locations and distances, and find the best route. For a larger project, students can try to improve the route taken by their school bus.

• Students study the role of apportionment in American history, focusing on the 1790 census (acting out the positions of the thirteen original states and discussing George Washington's first use of the presidential veto), and the disputed election of 1876, and discuss the relative merits of different systems of apportionment that have been proposed and used. (This activity provides an opportunity for mathematics and history teachers to work together.) They also devise a student government where the seats are fairly apportioned among all constituencies. (See the COMAP module *The Apportionment Problem* or Chapter 14 of *For All Practical Purposes.*)

• Students analyze mathematical methods for dividing an estate fairly among various heirs. (See Chapter 2 of *Discrete Mathematics Through Applications*, Chapter 3 of *Excursions in Modern Mathematics*, or Chapter 13 of *For All Practical Purposes.*)

• Students discuss various methods, such as preference schedules or approval voting, that can be used for determining the winner of an election involving three or more candidates (for example, the prom king or queen). With preference schedules, each voter ranks the candidates and the individual rankings are combined, using various techniques, to obtain a group ranking; preference schedules are used, for example, in ranking sports teams or determining entertainment awards. In approval voting, each voter can vote once for each candidate which she finds acceptable; the candidate who receives the most votes then wins the election. (See the COMAP module *The Mathematical Theory of Elections* or Chapter 11 of *For All Practical Purposes.*)

- Students find an efficient way of doing a complex project (like preparing an airplane for its next trip) given which tasks precede which and how much time each task will take. (See Chapter 8 of *Excursions of Modern Mathematics* or Chapter 3 of *For All Practical Purposes.*)
- Students find an efficient way of assigning songs of various lengths to the two sides of an audio tape so that the total times on the two sides are as close together as possible. Similarly, they determine the minimal number of sheets of plywood needed to build a cabinet with pieces of specified dimensions.
- Students apply algorithms for matching in graphs to schedule when contestants play each other in the different rounds of a tournament.
- Students devise a strategy for finding a "secret number" from 1 to 1000 using questions of the form *Is your number bigger than 837?* and determine the least number of questions needed to find the secret number.

References.

- Bennett, S., D. DeTemple, M. Dirks, B. Newell, J. Robertson, and B. Tyus. *The Apportionment Problem: The Search for the Perfect Democracy.* Consortium for Mathematics and Its Applications (COMAP), Module #18, 1986.
- Chavey, D. *Drawing Pictures with One Line.* Consortium for Mathematics and Its Applications (COMAP), Module #21, 1987.
- Consortium for Mathematics and Its Applications. *For All Practical Purposes: Introduction to Contemporary Mathematics.* W. H. Freeman and Company, Third Edition, 1993.
- Crisler, N., P. Fisher, and G. Froelich, *Discrete Mathematics Through Applications.* W. H. Freeman and Company, 1994.
- Francis, R. *The Mathematician's Coloring Book.* Consortium for Mathematics and Its Applications (COMAP), Module #13, 1989.
- Garland, T. H. *Fascinating Fibonaccis.* Palo Alto, CA: Dale Seymour Publications, 1987.
- Kenney, M. J., Ed. *Discrete Mathematics Across the Curriculum K-12,* 1991 Yearbook of the National Council of Teachers of Mathematics (NCTM). Reston, VA, 1991.

- Malkevitch, J. *The Mathematical Theory of Elections.* Consortium for Mathematics and Its Applications (COMAP). Module #1, 1985.
- Malkevitch, J., G. Froelich, and D. Froelich. *Codes Galore.* Consortium for Mathematics and Its Applications (COMAP). Module #18, 1991.
- Peitgen, Heinz-Otto, et al. *Fractals for the Classroom: Strategic Activities Volume One & Two.* Reston, VA: NCTM and New York: Springer-Verlag, 1992.
- Rosenstein, J. G., D. Franzblau, and F. Roberts, Eds. *Discrete Mathematics in the Schools.* Proceedings of a 1992 DIMACS Conference on "Discrete Mathematics in the Schools." DIMACS Series on Discrete Mathematics and Theoretical Computer Science. Providence, RI: American Mathematical Society (AMS), 1997.
- Seymour, D. *Patterns in Pascal's Triangle.* Poster. Palo Alto, CA: Dale Seymour Publications.
- Tannenbaum, P. and R. Arnold. *Excursions in Modern Mathematics.* Prentice-Hall, 1992.

Software.

- *The Chaos Game.* Minnesota Educational Computer Consortium (MECC).

DEPARTMENT OF MATHEMATICS, RUTGERS UNIVERSITY
E-mail address: joer@dimacs.rutgers.edu

Section 4

Integrating Discrete Mathematics into Existing Mathematics Curricula, Grades K–8

Discrete Mathematics in K–2 Classrooms
VALERIE A. DEBELLIS
Page 187

Rhythm and Pattern: Discrete Mathematics with an Artistic Connection for Elementary School Teachers
ROBERT E. JAMISON
Page 203

Discrete Mathematics Activities for Middle School
EVAN MALETSKY
Page 223

DIMACS Series in Discrete Mathematics
and Theoretical Computer Science
Volume **36**, 1997

Discrete Mathematics in K–2 Classrooms

Valerie A. DeBellis

Introduction

This article describes two K–2 classrooms that I have observed and/or taught during the 1996-97 school year. Critics have claimed that mathematics taught in primary grades (K–2) is nothing more than memorizing facts, contains little content beyond computation, and that topics in discrete mathematics cannot be thoughtfully discussed by children at these levels. I strongly disagree. For the past ten years, I have been involved with professional development projects for K–12 teachers of mathematics, including the Leadership Program in Discrete Mathematics (see Rosenstein and DeBellis [**7**]). This experience, coupled with my background in mathematics education, has provided many opportunities to collaborate with K–12 teachers who are implementing discrete mathematics in their classrooms. Based on these experiences, I have come to believe that not only is it important to incorporate discrete mathematics into existing curriculum, but that K–2 classrooms are a natural place to begin developing the rudiments of the subject.

The current K–2 curriculum

Traditional K–2 mathematics curricula include topics such as counting, writing numerals, whole number operations (addition, subtraction, multiplication), fractions, estimation, place value, measurement, geometry, and problem solving. Within the past ten years, some curriculum developers have also included topics in probability and statistics for K–2 children which typically focus on making predictions about experiments and on recording and interpreting data. The following general summary of grade level expectations in mathematics is based on my review of several current K–2 mathematics curriculum guides from New Jersey public schools.

By the end of kindergarten, children should be able to count and write numbers up to twenty, as well as add and subtract these numbers. They

1991 *Mathematics Subject Classification.* Primary 00A05, 00A35.

should be able to measure in ad hoc units — for example, a desk may be three pencils long — and understand spatial relationships such as over, under, top, bottom, middle, left, right, inside, and outside. They should be able to identify planar figures such as a circle, triangle, rectangle, and square; and sort or classify objects by attribute — color, shape, or size. Children in kindergarten should also begin flipping coins and recording outcomes.

By the end of first grade, children should be able to count and write numbers up to one hundred and add and subtract two-digit numbers. They should begin to have some part-whole understanding of fractions and be familiar with fractional amounts such as 1/2, 1/3, and 1/4. They should be able to identify spatial figures such as a ball, cube, cone, can, and box, and be able to acquire information from pictures, text, and charts. They should be able to identify and discuss notions of symmetry and perimeter in a square, rectangle, triangle, and circle. They should be able to solve two-step word problems.

By the end of second grade, these same children should be able to count and write numbers up to 999; add and subtract three-digit numbers; know multiplication facts with 0, 1, 2, 3, 4, and 5 as factors; and write fractions symbolically and work with mixed numbers. They should also know the place value system for ones, tens, and hundreds; be able to make and use charts, tables, and drawings to solve problems; identify three-dimensional geometric shapes — cube, cylinder, sphere, cone, and rectangular prism; and discuss area and volume. It is also during the primary school years that children learn about systems: coins, clocks, calendars, maps, metric system, standard measurement system, bar graphs, and pie graphs.

"Young children enter school with informal strategies for solving mathematical problems, communication skills, ideas about how number and shape connect to each other and to their world, and reasoning skills. In grades K–2, students should build upon these informal strategies" (see the *New Jersey Mathematics Curriculum Framework* [**6**], page 83). Cognitively, according to Piaget, this population acquires knowledge through thought and action (see Inhelder [**3**]). As a result, mathematical concepts are taught through the physical manipulation of objects, through role playing, through story telling, and through thematic teaching approaches.

Existing curricula for the primary grades already include natural connections to discrete mathematics topics. For example, during the first marking period, many K–2 grades spend time classifying and sorting, including pattern detection (identify the pattern) and pattern projection (what comes next in the sequence). In fact, several K–2 textbooks which claim to include discrete mathematics topics simply include sorting activities and nothing more. Second graders spend time learning the fundamentals of geometry. The curriculum usually includes topics on shape, size, what defines an object, and what makes two objects different from one another. But I have also observed second grade children explain what makes a triangle, circle, and square the same. These children are capable of doing far more complex mathematics than we have traditionally expected. The following accounts serve to demonstrate what can be done in K–2 classrooms.

A visit to Grade 2

The day was "math day" (an entire day devoted to learning mathematics) when I visited Sharon Heil's second grade classroom at the Kossmann School in Long Valley, New Jersey. The school has roughly five hundred children in grades K through 2. Ms. Heil teaches in a self-contained classroom of twenty-four students. She described this class as a truly heterogeneous group, comprising students from both farm families and middle-management families. Academically, the students have a wide range of abilities; some students receive academic support in the resource room, others receive basic skills assistance in mathematics, language, and/or reading, and others are high-achieving, articulate problem solvers. In general, she feels all her students are enthusiastic learners and very curious about the world around them. In the classroom descriptions that follow, the names of the children are fictitious so that they remain anonymous.

Until participating in the 1995 Leadership Program for Discrete Mathematics (see Rosenstein and DeBellis [**7**]), Ms. Heil had not taken any mathematics courses since graduating from college over 20 years ago. To her credit, she is among many elementary school teachers who recognize the need to upgrade their own mathematical learning. It was not easy for her, but I witnessed the benefits — a teacher who provides thoughtful, meaningful mathematical experiences to her students.

Sharon Heil sat on a chair near a carpeted open space in her classroom. The students systematically pushed their desks to the side of the room and lined up near the chalkboard, silently waiting for instructions. Ms. Heil asked the children to randomly sit on the floor in front of her without any parts of their body touching one another. They were excited because they saw her holding a kickball and thought the idea of playing with a ball inside the building was neat! She told them that this is an activity where everyone is silent. "I'm going to give the ball to Annie. You must pass the ball from student to student (without throwing it or moving from your seat) so that everyone touches it at least once and gets it back to Annie."

Ms. Heil was imagining the children as vertices in a graph, where two children were joined by an edge if they were close enough to hand the ball from one to the other without changing positions. She was asking them to find a circuit which included all the children; soon she would ask them to find a Hamilton circuit. Needless to say, she had initiated this activity without introducing any of these terms. The children began to pass the ball to each other without talking. When it got back to Annie, the teacher said, "raise your hand if you touched the ball once." Sixteen children raised their hands. "Raise your hand if you touched the ball twice." Eight children raised their hands and the teacher asked them to stand. These eight children are pictured as Frank, Charlie, Zachary, Lisa, Deanne, Michael, Anthony, and

Annie in Figure 1. The ball was given to Janet who was the last person to touch the ball once (and was sitting on the floor) before giving it to Frank who was the first person to touch the ball twice (and was standing). The teacher asked, "is there a shorter ..." and was interrupted by Daniel who suggested that there was another way to pass the ball. He said, "Janet gives it to Jackie. Jackie gives it to Lisa and Lisa gives it to Annie." The teacher asked these students to pass the ball in this fashion to show that such a path was possible. After doing so, Ms. Heil asked Jackie, Lisa, and Annie to stand and all others to sit. Figure 1 indicates the two paths proposed by the children; the original path consisting of eight children who touched the ball twice (Frank to Charlie to Zachary to Lisa to Deanne to Michael to Anthony to Annie) and a shorter path (Jackie to Lisa to Annie) introduced by Daniel.

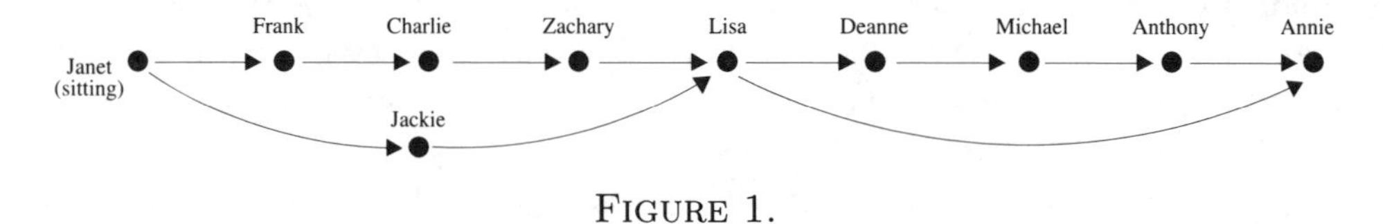

FIGURE 1.

A discussion ensued about how to make a shorter route. Lisa suggested that there is no route which leaves fewer than three people standing because, "how can you count one more person out? You'd have to throw the ball." Daniel insisted on a new proposal — Janet to Matt to Kenny to Maryann to Annie — then independently realized that this path was longer than his original three-person path. The group concluded that three was the fewest number of children who must touch the ball twice, until Daniel persisted that the ball can be passed with only two students touching it twice. He aggressively argued, "Janet to ..." but Cindy interrupted, "you can change the way you're passing the ball to only have Annie touch it twice." Daniel blurted, "you can just go in a circle." These suggestions happened simultaneously and the lesson that follows was crafted by a gifted teacher who encourages children to explain what they are thinking.

Ms. Heil interrupted to recognize appropriately the thoughtful comments that took place and asked, "Okay, let's consider individually what Cindy and Daniel each have said." The teacher stood and asked Cindy to exchange positions with her. All children were now sitting on the floor, waiting for their next instruction. Many children were laughing because they thought it was funny that the teacher was sitting on the floor, with her legs crossed like all the children, and Cindy was now in the teacher's position. Cindy began, "Okay, Annie gives to Lisa, Lisa gives to Robert, Robert gives to Sharon, Sharon gives to Mark, etc." Cindy orchestrated the movement of the ball in such a way that the only person who touched the ball twice was Annie. She worked outward from Annie, making sure that everyone touched the ball, but reserved a path of people along the front wall which she later used as the path that returned the ball to Annie. Her behavior was very similar to that of a mathematician as she or he works to find a Hamilton circuit in a graph; that is to say, each decision about to whom the ball should next be given is made keeping in mind that everyone needed to touch it once and "a last path" was needed in order to get back to the beginning. I thought to myself, "Am I really in a second grade classroom?"

The teacher stood after the task was completed and asked what just happened? The children explained, "If you do it the first way, the shortest way we could get is three people who touched the ball twice, but if you do it Cindy's way, you only get one person who touches the ball twice, Annie, so Cindy's way is shorter."

"Now, what about Daniel's comment. Daniel, what did you say before?" He replied, "You can just go in a circle." Ms. Heil suggested, "Okay everyone, let's get into a circle." From a theoretical perspective, the graph represented by the children has been changed, but from an educational perspective, Ms. Heil was presented with a valuable opportunity to take the lesson into uncharted territory, of which she quickly took advantage.

All the children sat in a large circle on the floor. The ball was given to Annie. "Now can you pass it so that everyone touches it once and it gets back to Annie?" The children passed the ball and when it was returned to Annie the teacher asked, "which way was easier?" They shouted, "circle!" Why? "Because you know where you're going." One child actually explained, "because you don't have to think about where to pass it next, you just get the ball from one side and pass it right to the next." This child was formulating a fundamental idea in computer science — that by arranging many individual units, each with a simple task, a large-scale, complex task can be performed. In computer-science terms, the children were simulating cellular automata.

Ms. Heil continued, "Is there any other way you could arrange yourselves so that ..." Another child shouted, "a square". The children arranged themselves into a square. They again passed the ball. "If I pass the ball along the square, is it similar or different if we pass it on a circle?" Several hands were raised immediately and the children responded, "similar." One child explained, "because we're still passing the ball to someone next to you." Another child shouted, "I think we should do a triangle because we could pass the ball there too." Ms. Heil said, "Good idea!" The class arranged itself into a triangle and passed the ball for a third time.

"So is the path in the triangle similar or different to the path in the square?" The class responded, "Similar." "What about the path in the triangle and the path in the circle?" "Similar." "What about the shape of the circle and the shape of the square?" "Different", they shouted. "What about the shape of the circle and the shape of the triangle?" "All their shapes are different." "Very good!" the teacher said as she looked at me in surprise. "So a path in a circle, square, or triangle is similar even though their shapes are different." This demonstrated that second graders are capable of understanding the rudiments of topological equivalence.

"What if we start with Annie, but don't end there? Could we arrange ourselves in such a way that the ball starts with Annie and everyone touches it exactly once but it doesn't have to end with Annie?" The students were still arranged in their triangle shape. They looked at each other as if this was too easy a question. One child said, "we don't have to move. Just pass it to Annie and end with Missy." Missy was the child who sat immediately to the left of Annie as the ball was passed to the right. Another child instantly shouted, "we could stand in a line." Ms. Heil began, "Okay, let's ..." and was interrupted by Daniel who said, "No, even if you stand in a line you get it back to the first person." The teacher and I were both confused. Did Daniel see a way for people to stand in a line and still make a circuit? Ms. Heil inquired, "What do you mean?" Daniel said, "You just have to give the ball to the first person, the first person gives it the third person, the third to the fifth, all the way to the end, and then that person just has to pass it back to the ones who didn't touch it yet." I was truly amazed at this second grader's insight.

Ms. Heil said, "Daniel thinks that you can get the ball back to the beginning if you stand in a line and everyone only touches it once except for the first person. Who agrees with Daniel?" A few hands were raised, tentatively. "Okay Daniel, show us what you mean." All twenty-four students stood in a straight line except Daniel. He gave the ball to Annie, who was standing at one end, and said, "Annie gives it to Frank, Frank gives it to Michael, Michael gives it ..." until the ball was passed back to Annie with everyone touching it exactly once. Figure 2 depicts a simplified version of the path that Daniel, a second-grade student, constructed in his mind; Daniel's path involved all twenty-three children.

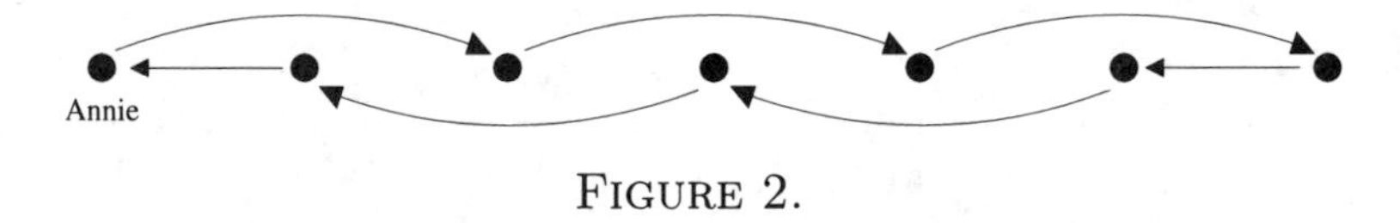

FIGURE 2.

"What just happened here?" the teacher asked. One child explained, "even though we're standing in a line you can get the ball back to Annie and only touch it once." The lesson concluded by introducing the words "path" and "circuit". When the teacher introduced the word circuit, Pete shouted, "is that like a circuit breaker?" Children make connections naturally if they're allowed to investigate their world. The words circuit and path were on the next's week spelling test.

I sat back in my chair in amazement. Second graders are quite capable of intuitively constructing paths and circuits in quite complex ways. They are able to recognize that a ball's path is the same in a circle, in a square, or in a triangle. Of course, they were unable to discuss graph isomorphism, but they found ways that a circle could be the same as a square and as a triangle. Further, they were able to maintain, at the same time, that these objects have different shapes in the Euclidean sense. They were able to identify and generate shorter paths — by finding a new way to pass the ball that would involve fewer children. A second grade classroom can fully engage in a dialogue which is filled with rich mathematical discourse.

This example shows how widely accessible topics in discrete mathematics can be. A young person's problem may be worded as follows: Given N children, randomly seated on the floor, how do we pass a ball so that each child touches the ball only once and so that it gets back to the first person who touched the ball? A similar challenge for a more mature problem solver may be worded a bit differently: Given N points in the plane — each connected by an edge with a few of its neighbors, find a Hamilton circuit. Essentially, both populations (children and adults) are able to discuss and solve these problems successfully. Having children think about such problems during their primary school years will provide a foundation for later mathematical development.

It might be said that this second grade classroom was full of gifted children, or at least Daniel (the child who generated many interesting paths during this lesson) was quite talented. Actually, none of the students in this class have been classified as "gifted", including Daniel. (In the Kossmann School, to be classified as "gifted" the child must score at least a 135 on the Wechsler Intelligence Scale for Children (W.I.S.C.).) Maybe educators need to evaluate how we determine if a child is mathematically talented. I saw a few children in this class who demonstrated powerful mathematical thinking and who I would classify as "gifted."

A visit to Kindergarten

I was invited into Michele Midura's classroom at the Irving Primary School in Highland Park, New Jersey to teach a discrete mathematics lesson. The school has roughly four hundred children in grades K through 2. She teaches a self-contained full-day kindergarten class with eighteen students, sixteen of whom were in attendance on the day of my lesson. She describes the class as developmentally, culturally, and economically diverse. Academically, the students have a wide range of abilities; some students have learning disabilities while others are reading and writing on a first grade level. Prior to my arrival, she informed me that the theme for the month was sports and nutrition and suggested that whatever math I did, I should somehow tie it to one of those themes.

In keeping with the "sports" theme, I decided to introduce the notion of a tournament by having each pair of children in a small group roll a giant die to determine a winner. I was uncertain how much of this topic kindergarten children would be able to understand, or even whether they would be able to determine if every player in their group competed against every other player exactly once. To see if they were capable of both enumerating all possibilities and knowing when they found all possibilities, I decided to begin with a combinatorics activity.

I entered her classroom with a large duffle bag filled with sports equipment, sneakers, and several two-foot long arrows made from poster-board paper. After their normal routine of hanging up coats, turning in homework, selecting hot or cold lunches, and telling their morning news (what Ms. Midura calls "show-and-tell"), I am able to introduce a math problem. The children were sitting on the floor arranged in a big square. I said, "Close your eyes! Keep them closed!" and reached into my bag. I pulled out a set of plastic bowling pins with two plastic bowling balls, still in their original wrapper. I asked, "What sport would you be playing if you needed these?" They simultaneously yelled, "Bowling!" Sixteen little people yelling an answer in unison caught me by surprise. Their level of excitement is infectious. This is nothing like teaching undergraduates! I placed the bowling pins on the floor in the middle of the square and said, "Close your eyes!" Several children began wiggling with anticipation. "Keep them closed!", I said. After I pulled out a tennis racket from the bag I asked, "What sport would you be playing if you needed this?" "Tennis!", they yelled. I placed the racket on the floor next to the bowling pins. We played the "close your eyes" routine two more times as I pulled out a pair of pink sneakers and a pair of Rebok sneakers and placed them on the floor.

"How many ways can you choose a pair of sneakers and a sport to play?", I asked. There was dead silence. I thought, "Uh, oh ... this is probably too hard." I regrouped and asked a different question, "Can anyone choose a pair of sneakers and a sport to play?" All sixteen children raised their hands. Anita chose the pink sneakers and bowling pins. I asked if anyone could find another way. Jimmy chose the Rebok sneakers and the tennis racket. Both children were standing in front of the class, wearing the sneakers they selected and holding their chosen piece of sports equipment. I pointed to each item and repeated, "Okay, Anita wears pink sneakers and bowls. Jimmy wears what?" The children together responded, "Rebok!", "and plays?", "Tennis!", they yelled. "Okay, can anyone find a different way to wear sneakers and play a sport?" Sean raised his hand, walked in front of the four items (now on the floor) and stared at them. After a few seconds, I asked if he would like a helper and I noticed Ms. Midura standing behind all the children nodding her head yes. Sean nodded his head up and down and picked the boy who was sitting next to him. Together they selected the Rebok sneakers (because pink sneakers were for girls) and the bowling pins. I repeated their choices, "Okay now we have a different way. Rebok sneakers and bowling pins. Can anybody find another way?" Cindy selected pink sneakers and the tennis racket. I asked, "What did Anita pick?" The students described her selection. "What did Jimmy pick?", "What did Sean pick?", "What did Cindy pick?" Each time the children described the choice of sneakers and sports equipment.

Now I returned to my original question, "How many ways can you choose a pair of sneakers and a sport to play?" "Four!", they yelled, continuing to respond in unison. "How did you know there were four?", I asked. One child explained, "because Anita's way is one, Jimmy is two, Sean is three, Cindy is four, and there's no other way to make a match that's different." "Okay, what are the different ways?" Together the children described each way of matching a pair of sneakers with a piece of equipment. Each time a new pair was mentioned, I placed a large fluorescent colored arrow on the floor to show the match. After all the pairs were found, we counted the number of arrow heads together and discovered four different ways. I said, "Close your eyes!" This was now a game for them. They're wiggling and getting excited because they know something else is coming out of the bag.

I placed a whiffle ball and bat on the floor next to the tennis racket, removed the arrows, and asked, "Now how many ways can we match a pair of sneakers with a piece of sports equipment?" The children enumerated all possibilities in a similar way described above. We again placed the arrows on the floor to show all six possibilities and count the arrow heads. "Close your eyes!" The children were now peeking (and telling me that they're peeking) and laughing as I pulled a pair of men's dress shoes from my bag. I asked them, "Do you know what these are?" No one responded. I said, "Geek sneakers!" and they all started laughing. "Now how many ways can we match a pair of sneakers with a piece of sports equipment?" A variety of children suggested simultaneously, "ten", "six", "nine", "twelve". I asked the young girl who responded "nine" to explain how she got her answer. After a bit of encouragement, she said that she counted every object on the floor (six individual sneakers and three pieces of equipment). Then one child yelled, "No, it's three plus three plus three." I am stunned by both responses.

I explained that mathematicians count all sorts of things. Sometimes we count individual things; for example, if we count all the sneakers and all the equipment, we find that there are nine things altogether. But sometimes we count groups of things, and that it is good to learn how to do both. Here we are counting how many ways there are of forming a group which has one pair of sneakers and one kind of equipment. I asked Carlo to use the arrows to show me what he meant by "three plus three plus three". He placed the arrows on the floor to show each possibility and then proceeded to count the arrow heads; altogether there are nine. He showed that each pair of sneakers can be matched with three different pieces of sports equipment and was able to describe this matching as "three plus three plus three". I reviewed this generalization with the children, and so ended the combinatorics activity.

Now convinced that kindergarten children will be able to determine if all players competed against one another in a tournament, I asked, "Find an X made from masking tape on the floor and sit on top of it." In advance, eighteen small X's were positioned in three circles (six to a circle) so that the fluorescent arrows could be placed between any two players to identify the winner and loser of that competition. All sixteen children, Ms. Midura, and her classroom assistant played in a tournament. Each group received one pair of giant dice and fifteen arrows of two lengths — nine long arrows (to be placed between players who are seated across the circle from one another) and six short arrows (to be placed between players who are seated next to one another). Player A rolled one die and Player B rolled the second die; together they determined the winner — the one who rolled the larger number — and placed an arrow on the floor pointing from the winner toward the loser. After all fifteen arrows were arranged according to the outcome of each competition, the children were asked to find a winning sequence — that is, a way of listing the six children in the group so that each one defeated the next one in the sequence; the winning sequence would be a Hamilton path in the directed graph.

In this lesson, it should be no surprise that the children were unable to find a winning sequence. I did several things wrong. First, the size of the groups (six) was too large for kindergarten children to even begin to look for a winning sequence. Having fifteen arrows pointing in many directions contained too much information for them to decipher. Second, the children were not comfortable with the arrows pointing to the loser; they wanted them to point to the winner and after several protests, the arrows pointed to each winner. Although this does not impede one from finding a Hamilton path, it does introduce another cognitive step in finding a winning sequence. Third, they really liked giant dice and I did not allow enough time for the children to "play" before I asked them to hold a tournament. Seasoned teachers of primary grades always allow time for play before they ask children to complete a task. Finally, I did not review the skill of finding a Hamilton path — that is to say, constructing a path along a series of directed edges so as to touch every vertex exactly once. I would certainly do this activity differently the next time, and expect that the children would find Hamilton paths; several kindergarten teachers in the Leadership Program have reported that they were able to do this with the children in their classes.

Although the lesson did not achieve what I initially intended, I learned that children in kindergarten are able to understand issues that arise in enumerating possibilities (sometimes an exercise in creative thinking) and determining that all possibilities have been exhausted. I also learned that children in kindergarten are quite capable of deciding who is a winner and how to position a directed edge to reflect that. I also believe that children at this age can identify a winning sequence in a tournament with four or five competitors, based on my observations of why they were unable to complete the task in the setting described here. In the long run, such activities help children develop strategies that are valuable for later use in problem solving, as well as for probability. Not to include such activities throughout the primary grades is a serious omission.

In this classroom I took a risk by attempting an activity which I had not tried before with children at this grade level. However, I have described this "failure" here so that teachers will understand that it is important for them to take similar risks in their classrooms. Teachers who attempt to bring discrete mathematics into their classrooms will find that sometimes their lessons are successful, and sometimes they are not. Nevertheless, they should continue to try out new activities, since ultimately their students will benefit, as they find ways of improving on their failed attempts, and even their present students will benefit, as the problem solving prepares them for future challenges.

A coloring example

A popular discrete mathematics topic among teachers who attend the Leadership Program is graph coloring. We introduce the topic by having groups of teachers work together to color a five-foot map of the United States. Each group is provided with two hundred circles cut from construction paper. There are ten different colors with twenty circles of each color. The initial question posed is to simply color the map. In a short time, some groups have nicely colored maps using all ten colors, other groups may be trying to use fewer colors, and yet others may define specific colors to represent characteristics of that state (i.e., all states that border the ocean, color orange or green). We now introduce the mapmaker problem. Imagine that you are a mapmaker and the cost to produce a map increases based on the number of different colors you use. Further, since every mapmaker wants individual regions to be clearly viewed on the map, no two regions that share a border may be colored with the same color. What are the fewest number of colors needed to color the map of the United States? Why? Can you identify areas of the map which cause problems? (If you haven't thought about this problem, take a few minutes to think about it before reading the rest of this article!)

When I walked into Sharon Heil's second grade classroom, I saw every piece of available wall space filled with students' work, from floor to ceiling. In the front of the classroom was a bulletin board dedicated to map coloring. There were several colored maps of the United States (partitioned into states) and several maps of Ohio (partitioned into counties, supplied by a Leadership Program participant from that state) which were colored by the students so that no two regions which share a border had the same color. I thought to myself, "I've seen that before ... but I wonder what second graders got from the experience."

Ms. Heil asked, "Does anyone remember what we did when we colored the maps?" Instantly, several hands were raised. One child explained that you "color two states with different colors if they're next to each other." "Anything else?", Ms. Heil asked. Another child explained, "We had to decide if corners counted or not." Several students proudly pointed to their map. Some groups decided that corners "counted", that is, they should be considered as shared borders, and some groups decided that they should not be considered as a common boundary. At the close of the discussion, Ms. Heil smiled and whispered, "we did that over one month ago."

This discussion was interesting because when we give the same exercise to K–8 teachers in our summer institutes, they too begin the activity with a similar struggle, namely, to decide if a point should be considered a shared boundary. An example of this occurs in the United States map at the point where Arizona, New Mexico, Colorado, and Utah meet. Most groups of teachers initially color these four states with four different colors. When they are asked to color their maps using the fewest colors, the issue arises as to what defines a border. Both populations (teachers and second graders) had to decide and define for themselves whether a point should be considered a shared border. All teachers were able to make the distinction that regions that meet at a point do not necessarily have to be considered a shared border; some second graders understood this, but others did not. When you make a decision that two regions that meet at a point may have the same color, the number of colors you will need to color a map may indeed be fewer. Hence some of the students' maps were colored with three colors, but others used more colors. Ms. Heil later explained that she did not focus on "fewest colors" and that the activity was intended to introduce coloring and discuss what makes something a border.

Map coloring is another example that shows how widely accessible topics in discrete mathematics can be; both adults and young children can engage in mathematical problem solving and experience similar difficulties. In the end they may resolve them quite differently, but they each have to define aspects of the problem that are not clear. Working through the "muck" of problem situation is one of the more difficult aspects of problem solving to teach; one must be willing to intimately engage the problem rather than passively perceive it (see Levine [**4**]). Introductory discrete mathematics topics seem to invite people from the non-mathematical community to think about their problems because difficult problems are easily understood and typically require little prerequisite knowledge; it may be that discrete mathematics can serve to attract under-represented persons into the mathematics field. It certainly attracts adults and young children alike!

Traditionally, coloring has been used in primary grade classrooms to help develop dexterity, creativity, and artistic talent. Now, with an introduction to discrete mathematics, teachers of K–2 classes can incorporate mathematics into their coloring book activities by asking the children to color the page so that no two regions that share a border have the same color. Once it is clear that everyone is able to color the page in this way, the teacher can introduce the question of using fewer colors. Before they begin, the children can be encouraged to talk about how they might develop a plan to use fewer colors. They can discuss why one color or two colors may not be enough to color the picture, or why N colors may be too many. These early conversations can help children begin to develop mathematical ideas about minimization and give practice in reasoning about lower and upper bounds. In addition, such lessons can also help develop powerful mathematical problem solving skills. For a good description of the value of coloring in K–4 classrooms, see Casey and Fellows [**1**].

Coloring books are also a good place to introduce children to constructing dual graphs, as part of an introduction to the topic of graphs. A typical page in a coloring book may look like the first bear in Figure 3. Primary

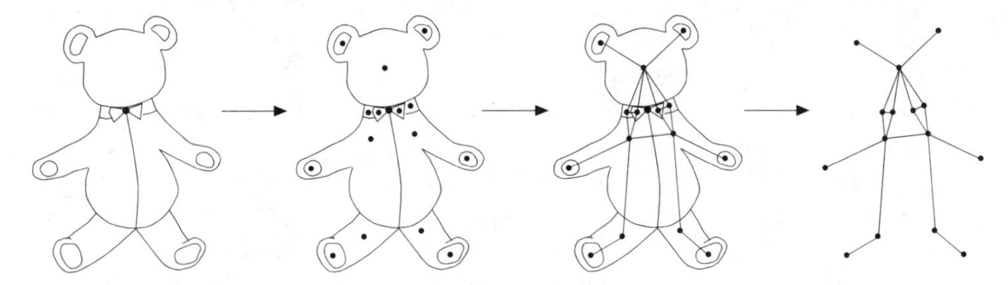

FIGURE 3.

grade children can be asked to place a dot inside each region and then connect two dots if the regions share a common border; the traditional kindergarten curriculum already includes learning notions like "inside/outside" and "next to" as geometry topics. These activities can provide a visual way to determine if children understand such notions. If the teacher chooses the picture wisely, she can then have further discussions with her students about the structure of the graphs. For example, in Figure 3, a teacher could ask the children to count the number of vertices and edges; to describe what parts of the graph look the same or what parts of the graph look different; to determine whether the graph is connected (together) or disconnected (in parts); or whether they can fold a piece of paper in such a way so that every vertex will lie on top of another vertex and every edge of the graph will lie on top of another edge. These experiences, if implemented in thoughtful ways, can help develop early notions of structure and symmetry.

The future of the K–2 curriculum

The K–2 curriculum for the twenty-first century needs to include technology topics and the mathematics that underlies computing. Some of these topics may be paths and circuits in graphs (consisting of vertices and edges), vertex coloring, Euler and Hamilton paths and circuits, shortest routes, counting, listing and sorting, and recognizing and using patterns in number and geometry. However, I am not suggesting schools should simply add more mathematical topics to an already packed curriculum. Rather, young children can learn to add numbers in the context of traveling along paths in weighted graphs (where each edge is assigned a "weight" which may, for example, be the distance between the sites represented by the vertices at the ends of the edge), or count "the number of ways" — an activity that can be done instead of just counting the natural numbers.

Primary grade students can establish efficient ways for dealing with their environment, and determine what makes something better (or shorter, or quicker) than something else. It is during this time that they can learn how to follow directions, follow classroom rules, follow a recipe, follow a map, and follow an algorithm. They can learn how to count the number of different ways to make change for one dollar, or how to systematically list the different ways to arrange three shirts with three pairs of pants. They can make flipchart storybooks to demonstrate the total number of outfits one can wear and, through such activities, come to know that mathematics is a way of thinking, not a way of memorizing. All of these topics are discussed in detail for the K–2 grade levels, as for other grade levels, in the discrete mathematics chapter of the *New Jersey Mathematics Curriculum Framework* (see Rosenstein, Caldwell, and Crown [**6**]).

At the primary grade levels, children can also be assisted and encouraged to come to understand what it means to be a powerful problem solver. A powerful problem solver is one who knows more than just a bunch of good strategies for solving a problem; it is a person who (among other things) uses intuition, generates conjectures, is creative, and perseveres. Young children can learn how to make a good prediction, how to remain comfortable even if a problem is left unsolved for several days, and that sometimes good problem solvers get wrong answers. They can also learn that working on a hard mathematical problem is sometimes frustrating, but that negative emotions can be regulated by the problem solver to a useful purpose (see DeBellis [**2**]). The ability to be successful at problem solving is no longer a higher order thinking skill that only mathematically talented children are expected to demonstrate; rather, all citizens of the twenty-first century will need this skill to function in a high-tech world. Today's kindergarten children will graduate in the year 2009.

Conclusions

I was quite surprised at the sophistication with which primary grade students can behave as scientists. As I walked around K–2 classrooms, observing other activities as well as those described in this article, I watched young children make conjectures, argue with team members for particular outcomes, demonstrate the ability to collect and record data accurately, verify that an experiment was run correctly by making sure the sum of each component equalled the total number of experiments conducted, and demonstrate the ability to make the distinction between a prediction and a best prediction. They also intuitively discussed fundamental notions of isomorphisms, algorithms, and topological equivalence. They were proud of themselves when each mathematical task was completed, just as the teachers were who worked on the same (or similar) problems in the Leadership Program.

Certain ideas in mathematics — such as "isomorphism", "enumeration" (systematic listing of possibilities), or the ability to generate global complex behaviors with simple local rules — are very important and should be developed in young children. It should not be that these discussions happen in K–2 classrooms by chance. Mathematicians, mathematics educators, and teachers need to collaborate to define what "big mathematical ideas" ought to be learned at each grade level. Technology will continue to evolve and new mathematical discoveries will unfold. Unless school systems allow for the constant infusion of new mathematical topics and information into their curriculum, they will forever be teaching archaic topics at inappropriate grade levels.

Finally, K–2 discrete mathematics topics, when introduced by good teaching methods, can serve not only to build the foundations for important mathematical ideas, but also can serve as a vehicle to help cover traditional curriculum topics. K–2 teachers need continued support from university and college faculty members who are both knowledgeable about the content and understand the mathematical development of young children. At the same time, teachers need to remain active in the learning of mathematics, at whatever level is appropriate for them. It is only when teachers themselves are active problem solvers who, for example, think about problems they cannot yet solve, that they can model the desired mathematical behaviors for the children in their classes. Such activities and collaborations can only benefit the children.

References

[1] Casey, Nancy, and Michael R. Fellows, "Implementing the Standards: Let's Focus on the First Four", this volume.

[2] DeBellis, Valerie A., *Interactions between affect and cognition during mathematical problem solving: A two year case study of four elementary school children.* Doctoral dissertation, Rutgers University, 1996. Ann Arbor, Michigan: University Microfilm 96-30716.

[3] Inhelder, Bärbel, "Some aspects of Piaget's genetic approach to cognition", in Hans G. Furth, *Piaget & Knowledge: Theoretical Foundations* (2nd edition), University of Chicago Press, 1981, p. 22.

[4] Levine, Marvin, *Effective Problem Solving*, Prentice Hall, 1994.

[5] Rosenstein, Joseph G., "A Comprehensive View of Discrete Mathematics: Chapter 14 of the New Jersey Mathematics Curriculum Framework", this volume.

[6] Rosenstein, Joseph G., Janet H. Caldwell, and Warren D. Crown, *New Jersey Mathematics Curriculum Framework*, New Jersey Mathematics Coalition, 1996.

[7] Rosenstein, Joseph G., and Valerie A. DeBellis, "The Leadership Program in Discrete Mathematics", this volume.

Center for Mathematics, Science, and Computer Education (CMSCE) and Center for Discrete Mathematics and Theoretical Computer Science (DIMACS), Rutgers University

E-mail address: debellis@dimacs.rutgers.edu

DIMACS Series in Discrete Mathematics
and Theoretical Computer Science
Volume **36**, 1997

Rhythm and Pattern: Discrete Mathematics with an Artistic Connection for Elementary School Teachers

Robert E. Jamison

[It is easy to] appreciate sunsets, and the ocean waves, and the march of the stars across the heavens. As we look into these things we get an aesthetic pleasure from them directly on observation. There is also a rhythm and a pattern between the phenomena of nature which is not apparent to the eye, but only to the eye of analysis; and it is these rhythms and patterns which we call Physical Laws.

—Richard Feynman, *The Character of Physical Law* [**11**]

1. Introduction

Over the past two years, I have had the privilege of offering a course entitled *Connecting Mathematics with Art, Music, and Nature* to two cadres of elementary school teachers participating in the Oconee County Lead Teacher program.[1] Each cadre of about twenty teachers dedicated every Monday night for two years to the project. Although similar in spirit to the elementary mathematics specialist program suggested by the NCTM (National Council of Teachers of Mathematics), the Oconee project focuses on *generalists*, rather than those who already have a special affinity or gift for mathematics, who can then become leaders in their schools for introducing new and more successful approaches to mathematics instruction. In this way the principle that "mathematics is for everyone, can be learned by everyone, and enjoyed by everyone" is emphasized. Quite naturally, many of the topics were in discrete mathematics. Some are fortunately becoming a common part of the curriculum: building polyhedra with *Polydron* shapes [**18, 20, 21**], making tessellations [**31, 33, 34, 38**], and classifying strip

1991 *Mathematics Subject Classification.* Primary 00A35.

[1]The Lead Teacher program was developed at the University of Chicago, and customized for the School District of Oconee County, South Carolina, by Ann Stafford, a district office staff professional, and Sybil Sevic, a classroom teacher.

patterns by symmetry types [**6, 9**]. The topics to be discussed here are less standard: drawing exercises for regular polygons, movement exercises to develop symmetry concepts, and connections between modular arithmetic and music.

Currently, I am using many of the same ideas and activities in an undergraduate geometry course for pre-service elementary school teachers. The course has four main goals:

1. to broaden the participants' view of mathematics;
2. to gradually stretch their level of comfort with mathematical ideas and abstractions;
3. to introduce the participants to a developmentally appropriate model of education; and
4. to give new meaning to mathematics by connecting it to subjects that have emotional content like art and music.

This article describes some of the activities that I use and the mathematics underlying them. It is addressed primarily to mathematicians and mathematics educators working with elementary school teachers or students; elementary school teachers may also find activities here to try in their classrooms.

As Peter Hilton [**17**] has said, "Geometry is a natural source of questions and algebra is a source of tools to answer them. When we teach algebra before geometry, we ask students to answer questions that no one would ever ask. And later in geometry, we give them problems that they have no hope of ever solving." This speaks to the idea that there is a preferred order in the introduction of mathematical concepts. I stand firmly with Rudolf Steiner [**30**], Jean Piaget [**8**], and the van Hieles [**10**] in believing this order is developmentally determined and that in rough outline it follows the historical development of the subject.

The impulse for many of the activities I describe here comes from the Montessori and Waldorf [**2, 3, 4, 5, 25, 26, 30, 37**] educational movements, which I believe have much of value to offer all schools. In the lead teacher courses as well as in other courses, I have not only used ideas from Montessori and Waldorf education but have also integrated workshops given by experienced Montessori and Waldorf teachers. The benefit of these ideas is that they stress the value of proper foundations for concept formation and learning.

In order to clarify my approach, for the sake of discussion, let me suggest five stages in the learning process:

1. encounter
2. observation
3. reflection
4. understanding
5. creativity

These terms have rather broad meanings in general use, so let me describe more specifically what I want them to mean here. Suppose that on my way to work I pass a particular old brick building. I see it everyday, perhaps, just out of the corner of my eye, without paying much attention. It is just there. Then I have encountered it. One day something causes me to pause and notice the building. "Oh," I think, "that is an interesting building, with rather attractive brick work." Now I have observed it. If I continue to notice the building, looking for different patterns in the brickwork, and wondering how they were made, then I have begun to reflect. In a garden walk behind my house, I successfully incorporate a brick pattern like one of those in the building. This demonstrates understanding. If I now choose to invent a pattern of my own, then I have become creative.

In this analysis, I see rule formation at the third stage and skill development at the fourth. These are the primary focus of our current educational system, reinforced by constant testing. But they depend very much on earlier experiences and observations to make them meaningful. These foundational experiences may lie several years back or may require repetition over a long period.[2]

Unfortunately our current educational system offers few incentives for, say, first grade teachers to provide the foundational experiences so essential to a fourth grade teacher's success. Nonetheless, I strongly encourage the teachers in my classes to consciously provide experiences for their students' later development even if it will not be tested in their classes. This is one of the main ideas underlying the activities presented here.

A second main idea is the value of kinesthetic and sensorial learning. This is a particularly strong feature of the Montessori materials and the morning "concentration exercises" in the Waldorf schools. The idea is that by actively involving our bodies and senses, the meaningfulness of learning is enhanced. For that reason, the activities which follow involve movement and color as essential features.

The third main idea behind the activities is the artistic element. The goal is not just to look for applications of mathematics in art. Nor is it to use art as a sugar coating for the bitter pill of mathematics. Rather I hope to capture the artistic spirit in mathematics as something beautiful and creative. Looking for symmetry patterns in medieval architecture should stimulate the students' esthetic sense and sense of history as much as pique their mathematical curiosity. Thus the goal is not a dominance of one subject over the others, but a balanced blend in which each is seen to offer something of value.

---

[2]This is one reason why both Montessori and Waldorf have the same teacher stay with a class for more than one year. Montessori teaches in three-year cycles with grades 1, 2, and 3 together followed by a cycle with grades 4, 5, and 6 together. The Waldorf teacher stays with the same class for eight years and so can enjoy in the eighth grade the blossoms from seeds planted in the first.

2. Drawing Polygons and Their Diagonals

Drawing Regular Polygons. The first task here is deceptively simple: to draw the regular n-gons (for $n \leq 10$). What makes it interesting is that the drawing is to be done freehand, without lifting the crayon from the paper and without turning the paper.[3] The goal is to develop an intuitive feel for the regular polygons in the hands and fingers so that the drawing comes with ease and freedom. Students must recall a mental image of the regular n-gon and use their understanding of the figure to reproduce it on paper. This requires students to actively reflect on their previous experience with the regular n-gon and can lead to some satisfying insights. Also, in our everyday experience, there are enough encounters with triangles, squares, hexagons, and octagons that most people can draw them rather well. The 5-, 7-, 9-, and 10- gons are less familiar and the missing experience with these figures needs to be provided.

For most people, drawing the pentagon and heptagon is already difficult, and it is best to approach the task in stages. You might begin by having students trace around templates[4] with (colored) pencil. This provides the student with models for later use as well as a valuable kinesthetic encounter with the shapes. The tracing should be done in one motion and if possible without lifting the pencil.

"Helping figures" should also be used in the beginning (see Figure 1). For example, in drawing an octagon, one can start with a square and cut off the corners (Figure 1a). Dually, one could "push out" the sides of a square by adding four new corners above the centers of sides of the square. The decagon can be derived from the pentagon in a similar way. To get the pentagon, it may be helpful to start with the 5-pointed star (the pentagram) and connect its points. The 5-pointed star resembles a person with legs spread and arms out-stretched, as in the celebrated da Vinci drawing—making the anthropomorphic connection explicit can strengthen the connection between the abstract world of mathematics and the direct experience of the child. The 9-gon can be built up by adding trapezoids to the sides of an equilateral triangle (Figure 1b), but getting the proportions just right is rather tricky. There is no way to reduce the 7-gon to a smaller polygon—since 7 is prime.

After a set of tracings is completed, and the students have created the shapes by modifying simpler polygons, the freehand drawing can begin. At first, allow the students to lift the crayon and turn the paper. As understanding, skill, and confidence grow, ask them to reduce the use of these aids. The accuracy of the final drawings can be visually checked by turning the page to see whether the figure looks the same no matter which side is chosen as base.

[3] I say "crayon" here because, as the students discover, crayon is more forgiving than pencil, and produces attractive, colorful drawings. I highly recommend the beeswax block crayons from Stockmar—lightly rubbing the paper produces a lovely, soft colored background. Of course, the use of rulers in this exercise is taboo!

[4] E.g., the Montessori "geometry cabinet" contains a tray of regular n-gons ($n \leq 10$).

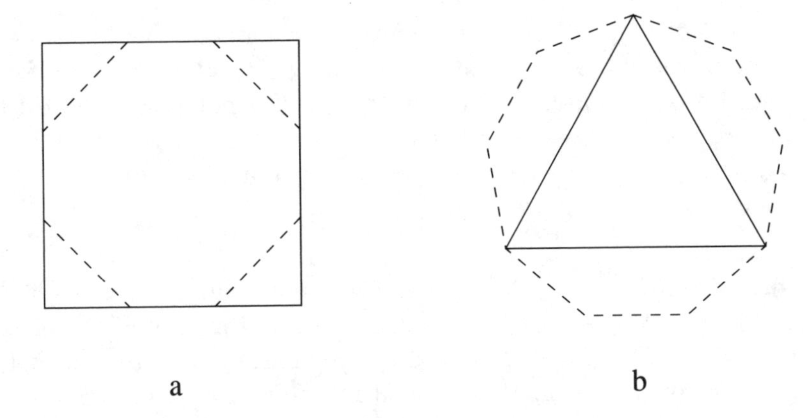

FIGURE 1. Using "helping figures" to draw regular n-gons: (a) deriving a regular octagon by cutting the corners of a square; (b) deriving a regular 9-gon by adding trapezoids to the sides of an equilateral triangle.

Several important mathematical ideas have crept into the exercise by this stage.[5]

- The drawing requires that segments have equal length and that angles be equal. That is, a polygon is regular if and only if it is both equilateral and equiangular. For triangles, these two properties are equivalent, but for quadrilaterals they define two different classes, the rhombi and the rectangles. This leads nicely into a discussion of the different classes of quadrilaterals, the nature of mathematical definitions, and that all-important question: "Is a square a rectangle?" (See [**7**, pp. 133–140] for a good discussion of the Aristotelian theory of definitions versus the modern theory.)
- The helping figures used along the way illustrate relationships among the regular n-gons based on the factors and divisibility properties of n. The geometry then becomes a visible expression of certain arithmetic relationships.
- Checking the accuracy by turning the drawing invokes a "transformational" definition of regularity (in mathematical terms, a figure is regular if and only if its group of symmetries acts transitively). It also prepares the way for a discussion of rotations and reflections in general.
- Certain pedagogical issues are also addressed. Holding the paper fixed requires the student to sense equal lengths and equal angles in a variety of orientations—not just in the natural horizontal-vertical frame of reference given by the bilateral symmetry of the student's

[5] I discuss these issues with the in-service teachers but urge them not to discuss them with their school children. They may serve as guides for observing a child's progress, but the proper qualities should be instilled by example, not by edict.

body. Not lifting the crayon requires concentration and forces steady adjustment of perception of length and angle. This requires effort and forces a very conscious encounter with the polygon. The drawings are also an exercise in neatness, patience, precision, and pride in workmanship, qualities which are valuable in mathematics as well as other endeavors.

Diagonals and Star Polygons. Once the students are comfortable with the freehand drawings, they can be asked to draw the diagonals of the regular n-gons, yielding the various star polygons, as in Figure 2. Color, of course, can be used to bring out special relationships. Students should be urged to use color not in a random way, but rather as an aid in revealing the inherent order.

In the Waldorf schools there is a lovely string exercise that complements this activity. Stand a group of n children in a circle. Number them 0 to $n-1$ (or 1 to n with younger children) and be sure they remember their numbers. Checking to see that students are standing on the vertices of a regular polygon involves recalling some interesting geometry: everyone should be the same distance from the center and everyone should be the same distance to their two neighbors. I also ask my students to check that each parallel class of diagonals is indeed parallel. Now, take a large ball of yarn and give it to the first child, that is, the child with number zero. Decide on a number k of "steps" to take in tossing the yarn around. Then have the children count off in turn and toss the yarn to every kth child. Each child holds onto the strand of yarn when received, so that in the end some star polygon is formed. (The yarn will be dropped occassionally, and there will be lots of laughter and giggling.) Holding onto the corners, the children can slowly lower the yarn to the floor to see the pattern emerge. (See Figure 2.) It is surprising, even for adults, what a different experience it is to "draw" these diagonals by tossing yarn rather than by using pencil and paper.

This exercise can be tailored to fit a variety of purposes and ages of children. For example, with younger children, I would have them count off the integers in their natural order. Have all the children count together out loud, as rhythmically as possible. This helps focus their attention and instills a kinesthetic sense of number. Since the yarn is tossed whenever a multiple of k is called out, this reinforces the multiples of k and is preparation for learning the multiplication tables. Here is how the pattern for $n = 12$ and $k = 5$ starts:

0 1 2 3 4 **5** 6 7 8 9 **10** 11 12 13 14 **15** 16 17 18 19 **20** etc.

In this case, the yarn is tossed when 5, 10, 15, 20, etc are called out. The yarn will be tossed from child 0 to child 5 to child 10 to child 3 to child 8, and so on.

Implicit in this version is an indirect encounter with division and remainders. With older children, you may wish to make this more explicit. Before tossing the yarn, have each child write his number on a name tag. Again

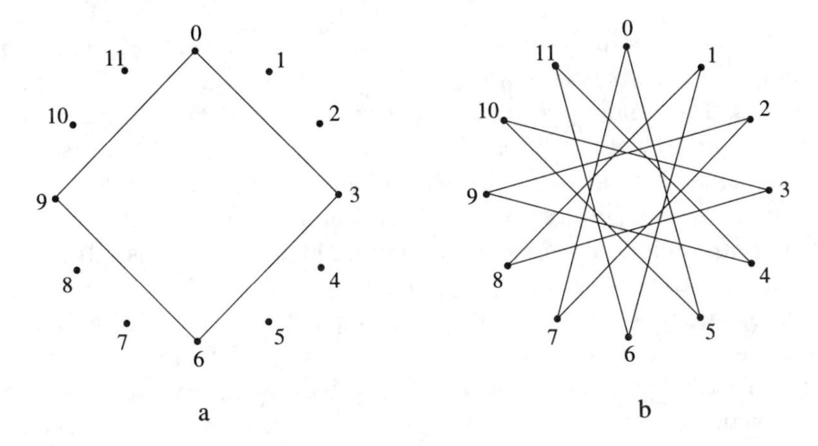

FIGURE 2. Star polygons created by yarn-tossing for $n = 12$; (a) $k = 3$; (b) $k = 5$.

have the children count off the integers in their natural order. This time, however, have them call out their numbers individually, one after the other, rather than in chorus. Thus for $n = 12$, child number 3 will call out 3 the first time around, 15 the second time around, 27 the third time around, and so on. These are precisely the positive integers that leave a remainder of 3 on division by 12. This version of the exercise is a preparation for modular arithmetic, discussed below in Section 5.

When you have chosen a number n of children to work with, it is best to systematically go through all values of k from 1 to $n - 1$. Children should note that k and $n-k$ always give the same shape, but traced out in opposite orders. The value $n = 12$ seems to be a good one to start with for several reasons. For younger children it is related to the familiar clock face. For older children who are learning geometric constructions, it is possible with only moderate difficulty to construct a regular dodecagon with straightedge and compass. Hence they can construct accurate diagrams of the patterns they first formed with the string. Moreover, for $n = 12$, several familiar geometric shapes appear: a hexagon for $k = 2$ and 10, a square for $k = 3$ and 9, and a triangle for $k = 4$ and 8.

The exercise can be further modified to investigate many other mathematically significant questions.

1. Which polygons are the same?
2. Does the yarn always come back to where it started?
3. For which k will everyone get the yarn?
4. If only some children get the yarn, can you predict how many?
5. If only some children get the yarn, can you predict which ones?
6. Suppose we start tossing the yarn from a child other than 0. Who will get the yarn then?
7. For which n will everyone *always* get the yarn?
8. For which k do you use up the most yarn?

All of these are foundations for more advanced concepts. Question 1 is a question about *isomorphism* of figures that are the same but look different; Questions 2, 3, 4, and 7 are related to questions of divisibility and primes. Questions 5 and 6 are the basis of the idea of *subgroups* and *cosets.*

I have done this yarn exercise with elementary school children, parents of elementary school children, elementary school teachers, and most recently with Computer Science majors in a Discrete Mathematics class. It is fun and stimulating for all these groups, but obviously I emphasize different stages of the learning process with the different groups. With the children and their parents, I leave it mostly at the "encounter" stage. We do the exercise and admire the patterns. I ask a few leading questions about what they expect to happen and generally get some good answers back. With the teachers I have worked more at the "observation" level — explicitly calling their attention to the patterns that are developing. In my Discrete Mathematics class, I was aiming at reflection and understanding. We openly discussed modular addition and multiplication, congruence modulo k, and equivalence classes, all topics in the course. My point is that for all of these groups, the same mathematical ideas were being presented — only at different levels of explicitness as appropriate to the group's degree of mathematical awareness.

3. Polyhedra and Schlegel diagrams

Three dimensions are really more concrete than two dimensions since the physical space we live in is three dimensional. Young children, having only recently mastered the difficult tasks of holding themselves upright and walking, have an intuitive feel for three dimensions that surprises many adults. Thus encounters with polyhedra can begin very early and naturally come before the study of polygons. In the pre-school and early grades teachers should encourage children to explore on their own with *Polydrons* (see [**20, 21**]) or other building materials. Teachers should keep colorful models of polyhedra on the shelves of their classrooms just as decoration, maybe for occassional discussion. In the early stages it is the encounters with polyhedra rather than their formal study which matters. Young children are just as fond of learning impressive words like "icosahedron" as they are of learning "brontosaurus" and they should be casually introduced to the names and models of the regular solids. (But please don't quiz them on it!)

A good task for children of all ages is to ask them to count the number of faces (or edges or vertices) of some polyhedral model. It instills an appreciation for systematic counting and can be used to teach quite a bit of combinatorics. In order to encourage and guide the children's explorations, the teachers need a fairly sound knowledge of polyhedra themselves.

There are numerous excellent books on building polyhedra [**18, 20, 21, 22, 24, 28**], so I need not go into that here. However, I would like to discuss drawing the *Schlegel diagrams* of polyhedra (see Figure 3). This exercise stretches the visual imagination, provides an occasion for the artistic use

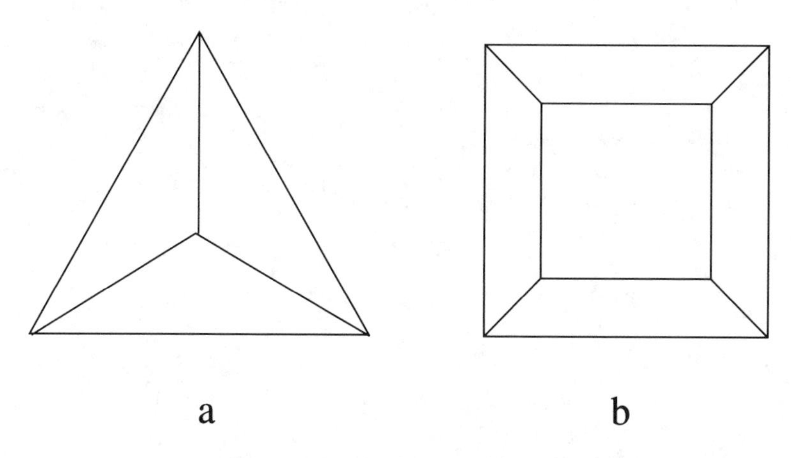

FIGURE 3. Schlegel diagrams for the tetrahedron (a), and the cube (b).

of color, and lays a experiential background for two important subjects: topology and its step-child—graph theory.

The *Schlegel diagram* of a polyhedron is what you would see if you have very good peripheral vision and you put your eye right in the middle of one face of a wire frame model of the polyhedron. In other words, it is a *perspective projection* of the vertices and connecting edges onto the plane of the face. In practice, it is hard to view the Schlegel diagram this way, because the outside faces appear so skinny. There is another way to think of the Schlegel diagram which is often more useful. Imagine the polyhedron is made out of rubber. Now pump air into the polyhedron until it bulges and rounds out like a sphere. Now imagine the faces disappearing so that only the edges and vertices are left. Put your hands through one face and pull outward on the edge-vertex skeleton until it lies flat in a plane. That is the Schlegel diagram. This approach suggests topology—"rubber sheet geometry." It also brings out two important principles of discrete mathematics. First, in the graph of a polyhedron, it is only the incidence relations between vertices and edges that matters, not the exact lengths of the edges and the angles betwen them. Second, a polyhedron can be thought of as a tessellation of the sphere.

There are actually two kinds of Schlegel diagrams: those with all vertices "finite" and those with one vertex "at infinity." (See Figure 4.) The vertex at infinity can be visualized using a spherical model. Imagine the polyhedron again as vertices and edges stretched out on a sphere. Pull all the vertices except one into the single hemisphere you are looking at. The other vertex remains on the other side—at infinity, so to speak—with the edges going to it wrapping around the sphere. Now cut these edges and take what is in the hemisphere you can see and flatten it out into the plane. The cut edges dangle off into the outside region of the diagram, running off to meet at

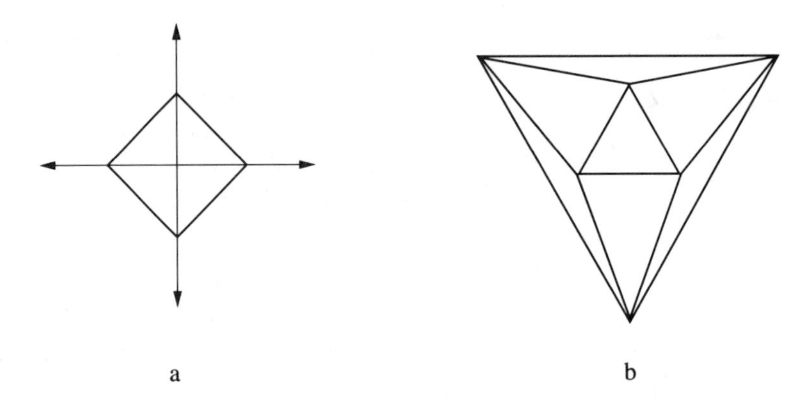

FIGURE 4. Schlegel diagrams for the octahedron. (a) Vertex at infinity; (b) all vertices finite.

infinity at the missing vertex. I have recently illustrated this for pre-service elementary teachers by having them work in groups making drawings on transparent plastic spheres[6] with non-permanent overhead markers. It was time consuming, but very instructive.

The drawings of Schlegel diagrams present numerous challenges, and the time spent drawing the regular polygons first is very helpful. Because perspective projection distorts angles and lengths, most of the polygons in a Schlegel diagram will not be regular. However, there are usually several key regular polygons in each diagram. Drawing these first in the correct position makes the rest of the diagram easier to draw. For example, to draw the octahedron in Figure 4b, start with two equilateral triangles, oppositely oriented, with one inside the other. Now "stitch" back and forth between these two triangles creating a "seam" of edges that alternates between the vertices of the inner and outer triangle. This "seam" is what is known in graph theory as a Hamiltonian cycle. This is an excellent place to use color to make the "seam" stand out.

Two common stumbling blocks are 1) drawing faces that are not convex, and 2) leaving some vertices with only two edges. For example, the regular dodecahedron has twelve pentagonal faces. Students drawing the Schlegel diagram for the first time often end up with some of these faces nonconvex. Although such drawings may be correct as graphs, they are not true Schlegel diagrams since perspective projection does preserve convexity. I encourage the students to think about how they would have to move the vertices and edges in their drawings to make all the faces convex. This provides an excellent opportunity to emphasize visualization and the "rubber sheet" nature of the drawing as well as the importance of an aesthetically appealing product.

[6]These *Lénárt* Spheres were originally produced in Hungary, but Key Curriculum Press has taken over their production in the USA. *Editors' note*: You may be able to find inexpensive Christmas ornaments or toys for this purpose.

The second stumbling block is more serious because it is geometrically impossible for a three dimensional figure to have a vertex attached to only two edges. In fact, counting the number of edges at each vertex is a good way to detect errors in a drawing. In a regular polyhedron, each vertex will have the same degree – that is, number of edges at that vertex. In a pyramid, there will be vertices of different degrees. Exploring the degrees of vertices as well as the number of sides or faces in a variety of polyhedra can lead to the discovery of many beautiful relationships, such as the celebrated Euler formula: $V - E + F = 2$, where V is the number of vertices, E the number of edges, and F the number of faces. In keeping with the philosophy of this paper that even introductory material should provide an encounter with deep and significant mathematics, I want to point out that I always write the Euler formula as $V - E + F = 2$ because it is this form which generalizes to higher dimensions and best displays the role of 2 (the Euler characteristic of the plane) as a topological invariant.

4. Movement and Symmetry

The following activity was inspired by a highly geometrical form of dance-movement known as *eurythmy*, taught only in the Waldorf schools. These exercises are intended to give a kinesthetic sense of the symmetries of the regular polygons.

Divide the class into groups of six, each consisting of a leader and five children who will form the vertices of a regular pentagon. As in the yarn-tossing exercise, it is valuable to review the properties of regular polygons: equal distance to neighbors and parallels in order. The first exercise is quite simple: each person in the pentagon is to walk (counter-clockwise) along an edge of the pentagon until he or she reaches the position previously occupied by his or her neighbor. All five children are to walk at once, to a steady rhythm, say of three beats, clapped by the leader (or teacher). Before beginning, ask the children to point to where they are going—this helps to avoid mishaps and clear up misunderstandings. The end effect is a rotation 72 degrees counter-clockwise. Repeating this exercise yields a rotation 144 degrees counter-clockwise.

Now comes an exercise that ties in with the drawing of diagonals. This time the children are to walk a *diagonal* to the position of the *second* person from them counter-clockwise around the pentagon. Again they are to walk in a straight line to a steady rhythm. More beats are required, because the distance is further. The end result is much less clear! Ask the children first if they think they will collide. Some will think they will collide at the center of the pentagon. But if they have drawn the diagonals of a pentagon beforehand, some will remember that the diagonals do not meet at the center of the pentagon, but rather pass around it. Thus, if the pentagon is large enough, they will not collide, but steadily move past each other. In fact, the whole pentagon contracts spirally and then expands again. This is quite a beautiful form. The steady rhythm is essential to insure a smooth transformation of the pentagon.

Let us move on to reflections. Have the leader choose a person in the pentagon and stand on the edge opposite to form a mirror line. Now the pentagon is to be *reflected* through the mirror line. Again ask the children to point to where they are going. This will invariably lead to considerable confusion, which is important to straighten out. In particular, the vertex-person on the mirror line often wants to know: "Where do I go? Do I exchange with the leader?" The answer, of course, is "No". A point on the mirror stays fixed and does not move. The leader is needed because it takes two points to determine the mirror line. Doing the reflection is slightly more tricky than the rotation because (1) the distances to walk are not the same, and (2) exchanging children *will* collide unless they walk around each other.

Having practiced both rotations and reflections, it is now possible to explore them a little more. Ask the children to note whom they are standing next to. Now do a reflection and look again. They will have the same neighbors, but left and right will be reversed. This illustrates that reflections reverse orientation whereas rotations preserve orientation.

Now ask the children to take note of the spot where they are standing. Do two consecutive reflections around different mirror lines and ask the children how the pentagon has moved from its original position. It will have rotated, illustrating the fact that the product of two reflections (whose mirrors intersect) is a rotation.

These activities can be repeated with groups of more than six children, of course. The smaller groups, however, are more manageable, and an odd number works best for walking the diagonals.[7]

I use these exercises as a precursor to the study of symmetry groups. This usually includes a full study of the symmetries of the equilateral triangle and the square, the classification of the symmetries of frieze patterns, and a brief discussion of the symmetries of regular polyhedra and wallpaper patterns. In fact, the two kinds of Schlegel diagrams of the regular polyhedra help to illustrate two kinds of rotational symmetries:

1. diagrams with all vertices finite illustrate rotational symmetry around a face-to-face axis;
2. diagrams with one vertex at infinity illustrate rotational symmetry around a vertex-to-vertex axis.

Symmetry abounds in art in the form of ornaments [**9, 31, 33**]. I usually go through a series of slides of medieval architectural monuments with my class, asking them to identify the symmetry groups of frieze patterns [**9**]. Round singing or clapping rhythmic patterns is an excellent way to illustrate

[7]Needless to say, these exercises need to be done outdoors where there is plenty of room. I have done these exercises with in-service and pre-service elementary teachers, all women, who enjoyed them immensely. I have also tried them with a mix of male and female math and math education majors in a Modern Algebra course; this was less successful, because the men especially were much more resistant and self-conscious.

translational symmetry aurally. Baroque music is full of sequences [**27**, pp. 230–242] which can also be used to illustrate the concept of translational and even reflectional symmetry in a non-geometric space having time and pitch as its "dimensions."

My intention is not to force a complete understanding of these rather subtle ideas. Rather the intent is to slowly and carefully prepare the intellectual ground in the student for the planting of an intellectual seed which may take years to ripen. When it does, it grows with the strength of a self-discovered idea rather than as an idea imposed from outside.

5. Modular Arithmetic

Many elementary-school mathematics curricula include a section on arithmetic in other bases.

A much more meaningful alternative that is closely related to base arithmetic is modular arithmetic. In fact, I have introduced it to my in-service teachers as the study of what happens to the last digit in computations in other bases. But this is only in passing, because there are many much more important and serious applications of modular arithmetic—especially in coding theory [**12, 13, 14, 16**]. Moreover, there are many immediate examples in everyday life with which students are familiar. Here are some sample problems which convince beginners that they already know some modular arithmetic:

1. You leave on a 5 hour trip at 10:00 am. What time will you arrive?
2. What day of the week will it be 10 days from today?
3. October 4 is a Monday: what day of the week is October 23?
4. What is the date exactly 3 weeks after June 20?

These examples make it easy to grasp the idea of "clock" arithmetic. However, they all involve addition. In introducing multiplication, it is helpful to recall that multiplication of integers is just repeated addition. This makes the definition appear less arbitrary. I also tie this in with the diagonals of a polygon. For example, the "multiples of 5 mod 12" are obtained by taking the sequence of "5-step diagonals" of a dodecagon (shown in Figure 2b). At this stage, students spontaneously ask about subtraction and division. And they discover that while subtraction always makes perfect sense, division is far more problematic.

At this point, it is helpful to systematize the study by having the students write out tables for addition and multiplication modulo m for small m—say, addition for $m \leq 6$ and multiplication for $m \leq 13$. The students are quick to spot many patterns. In the addition tables, rows and columns are obtained just by cyclic permutaton and hence each element appears exactly once in each row and column. The situation is more complex for multiplication, and hence the need to write out a larger number of tables. In order to help bring out the patterns, I ask my students to write each "0" in red and each "1" in blue.

There is always a red border of zeroes. Even if this is discarded, some elements can occur more than once in a row or column. This, of course, spells trouble for division. Students are led to guess that the moduli for which this bad situation does not occur are precisely the primes. Hence for prime moduli, modular arithmetic is very similar to regular rational arithmetic with all four operations defined. Other patterns that can be elicited are as follows.

1. The tables are symmetric about the main diagonal and this means the operations are commutative;
2. The rows and columns with "extra" zeroes all correspond to numbers which occur more than once in some rows and columns;
3. A row or column contains a (blue) one if and only if it does not contain an "extra" (red) zero.

A lovely illustration of a practical use of these ideas is the International Standard Book Number (ISBN) which every book possesses [**23, 16**, pp. 36–39]. This consists of a ten-digit code $a_1, a_2, a_3, \ldots, a_{10}$ of which the first nine digits identify the language, publisher, and catalogue number of the book. The last digit a_{10} is a "check digit" which allows single-digit errors and even the transposition (reversal) of two adjacent digits to be detected. It is also possible to determine a missing digit if its position is known. This all follows from the fact that a_{10} is chosen so that the equation

$$10a_1 + 9a_2 + 8a_3 + \ldots + 2a_9 + a_{10} = 0 \pmod{11}$$

holds. Notice that a_{10} may be required to take the value 10; if this happens, it is represented by an X in the ISBN number. In introducing modular arithmetic, I illustrate the above equality with several ISBN numbers. Later, we actually solve for missing digits. This involves solving a linear equation $kx + b = 0 \pmod{11}$, and inevitably leads to a healthy discussion of what it means to solve such an equation.

6. Modular Arithmetic in Music

I will give here a sketchy account of connections between music and modular arithmetic, all possible to describe within the realm of elementary mathematics. A detailed discussion can be found in [**1**] or [**37**, pp. 88–103]. Unfortunately, this material takes a long time to cover, and only at a very slow pace, because most of my students have a poor musical background, and we must start from scratch. However, even those with an extensive musical background find that the careful mathematical treatment puts the theory into a new and clearer light. In either case, I consider the time well spent because music is such an important background for mathematics.

The simplest way that modular arithmethic enters music is through the cyclic naming of the notes as A,B,C,D,E,F,G. When G is passed, we start over again with A. Thus stating that C (note 3) is four notes above F (note 6) corresponds to the equation $3 = 4 + 6 \pmod 7$. The white keys on the piano get the names A,B,C,D,E,F,G in cyclic order. The black keys are

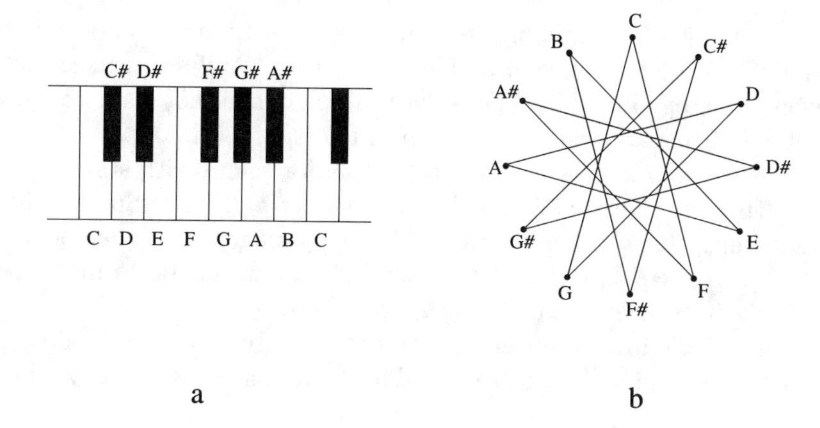

FIGURE 5. (a) A standard octave on the piano; each white or black key represents a half step. Recall that the C-major scale is the sequence of tones played on the white keys only. (b) The *Circle of Fifths* shown as diagonals of a dodecagon.

named using sharps (or flats). (See Figure 5a.) A *scale* is a sequence of notes, beginning at a fundamental note and ascending to the next occurrence of the same note name. Thus a C major scale begins on a C and ends at the next higher C. Because Western scales typically have 8 notes, the interval between the beginning and ending note of a scale is called an *octave*. Another important interval is that between the fundamental and the fifth note of the scale. This is the fifth interval of the scale and is usually simply called a *fifth*. For example, the strings of a violin are tuned in fifths: G, D, A, E. Notice that "fifth" is used here as an ordinal number, not as a fraction. Since this is a frequent source of confusion, it provides an opportunity to clarify the distinctions between different types of numbers and our words for them.

All this may appear quite arbitrary, but it is not. The octave is the first harmonic (or overtone) and the fifth is the second harmonic over the fundamental. Vibrational energy is easily passed between a tone and its harmonics, and this sounds pleasant (or harmonious!) to our ears. Pitch is determined by rate of vibration. The octave vibrates twice as fast as the fundamental whereas the fifth vibrates 3/2 as fast. Standard pitch for a violin A string is 440 vibrations per second. Thus the violin E string a fifth above is 660 vibrations per second, and the 'cello A string an octave below is 220 vibrations per second.

The characteristic sound of a scale is determined by the frequency ratios between the notes of the scale. If we wish to build a major scale starting on D instead of C, we are forced to add sharps in order to keep the same frequency ratios as in C major. In music theory, the basic frequency ratios (intervals) are expressed in terms of whole steps and half steps. On the piano, a half step is the interval between a note and the note immediately next to it. Thus B-C and E-F are half steps because there are no black notes in between, but G-A is a whole step since it consists of the two half steps G-G♯ and G♯-A. (See Figure 5a). There are twelve half steps in an octave and seven half steps in a fifth. Thus determining which note is a fifth above a given fundamental is really a problem in addition modulo 12.

There is a systematic way of listing the major scales with starting notes rising a fifth each time. This is called the *Circle of Fifths*. Since a fifth is seven half steps, this can be viewed as the 7-step diagonals of the dodecagon (Figure 5b). Notice that this is the pattern that would be formed by the yarn exercise in Section 2 with $n = 12$ and $k = 7$.

Writing out the first eight scales with successively more sharps illustrates cyclic rotation again, and reveals several beautiful patterns, as shown below:

C	D	E	F	G	A	B	C
G	A	B	C	D	E	F♯	G
D	E	F♯	G	A	B	C♯	D
A	B	C♯	D	E	F♯	G♯	A
E	F♯	G♯	A	B	C♯	D♯	E
B	C♯	D♯	E	F♯	G♯	A♯	B
F♯	G♯	A♯	B	C♯	D♯	E♯	F♯
C♯	D♯	E♯	F♯	G♯	A♯	B♯	C♯

The new sharps always appear in the table in the same (seventh) position. Visually, the pattern of sharps in the table is essentially a pair of triangles. There are also several diagonal relationships which can be brought out nicely with the use of color, say by making all the F♯'s the same color. Notice also that the last four notes of each scale are the same as the first four notes of the next scale.

A more involved connection, both mathematically and musically, is the calculation of frequency intervals from basic harmonics. As noted above, the frequency ratios for the octave and the fifth are 2:1 and 3:2. In fact, every musical interval can be expressed as a simple ratio of two small integers [**15, 35**]. This discovery really dates back to Pythagoras [**7**, p. 72] in the 5th century BC. It represents the first expression of a physical law in mathematical terms. Mathematically, following one interval by another involves the multiplication of fractions giving their frequency ratios. This leads to certain rather surprising musical consequences of the unique factorization of integers into primes. For example, if we go all the way around the Circle of Fifths, we will go through twelve fifths and end up seven octaves higher. Since the frequency ratio for a fifth is 3:2 and for an octave is 2:1, we should have

$$(3/2)^{12} = 2^7,$$

or

$$3^{12} = 2^{(12+7)} = 2^{19}.$$

The first equation may seem plausible; indeed, $(3/2)^{12}$ is 129.746..., which is convincingly close to $128 = 2^7$. The second equation, however, says that some large number has two different prime factorizations, an impossibility. Musically, this means that going around the Circle of Fifths will not return us to the original C, but to a tone just a shade higher. This mathematically inescapable fact says that it is impossible to tune a piano so that all octaves are true and all fifths are true.[8]

7. The Mathematics Underlying the Activities

The activities described in this paper are not intended merely as enrichment exercises or as entertainment to make mathematics more palatable. They are designed to provide kinesthetic experiences leading to deep and significant mathematics. Even for the majority of students who will not go on to higher mathematics, I think it is still advisable that their limited mathematical experience be based on sound and significant mathematics. In this section, I would like to indicate some of the deeper mathematical concepts underlying the activities.

The basic mathematical idea underlying these activities is that of a *group of transformations.* The concept of group is, of course, the formalization of the general notion of symmetry. The cyclic groups capture rotational symmetry and of course periodic behavior which is commonly known as rhythm. Part of my goal is to make my students aware of the many diverse contexts in which periodic behaviour and rhythm occur. There is rhythm in the columns along the nave of a Romanesque church as well as in music and in the seasons of the year. There is also a rhythm in the way the numbers are arranged in the (Cayley) tables for addition and multiplication modulo n. We also see rhythm in the rotational symmetry of many flowers.

Polyhedra and Schlegel diagrams lead naturally to graphs and important notions in graph theory: the "fundamental theorem" that the sum of the degrees is twice the number of edges, planarity, and Euler's formula.

The introduction of modular arithmetic naturally lays the foundation for a whole host of algebraic and number theoretic notions: finite fields, coding theory, and prime factorization among them. Scales reveal many patterns related to arithmetic modulo 7 and 12. And the study of frequency ratios leads to prime factorization, logarithms, and the arithmetic of the rationals modulo 1.

It is important to point out that I do not go into a prolonged discussion of the underlying mathematics with the elementary school teachers in my class. I only wish to give them a hint of the broader significance that lies down the road. I expect that they will say even less to their school children. What is important is that a host of meaningful experiences should be a part

[8]This fact and others related to it involving the tuning of thirds led to a host of compromise tunings or temperaments in the baroque period. The modern solution is the "equal temperament" system, in which all intervals except the octave are just slightly out of tune, but all equally so. All half-steps on the piano are tuned in the frequency ratio of the twelfth root of 2, an irrational number, and intervals are sometimes measured in a logarithmic scale giving 1200 "cents" to each octave.

of each student's background so that as more advanced mathematical topics are introduced in middle school, high school, or college, the student will have some personal experience to connect with them.

8. Conclusion

There is really a curious paradox here and a serious lopsidedness in our educational system. We are now expecting of all sixth graders a deeper understanding of arithmetic than what the most learned men in Europe possessed in 1500! Only after 1494 did Hindu-Arabic numerals finally replace Roman numerals in all Medici account books [**36**, p. 81]. Simon Stevin's *La Disme* introducing decimal notation appeared in only 1585 [**36**, p. 89]. And in 1637 Descartes still referred suspiciously to the negative roots of an equation as "false roots" [**36**, p. 96].

Our expectation may be reasonable, but it is by no means trivial. And unfortunately it is not balanced by a supporting expectation in geometry. Even more unfortunately it is not balanced by a supporting expectation in music. The Greeks in establishing the quadrivium understood the vital connections of these areas:

arithmetic — numbers at rest	music — numbers in motion
geometry — figures at rest	astronomy — figures in motion

It is sad that the routine computational aspects of arithmetic have come to dominate our elementary mathematics curriculum and that musical skill has come to be regarded as a special talent. The Suzuki method of violin instruction has given the lie to the limiting idea that it takes special inborn musical skill to play an instrument. Suzuki's philosophy is that talent can be trained and the success of his instructional "mother tongue" method, based on slow and careful steps, imitation, and positive reinforcement gives evidence that he is right. It is important to remember that the goal of the Suzuki method is not to produce musical specialists (i.e., concert violinists) or musical consumers (i.e., music appreciators) but "beautiful human beings." The goal is to open up an avenue of enjoyment and expression for the child, to develop a skill that can enrich a whole life.

I would like to suggest that the same should be the goal of mathematical instruction. The real goal of the educational system should be to help students develop the intellectual and emotional skills necessary to have the freedom to choose their own futures wisely.

Acknowledgments

I would like to thank Furman University for continuing to support the development of this course in Ann Stafford's Lead Teacher program after Clemson University eliminated its in-service mathematics offerings due to budgetary constraints. Thanks are also due to Marjorie Senechal and the NSF Regional Geometry Institute of 1993 at Smith College, where the idea for this paper first took shape. I am also grateful for the hospitality of the

Mathematics Department at Cornell University during my sabbatical year there and for stimulating interactions with David Henderson, Bob Connelley, Tom Rishel, and Maria Terrell at an NSF Institute held there in 1994. Special thanks are due to two of the editors, Deborah Franzblau and Joseph Rosenstein, for their patient encouragement of the writing of this report and for their editorial assistance in putting this paper into final form.

References

[1] Willi Apel, *Harvard Dictionary of Music*, The Belknap Press of Harvard University Press, Cambridge MA, 1969. See articles on Acoustics (p. 9), Intervals, Calculation of (p. 419), Pitch (p. 679), Temperaments (p. 835).

[2] Hermann von Baravalle, *Geometric Drawing and the Waldorf School Plan*, Waldorf School Monographs, 1967.

[3] Hermann von Baravalle, *The Teaching of Arithmetic and the Waldorf School Plan*, Waldorf School Monographs, 1967.

[4] Henry Barnes, "Learning that Grows with the Learner: An Introduction to Waldorf Education", *Educational Leadership*, October, 1991, 52-54.

[5] Henry Barnes et al., "Waldorf Education: a Symposium," *Teachers College Record*, Vol. 81, Nr. 3, Spring 1980, 322-370.

[6] Richard G. Brown , *Transformational Geometry*, originally published by Silver, Burdett, & Ginn Inc., 1973. Reprinted by permission by Dale Seymour, Palo Alto, CA.

[7] Lucas Bunt, Phillip S. Jones, Jack D. Bedient, *The Historical Roots of Elementary Mathematics*, Dover, Mineola NY, 1988.

[8] Richard W. Copeland, *How Children Learn Mathematics: Teaching Implications of Piaget's Research*, Macmillan, New York, 1974.

[9] Donald Crowe, "Symmetry, Rigid Motions, and Patterns," HiMAP Module 4, COMAP, Arlington MA, 1986.

[10] Mary L. Crowley, "The van Hiele Model of the Development of Geometric Thought", in *Learning And Teaching Geometry K–12*, NCTM 1987 Yearbook, NCTM, Reston VA, 1987, pp. 1-16.

[11] Richard Feynman, *The Character of Physical Law*, M.I.T. Press, Cambridge MA, 1965, p. 13.

[12] Joseph A. Gallian, "How Computers can Read and Correct ID Numbers", *Math Horizons*, Winter 1993, pp. 14-15.

[13] Joseph A. Gallian, "The Mathematics of Identification Numbers", *The College Mathematics Journal*, 22 (1991), 194-202.

[14] Joseph A. Gallian, "Assigning Driver's License Numbers," *Mathematics Magazine*, 64 (1991), 13-22.

[15] G. D. Halsey and Edwin Hewitt, "More on the Superparticular Ratios in Music", *Am. Math Monthly* 79 (1972), 1096 -1100.

[16] Raymond Hill, *A First Course in Coding Theory*, Oxford Applied Mathematics and Computing Science Series, Clarendon Press, Oxford, 1986, pp. 36 - 39.

[17] Peter Hilton, Lecture at the Howard Eves 80th Birthday Conference, University of Central Florida, May, 1991, Orlando, Florida.

[18] Peter Hilton and Jean Pederson, *Build Your Own Polyhedra*, Addison-Wesley, New York, 1988.

[19] Jay Kappraff, *Connections: The Geometric Bridge Between Art and Science*, McGraw-Hill, New York, 1991.

[20] Marilyn Komarc and Gwen Clay, *Exploring with Polydron: Book 1 (Grades 3-9)*, Cuisenaire, New Rochelle NY, 1991.

[21] Marilyn Komarc and Gwen Clay, *Exploring with Polydron: Book 2 (Grades 3-9)* Cuisenaire, New Rochelle, NY, 1991.

[22] Mary Laycock, *Dual Discovery Through Straw Polyhedra*, Creative Publications, Palo Alto CA 1970.

[23] Joseph Malkevitch, Gary Froelich, and D. Froelich, *Codes Galore*, Consortium for Mathematics and its Applications (COMAP). Module #18, 1991.

[24] David Mollet, "How the Waldorf Approach Changed a Difficult Class," *Educational Leadership*, October, 1991, 55-56.

[25] Hans R. Niederhauser and Margaret Frohlich, *Form Drawing*, Mercury Press of Rudolf Steiner College, Sacramento CA, 1974.

[26] Peter and Susan Pearce, *Polyhedra Primer*, Dale Seymour, Palo Alto CA, 1978.

[27] Walter Piston, *Harmony*, W.W. Norton & Co., New York, 1962.

[28] Anthony Pugh, *Polyhedra: A Visual Approach*, Dale Seymour, Palo Alto CA, 1990.

[29] Victoria Pohl, *How to Enrich Geometry Using String Designs*, NCTM, Reston VA, 1986.

[30] Rene M. Querido, *Creativity in Education: The Waldorf Approach*, H. S. Dakin Co., San Francisco, 1984.

[31] Issam El-Said and Ayse Parman, *Geometric Concepts in Islamic Art*, Dale Seymour, Palo Alto CA, 1976.

[32] Doris Schattschneider, *Visions of Symmetry (Notebooks, Periodic Drawings, and Related Work of M.C. Escher)*, W. H. Freeman, New York, 1990.

[33] Dale Seymour, *Geometric Design – Step by Step*, Dale Seymour, Palo Alto CA, 1988.

[34] Dale Seymour and Jill Britton, *Introduction to Tessellations*, Dale Seymour, Palo Alto CA, 1989.

[35] A. L. Leigh Silver, "Musimatics or the Nun's Fiddle," *Am. Math Monthly* 78(1971), 351-357.

[36] Dirk J. Struik, *A Concise History of Mathematics*, Dover, Mineola NY, 1987.

[37] Bengt Ulin, *Finding the Path: Themes and Methods for the Teaching of Mathematics in a Waldorf School*, The Association of Waldorf Schools of North America, Wilton, N.H., 1991.

[38] John Willson, *Mosaic and Tessellated Patterns: How to Create Them*, Dover, New York, 1983.

Department of Mathematical Sciences, Clemson University, Clemson, SC 29634-1907

E-mail address: `rejam@clemson.edu`

DIMACS Series in Discrete Mathematics
and Theoretical Computer Science
Volume **36**, 1997

Discrete Mathematics Activities for Middle School

Evan Maletsky

There is a body of knowledge that has come to be known as discrete mathematics and much of it is accessible to middle-school students. Many related topics can already be found in the existing curriculum and others can be readily integrated into it. Discrete mathematics problems tend to be simply stated and easily motivated. They offer a rich, new source of diversified problem-solving experiences that range across all ability levels. Furthermore, they serve to portray mathematics from a broader perspective than many typical practice exercises.

It is equally important to note that problems in discrete mathematics can be incorporated into many of the hands-on activities that already are part of the established classroom scene. This article focuses on that connection through the two central ideas of *counting* and *change*. Counting is viewed through number patterns, computation, manipulation, and visualization, and these are connected to change through the mathematical idea of *iteration*. It is the notion of iteration — arithmetic, algebraic, and geometric — that brings alive the subject of mathematics, and it is through hands-on activities that it is made real. Emphasizing this combination when we teach offers a dynamic view of the discipline so greatly needed by today's middle school students.

This article begins with a sampling of discrete mathematics activities arising from a simple counting problem involving paper folding, then moves through others that can be analyzed by graphs, and ends with some applications of iteration through geometric transformations. The examples illustrate the importance of both content and pedagogy and show how discrete mathematics can be designed and woven into the broad fabric of middle-school mathematics.

Counting

Almost every middle school student and teacher has, at one time or another, used the folding of paper to explore a mathematical relationship.

1991 *Mathematics Subject Classification.* Primary 00A05, 00A35.

This first illustration shows some different ways one simple paper model can be tied into the arena of discrete mathematics.

Cut out some 2x8-inch strips of paper, one for each student. Have them fold the strips in half and in half again as shown in Figure 1. Let them visualize in their mind what the strip would look like unfolded.

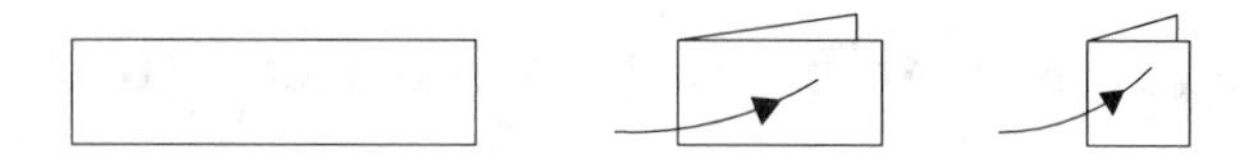

FIGURE 1. Folding a strip of paper

Ask the students to mentally count all the rectangles that they visualize, including the squares. After writing their individual answers, let them compare and discuss their answers with other students. Once an agreement is reached in their groups, they can unfold the strips and check their answers by actually counting from the model. Finally, as a writing activity, have your students describe the algorithms they used for their counting, both in the abstract and in the concrete case.

This activity is much more than just one of visualization. It involves analysis and systematic counting. One approach might be to letter the squares (as in Figure 2a) and make a list of the 10 different rectangles using successive letters, four, three, two, and one at a time (Figure 2b). Another approach might be to show the solution in a graph with 4 vertices and 10 edges (Figure 2c). Six edges connect different vertices, denoting different starting and ending squares. Four edges connect vertices to themselves, indicating the same starting and ending square.

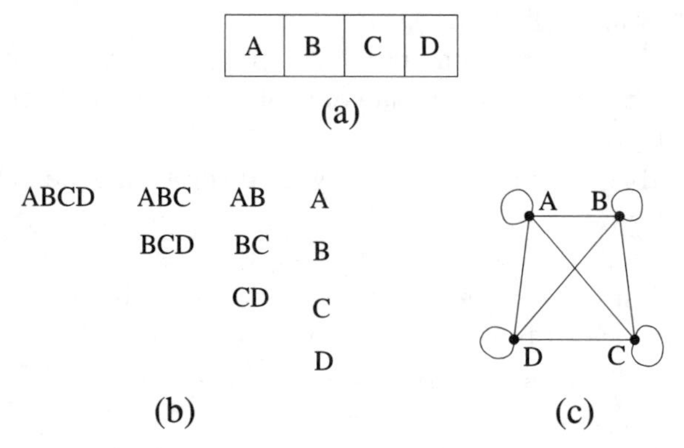

FIGURE 2. (a) An unfolded and labeled piece of paper. (b) Systematic listing of rectangles. (c) Using the edges of a graph to represent starting and ending squares of each rectangle.

The list reveals that, for two successive folds, the answer is 10, the sum of the first four counting numbers. Compare the number 10 for two folds to the number 3 for one fold:

Folded once: $1 + 2 = 3$.
Folded twice: $1 + 2 + 3 + 4 = 10$.

Ask your students to fold the strip in half a third time and ask for an educated guess as to how many rectangles will be in the unfolded strip now. See how many students can find and extend the pattern.

Folded three times: $1 + 2 + 3 + 4 + 5 + 6 + 7 + 8 = 36$.

Some middle school students may want to explore this problem further and look for a general solution. For n successive folds, the number of rectangles is the sum of the first 2^n counting numbers.

Folded n times: $1 + 2 + 3 + 4 + \ldots + 2^n = 2^{n-1}(2^n + 1)$.

Given the formula, this paper-folding activity now offers students an additional important experience with exponents and other algebraic symbolism. For example, with four successive folds, there are 136 different rectangles, since with $n = 4$,

$$2^{n-1}(2^n + 1) = 2^3(2^4 + 1) = 8(16 + 1) = 8(17) = 136.$$

Are the counting numbers that come from this paper-folding activity, such as 3, 10, 36, and 136, special in any other way? You may recognize them from another discrete mathematics topic already in the middle school mathematics curriculum. They are members of the set of triangular numbers.

Figure 3a shows the triangular arrays which account for the name "triangular numbers". Triangular arrays such as these can be easily built and vividly displayed on an overhead projector. Figure 3b shows how the triangular numbers are calculated by summing the rows of the triangular arrays.

Another feature of the triangular numbers emerges if we look at a difference table. In a difference table, we first record the differences between successive triangular numbers — these are called "first differences". Then we record the difference between successive first differences — these are called "second differences." For the triangular numbers, second differences are all 1, as in Figure 3c.

Compare this to the familiar square numbers where the second differences are all 2, as in Figure 4. Here we see another topic from discrete mathematics, finite differences, closely connecting to the existing middle school curriculum.

We can also look at other geometric arrays — squares, pentagons, hexagons, etc. — and introduce other sequences of "figurate numbers" — square numbers, pentagonal numbers, hexagonal numbers, etc. These geometric arrays lead to counting activities, number patterns to explore, discoveries

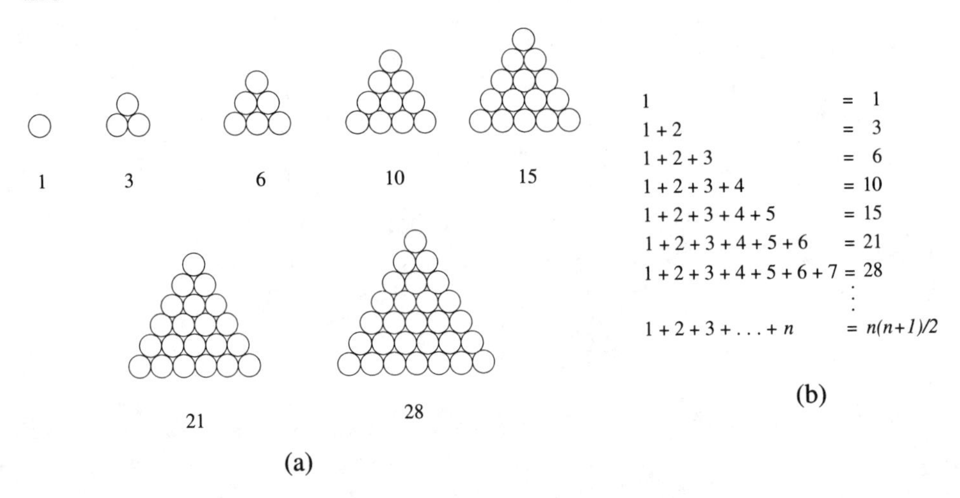

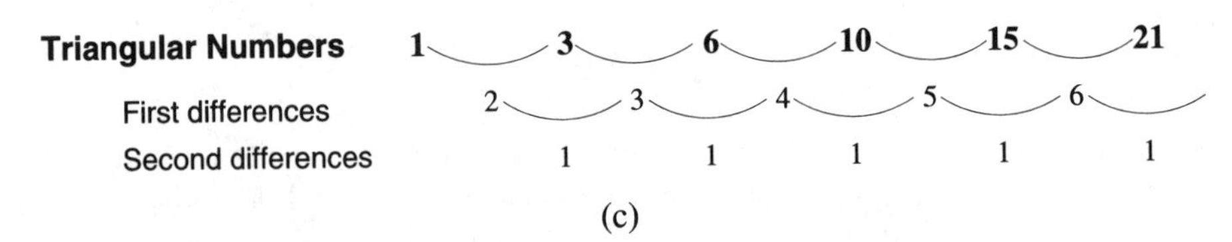

FIGURE 3. (a) The triangular numbers. (b) Calculating the triangular numbers. (c) Difference table for triangular numbers.

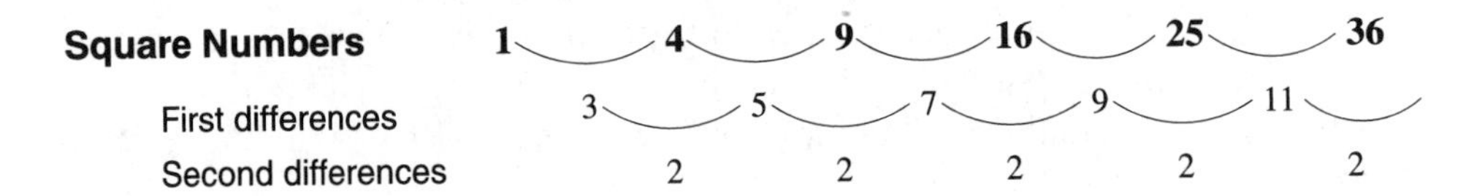

FIGURE 4. Difference table for the square numbers.

to make and test, and more questions worth investigating. For example, will pentagonal numbers have successive second differences that are all 3? For hexagonal numbers, will the successive second differences all be 4? The answer is yes for all figurate numbers of this type. In fact, any second degree, quadratic expression such as $n(n+1)/2$ must have constant second differences, an idea worth challenging your students to explore as a calculator activity.

Let us go back to the folded strip of paper for some more counting activities. Have your students label the squares on one side with the digits 1, 2, 3, and 4. Tear apart the four squares and the students have a nice model for some counting problems.

One good question is the following: how many different four-digit numbers can be formed using the squares? Let students work in teams arranging the digits, making lists, finding and applying counting procedures, and writing about their methods. This can lead nicely into the topics of permutations and factorials, since many students will discover that the answer 24 is expressed as $4 \times 3 \times 2 \times 1$. You can also ask how many numbers with 4, 3, 2, and 1 digits can be formed. The answer here is

$$(4 \times 3 \times 2 \times 1) + (4 \times 3 \times 2) + (4 \times 3) + 4 = 24 + 24 + 12 + 4 = 64.$$

For students at a higher level, ask them to turn the strip over and put the digits 5, 6, 7, and 8 on the back, with the 8 behind the 1, before tearing the squares apart. Ask the same two questions noted above. Here the algorithmic thinking is every bit as important as the numerical answers of

$$8 \times 6 \times 4 \times 2 = 384 \text{ and } (8 \times 6 \times 4 \times 2) + (8 \times 6 \times 4) + (8 \times 6) + 8 = 632.$$

For those who want a real challenge, label both sides of the strip 1 through 8 as noted above, but don't tear the strips apart. Folding only on existing creases, how many different numbers with 1, 2, 3, and 4 digits can be formed? At this level, some difficult analysis is called for by the students. Let them discuss and explore the problem mentally before they start folding and forming numbers in their hands with their paper strips.

On another day, review these results and then offer a variation. Mark the four squares of a newly folded strip of paper with the digits 1, 2, 3, and a decimal point (as in Figure 5). Separate the squares and think about possible arrangements using one or more of the squares. What are the different decimal numbers that can be formed?

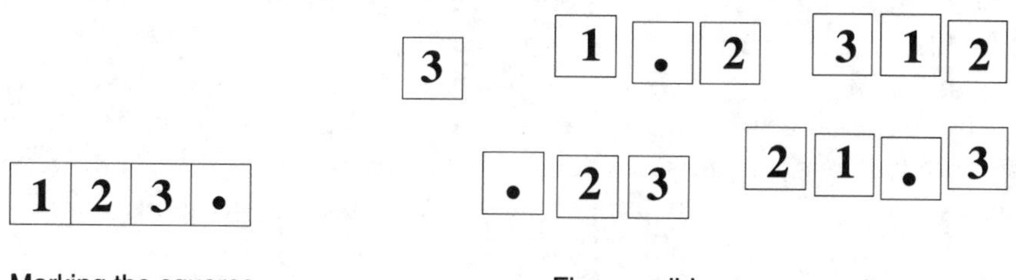

FIGURE 5. Creating decimal numbers with digits 1, 2, 3, and ".".

Counting the different possibilities can be an interesting and challenging activity for the middle-school student. But couched in the form of a class game or competition, much more classroom excitement and enthusiasm can be generated. Middle school teachers from the Rutgers University Leadership Program in Discrete Mathematics and others have related back to me several game variations they have used in their classes with great success. One of the more interesting formats uses no paper other than the strips used to introduce the activity. Every day from there on, begin the class with a number from the set, say 2. See how far you can get around the class, asking each student for the next larger decimal that can be formed from the set, before a mistake is made. When one occurs, stop. The following day, try again. Challenge the students to get through to the largest decimal without any mistakes. This may seem easy, but experience proves otherwise. The first few correct choices, in order starting with 2, are $2, 2.1, 2.13, 2.3, 2.31, 3, 3.1, 3.12, \ldots$

From the point of view of mathematical content, this activity deals with the important skill of ordering decimals. But even more important, students must create them, and to do so requires the ability to play freely and imaginatively with numbers and shapes in situations involving discrete choices. This skill needs to be developed and nurtured thoroughly in the middle grades by embedding it within the existing curriculum and around familiar classroom experiences. These kinds of simple exercises, while both fun and challenging for students at this age, lay the foundation that will enable them, in later years, to approach more profound and intriguing applications.

Graphs

Many problems can best be approached through models in the form of graphs. Graph models offer a kind of organizational structure that can be utilized in many problem-solving experiences involving both manipulatives and counting. Let us look at an example.

> *Five cubes of different colors are arranged in a row. How many different arrangements are possible?*

Many students familiar with counting know this is a permutation problem and know the answer to be $5! = 120$. But, when asked for an explanation or meaning, they have little to say because they really see nothing. Early counting experiences of this type need to be done with concrete materials and modeled in diagram form for better understanding. In the following example, we use five blocks, one of each of the colors green (G), orange (O), red (R), yellow (Y), and blue (B).

You might begin by arranging the cubes in a row and discussing their order. Have students suggest and show other orderings. Put the cubes in your hand and ask how many choices there are for the first position. How many choices remain for the second, and then the third, and the fourth, and the fifth positions? Connect these questions to the blocks and to the diagram in which the numbers are entered one at a time (see Figure 6), and to their product.

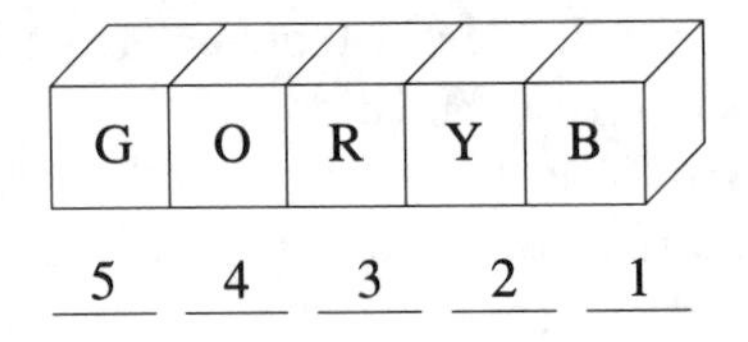

FIGURE 6.

A systematic listing of all solutions is often accessible and useful in solving many counting problems. However, a listing of the 120 choices here seems a bit tedious. This is one place where a graph can be useful. The vertices represent the cubes and the edges show all the possible connections (see Figure 7a). Every one of the 120 possible arrangement of the cubes is a distinct, directed path of four edges connecting the five vertices. The arrangement GORYB can be represented by a path as shown in Figure 7b.

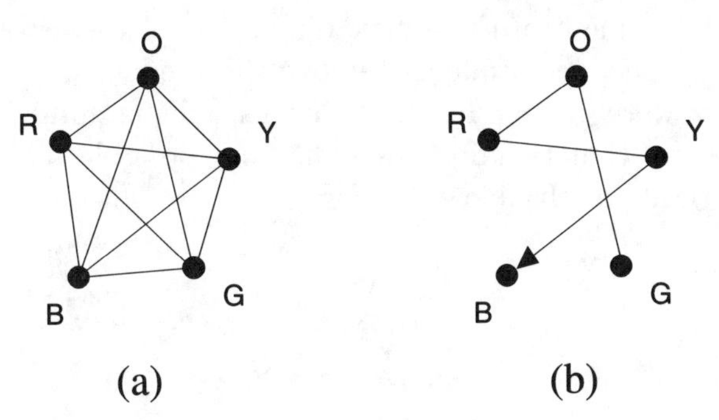

FIGURE 7. (a) Graph with vertices representing blocks. (b) Directed path representing the linear arrangement of blocks, GORYB.

Many good counting questions can be asked. How many of these paths start at G? How many start at G and end at B? How many have G next to B? How many do not have G next to B?

Situations can be analyzed and answers can be found from the graph. Have students trace out paths for given arrangements and arrange the cubes for given paths. (These require very different skills.) Have students count the number of edges in the complete graph and explain what the number means. Connect the answer to the problem of choosing two cubes from the set of five. See if they recognize the answer as a triangular number.

Many discrete mathematics problems are already in the textbooks and other available literature as examples addressing teaching methods or classroom issues. The EQUALS project at the Lawrence Hall of Science at the University of California at Berkeley, through its publications, *Get It Together*, suggests an interesting cooperative learning activity similar to the one just described. It is an arrangement problem involving six colored cubes.

Four students independently receive critical information, that they alone possess, about the arrangement. All students must participate because each student has information to contribute and needs to do so at the right time. The task is to arrange the six colored cubes in a row in the correct order.

a: Green is not next to yellow and purple is not next to green.
b: Orange is not next to yellow and green is not next to blue.
c: Yellow is not next to red, blue not next to purple, and red not next to orange.
d: Purple is not next to yellow, blue not next to orange, and green not next to red.

The problem offers an excellent example of a cooperative learning situation in the arena of discrete mathematics. One approach is hands-on, with the solution emerging through the arranging and rearranging of the colored cubes. Another approach is to draw a complete graph with 6 vertices representing the colors and 15 edges representing all possible ways any two colored cubes might touch each other, when arranged in a row. Clearly, in any given arrangement of the cubes, only some of these connections will be made. One by one, the students remove those edges not allowed by the restrictions they were given. In all, 10 edges will be eliminated from the graph. The 5 edges that remain reveal the only possible sequence, ordered left-to-right or right-to-left, shown in Figure 8.

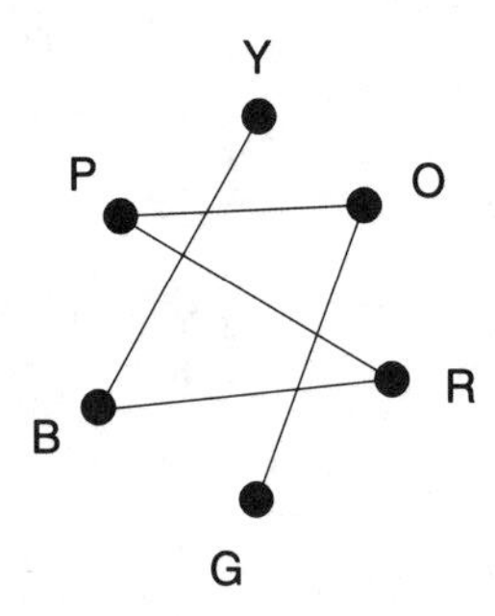

FIGURE 8. This path shows the only two possible arrangements of the six cubes when placed in a row: YBRPOG or GOPRBY.

Encourage students to make up similar sets of conditions on their own. Let them check one another's suggestions. Have them describe algorithms for creating problems that will ensure unique solutions. These are some of the important components of the critical thinking required for doing mathematics.

How can discrete math problems such as these, involving the ordering of colored cubes, be modified to assign lengths to the edges? Suppose, instead of having five colored cubes, teams of students select five whole numbers in the range 0 to 100. Imagine the numbers as the names of cities which are connected by airplane flights.

Begin by having the teams arbitrarily place their five vertices, identified by their choice of numbers. Next, have them assign distances to the edges corresponding to the differences between the numbers on the connected vertices representing cities. Here are some possible investigations to consider.

- Try to find the shortest route connecting all five cities. Where would you start and where would you end? What about a round trip that takes you through all five cities?
- Where would you start and end for the longest route, without repeating any connections? Is the same sequence the best for the longest round trip?

Have the teams try to find algorithms for solving these problems. Ask whether their procedures would change for an even instead of an odd number of cities. In middle school, students need the experience of exploring, trying, testing, and expressing their ideas in situations like these as much as they need to learn and apply known algorithms from discrete mathematics.

Figure 9 shows a complete graph, weighted on the edges by the distances for the five cities numbered 6, 32, 19, 84, and 61.

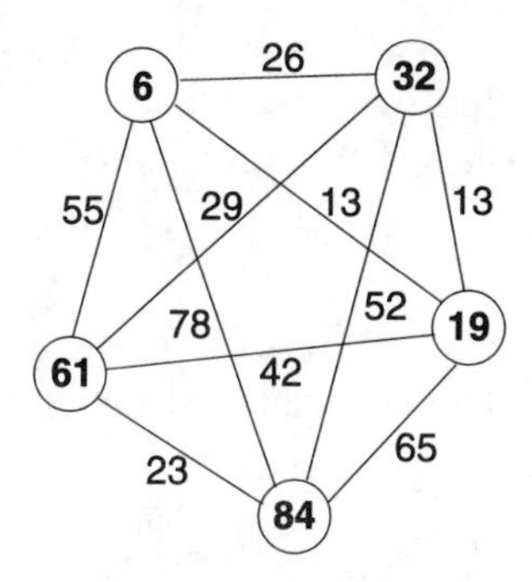

FIGURE 9. Graph showing five cities with distances.

There are $5! = 120$ directed paths that connect the five vertices and $4! = 24$ directed trips through them back to the starting point.

Finding the shortest paths and circuits through the five vertices in this situation does not require a great effort, especially if one realizes that tours among the points on the graph correspond to routes along the real number line. Finding the longest paths and circuits requires more thinking and testing. Searching for appropriate algorithms for any set of vertex values poses some interesting challenges.

Iteration

When the dynamics of change is built into a hands-on activity for the mathematics classroom through some iterative process, the experience becomes all the more powerful. One reason is that numerical, geometric, and algebraic relationships and connections often emerge from a single experience, as in the following activity.

Start with an equilateral triangle cut from paper. Mark a vertex P and repeat the following folding procedure through several stages:

> *When the vertex P appears in a triangle, fold it to the midpoint of the opposite side and then unfold.* (See Figure 10).

The outline of the folded paper at each stage is a trapezoid, but these trapezoids change through successive stages. How are they changing? What do you see?

From a measurement point of view, the trapezoids are growing in height. Start with a triangle whose area is 1 square unit and watch the areas of the trapezoids change.

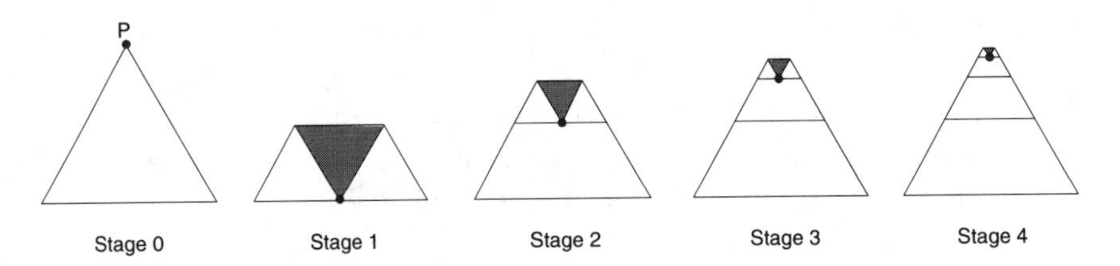

FIGURE 10. A trapezoid folding activity

The first triangle folded over has an area of $1/4$.
The second folded triangle has an area of $(1/4)^2$ or $1/16$.
The third has an area of $(1/4)^3$ or $1/64$, and so on.

Subtract these successive powers of $1/4$ from the original area of 1 to find what area remains for the trapezoid at each stage:

$$\underset{\text{Stage 1}}{\tfrac{3}{4}} \rightarrow \underset{\text{Stage 2}}{\tfrac{15}{16}} \rightarrow \underset{\text{Stage 3}}{\tfrac{63}{64}} \rightarrow \underset{\text{Stage 4}}{\tfrac{255}{256}}$$

What else is changing as the process is repeated over and over? The unfolded stages reveal other interesting patterns of a discrete nature, as in Figure 11.

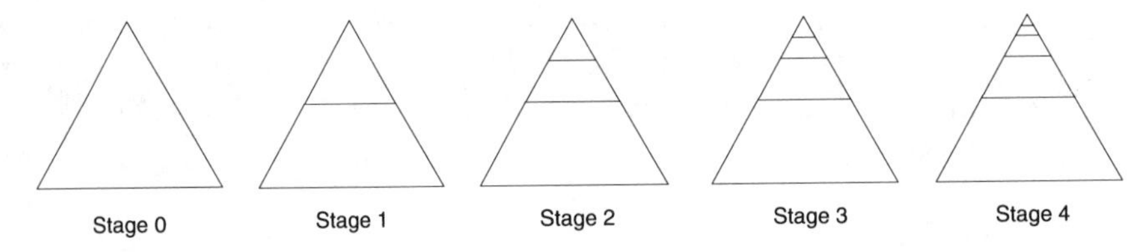

FIGURE 11.

In this form, we can view triangles and trapezoids in quite a different way, as shown in this table:

Stage	0	1	2	3	4	n
Number of triangles	1	2	3	4	5	$n+1$
Number of trapezoids	0	1	3	6	10	$n(n+1)/2$

Here again, we find the triangular numbers embedded in a counting problem centered around a geometric activity. Looking at the folds themselves, still another vision may appear. Let your students describe what they see.

One image is that of a strangely distorted ladder. When you climb it, each successive step is half as high and each successive rung half as wide. When you look up, you forever see reduced versions of exactly what you saw before. And the climb, step-by-step, is endless!

You can quickly see how some more powerful notions, such as perspective in art and limits in mathematics, can be brought into play. Students need to see, think, and talk about concepts such as these from an intuitive point of view during the middle school years. By choosing a good visual model and asking the right questions, one can bring together a host of related mathematical ideas in a single activity. And it is not surprising that many of these turn out to be discrete in nature.

Suppose the folding process is changed a bit, as described in the following algorithm.

> *Every time you have a triangle, fold each vertex to the midpoint of the opposite side. Cut off the corners and keep only the middle triangular piece at each stage.* (See Figure 12.)

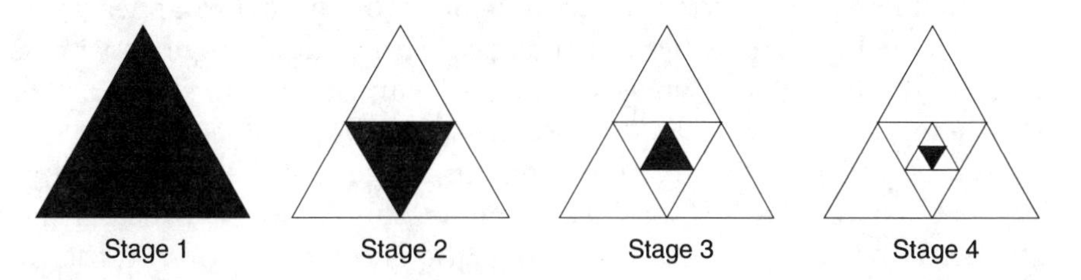

FIGURE 12.

A new set of figures is generated and new sets of number patterns emerge.

Stage	0	1	2	3	4	n
Number of triangles	1	1	1	1	1	1
Area	1	1/4	1/16	1/64	1/256	$(1/4)^n$
Perimeter	1	1/2	1/4	1/8	1/16	$(1/2)^n$

By interchanging what is kept and discarded in the folding and cutting process, an entirely different sequence of figures is created, as shown in Figure 13. This time, keep the corner pieces and discard the middle piece at each stage with each triangle. Now the process leads to an entirely different structure, a *fractal* called the Sierpinski triangle.

Stage	0	1	2	3	4	n
Number of triangles	1	3	9	27	81	3^n
Area	1	3/4	9/16	27/64	81/256	$(3/4)^n$
Perimeter	1	3/2	9/4	27/8	81/16	$(3/2)^n$

As an alternative approach in the classroom, have your students draw these two sets of figures on triangular dot paper. Choose a large triangle where the dots divide the sides into units that number a power of 2. This way the spacing of the dots will facilitate drawing several repeated reductions by one-half. For many students, both types of activities would be worthwhile. Indeed, seeing, drawing, and visualizing experiences all need to occur more

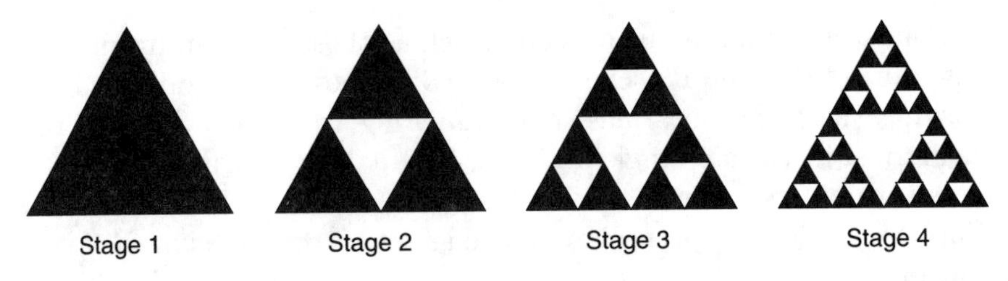

FIGURE 13.

often in the mathematics classroom to improve our students' abilities in visual literacy.

The distinct, discrete stages of growth clearly show an underlying property of fractals, that of *self-similarity.* Copies of the figure appear within itself at all scales. Three reduced images of the initial stage can be seen in stage 1. Three reduced images of stage 1 can be seen in stage 2. Three reduced images of stage 2 can be seen in stage 3, and so on.

The intricate structure of the emerging fractal can be measured by its *fractal dimension.* For the Sierpinski triangle, this complexity measurement is approximately 1.58. See Volume 1 of [**2**] for an introduction to the topic of fractals.

Is there an underlying structure here that is independent of the shape of the initial figure? That is, if we start with a different figure and repeatedly put together three copies of the figure, scaled to one-half, what do we get? Have students explore this question starting with other figures, such a right triangle, a scalene triangle, or even a square, rather than an equilateral triangle.

Start with a square cut from paper. Cut it in half vertically and horizontally. Use the rebuilding process shown in Figure 14 with three reduced copies at each stage placed in the shaded cells.

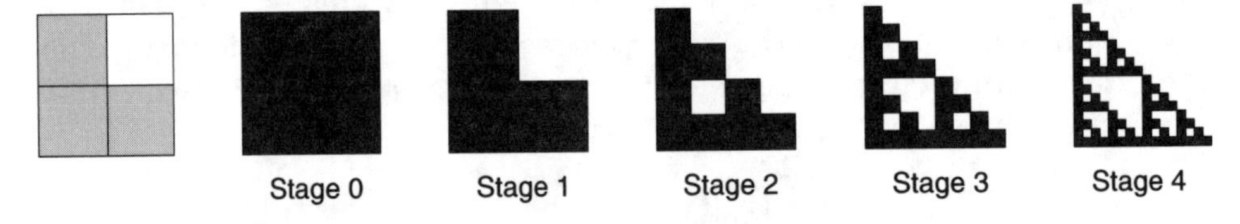

FIGURE 14. Iteration based on a square

It does not take many stages to see a familiar shape emerging. Have your students think, talk, and write about the similarities and the differences between the changing structures being generated from squares and those that were generated above from triangles. In both cases, of course, the limit structure is the Sierpinski triangle.

As a final activity, have students put their own personal twist to the rebuilding step in the iteration process, which can be abbreviated as *Reduce, Replicate, and Rebuild.*

Mentally label the three cells A, B, and C, as shown in Figure 15. When the reduced images are dropped back into the appropriate cells, consider possible rotations. In the sequence of figures shown in Figure 15, the reduced copy in cell A is always rotated 270° clockwise at each stage. Those copies placed in cells B and C always remain in their original orientation, which, for convenience, can be called a rotation of 0°.

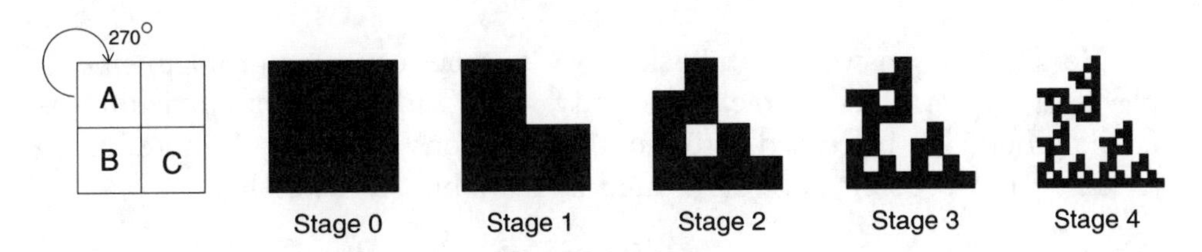

FIGURE 15.

Four choices of rotation are possible for each of the three cells. That gives $4 \times 4 \times 4 = 64$ different rebuilding codes using rotations. This can lead to the exploration of a whole family of related fractals with many different structures. Have students create their own personal fractals by making individual choices of rotations for cells A, B, and C. They can cut out and tape together their images or draw the first few stages on graph paper. The first four stages can be readily drawn using 2×2-inch initial squares on 1/8-inch graph paper.

When reflections are considered, another four transformations of the square can be explored. See Figure 16.

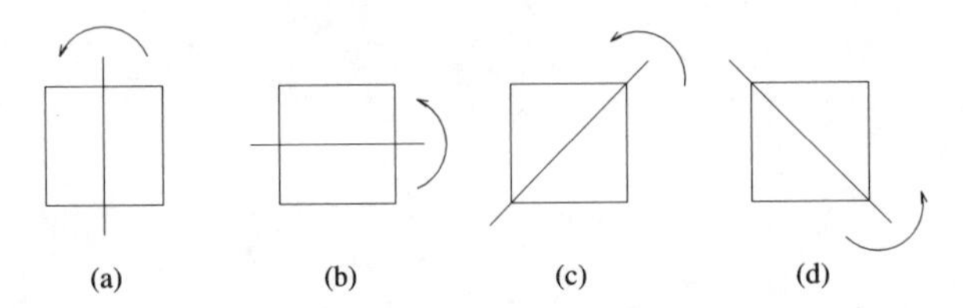

FIGURE 16. (a) Horizontal reflection about the vertical axis. (b) Vertical reflection about the horizontal axis. (c) Reflection about the lower-left, upper-right diagonal. (d) Reflection about the upper-left, lower-right diagonal.

In the sequence of iterations shown in Figure 17, the reduced copy in cell A is reflected about the upper-left, lower-right diagonal at each stage. Those copies in cells B and C remain in their original orientation.

In all, four rotations and four reflections can be made in each of the three square cells. With eight transformations possible in each cell, there must be $8 \times 8 \times 8 = 512$ different rebuilding codes. Will all 512 different building codes produce different fractals? The answer is no. Because of symmetry, some images will be duplicated. How many distinct fractal images will there be? The question is left for the reader to investigate and answer.

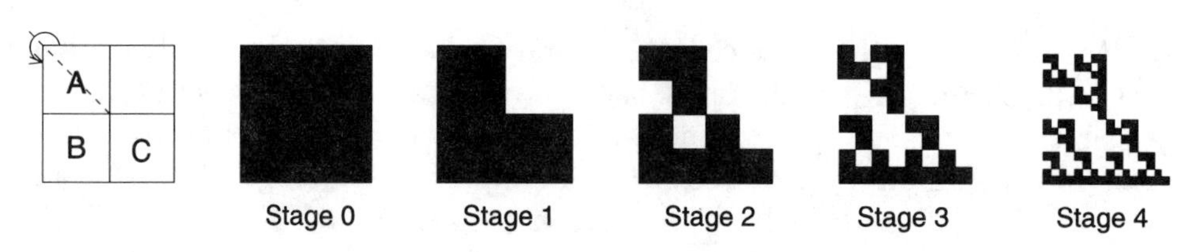

FIGURE 17.

Here we are, answering and asking yet another counting problem emerging from an iterative, geometric activity. The middle school curriculum is fertile ground for increased attention to situations involving discrete mathematics. The problems are all around us if we but look for them.

References

[1] Erickson, T., *Get It Together: Math Problems for Groups — Grades 4-12*, Lawrence Hall of Science, Berkeley CA, 1989.

[2] Pietgen, H-O., Jurgens, H., Saupe, D., Maletsky, E., Perciante, T., and Yunker, L. *Fractals for the Classroom: Strategic Activities, Volumes One and Two*, Springer-Verlag, New York, 1991.

[3] Sobel, M., and Maletsky, E., *Teaching Mathematics: A Sourcebook of Aids, Activities, and Strategies*, Allyn & Bacon, Needham Heights MA, 1988.

MONTCLAIR STATE UNIVERSITY, UPPER MONTCLAIR, NEW JERSEY 07043
E-mail address: maletskye@alpha.montclair.edu

Section 5

Integrating Discrete Mathematics into Existing Mathematics Curricula, Grades 9–12

Putting Chaos into Calculus Courses
ROBERT L. DEVANEY
Page 239

Making a Difference with Difference Equations
JOHN A. DOSSEY
Page 255

Discrete Mathematical Modeling in the Secondary Curriculum: Rationale and Examples from The Core-Plus Mathematics Project
ERIC W. HART
Page 265

A Discrete Mathematics Experience with General Mathematics Students
BRET HOYER
Page 281

Algorithms, Algebra, and the Computer Lab
PHILIP G. LEWIS
Page 289

Discrete Mathematics Is Already in the Classroom — But It's Hiding
JOAN REINTHALER
Page 295

Integrating Discrete Mathematics into the Curriculum: An Example
JAMES T. SANDEFUR
Page 301

DIMACS Series in Discrete Mathematics
and Theoretical Computer Science
Volume **36**, 1997

Putting Chaos into Calculus Courses

Robert L. Devaney

Our goal in this paper is to give a brief description of how some elementary ideas which typically belong to the realm of discrete mathematics may be easily and beneficially incorporated into the standard calculus course. These ideas come from dynamical systems theory. They form a unified thread that begins with the basic topics in the calculus and culminates in a modern treatment of Newton's method.

There are many reasons for incorporating ideas from dynamical systems theory in the calculus curriculum. One reason is the fact that it is becoming increasingly important for mathematics and science students to understand and be able to use such numerical algorithms as Newton's method. In the same vein, it is important that these students understand the limitations of computer implementations of these algorithms (e.g., they may fail to converge, round-off error may affect results, etc.). Another reason for including dynamical ideas is the ease with which students may be exposed to topics of contemporary research interest in mathematics (iteration of quadratic functions of a real variable remains an active field of research interest!) A third reason is that dynamics provides a natural arena in which to couple theoretical results from calculus with computer experimentation.

In this paper we will present a thread of ideas which runs through such topics as iteration, graphical analysis, attracting and repelling periodic points, and chaos. This thread terminates with Newton's method, where we show that all of these dynamical ideas may be combined to present a coherent treatment of this algorithm, its results as well as its limitations. This can also serve as a "jumping-off point" for student computer experiments or research projects involving Newton's method in the plane and the related concept of Julia sets or the Mandelbrot set.

1. Iteration

Iteration is one of the basic operations of dynamical systems. Given a function F, the basic question is what happens when we compose F with

1991 *Mathematics Subject Classification.* Primary 00A05, 00A35.
Partially supported by NSF Grant ESI-9255724.

itself many times in succession. That is, given a number x_0 called the *seed*, the basic question is what happens to the following sequence:

$$\begin{aligned} x_1 &= F(x_0) \\ x_2 &= F(x_1) = F(F(x_0)) \\ x_3 &= F(x_2) = F(F(F(x_0))) \\ &\vdots \\ x_n &= F(x_{n-1}). \end{aligned}$$

This sequence is called the *orbit* of x_0. The question is then: What is the fate of orbits?

For example, if $F(x) = x^2$, the orbit of $1/2$ is

$$\begin{aligned} x_0 &= \frac{1}{2} \\ x_1 &= \frac{1}{4} \\ x_2 &= \frac{1}{16} \\ x_3 &= \frac{1}{256} \\ &\vdots \\ x_n &= \frac{1}{2^{2^n}}. \end{aligned}$$

Thus we see that the orbit of the seed $x_0 = 1/2$ under $F(x) = x^2$ tends to 0. On the other hand, the fate of the orbit of $x_0 = 2$ is much different: It tends to infinity.

$$\begin{aligned} x_0 &= 2 \\ x_1 &= 4 \\ x_2 &= 16 \\ x_3 &= 256 \\ &\vdots \\ x_n &= 2^{2^n}. \end{aligned}$$

For simplicity, we often write F^n to mean the n-fold composition of F with itself. That is, $F^2(x_0) = F(F(x_0))$ and $F^3(x_0) = F(F(F(x_0)))$.

There are all sorts of possibilities for the behavior of the orbit of a given seed. The point x may be a *fixed point*, i.e., $F(x_0) = x_0$, or a *periodic point*, i.e., $F^n(x_0) = x_0$. In the latter case, n is the *period* of x_0 and the orbit of x_0 is called a cycle of order n. For example, 0 is a fixed point for $F(x) = x^2$, since its orbit is the constant sequence $0, 0, 0, \ldots$. On the other hand, 0 is periodic with period 2 for the function $F(x) = x^2 - 1$. The orbit of 0 in this case is the repeating sequence $0, -1, 0, -1, 0, -1, \ldots$.

Cycles and fixed points are among the most important types of orbits in any dynamical system. In applications, such orbits correspond to cyclic or periodic behavior or to equilibrium points.

Other types of orbits include those that are asymptotic to fixed or periodic points and orbits which behave randomly or chaotically. All of these orbits are often present in even the simplest of dynamical systems, including Newton's method.

With this in mind, it is natural to introduce the concept of iteration whenever composition of functions is defined. One may introduce the concept of orbit at this time and then begin to hint about some of the contemporary research topics the class is about to experience. This is nothing but advertising at this point, but it serves the purpose of convincing the students that something exciting will occur in this course.

Iteration is particularly easy to illustrate using technology. Armed with a computer with a simple program, a spreadsheet, or a calculator, students may easily experiment with a variety of iterations. The "sequence" mode and web plot of the Texas Instruments TI-82 graphing calculator is particularly useful for iteration. Other calculators can be programmed to produce similar results.

2. Graphical Analysis

Most students who enter calculus are quite familiar with the concept of the graph of a function. Nevertheless, most instructors find it useful to review a number of basic graphs that students must know early in the course. At this juncture, it is natural to introduce the concept of graphical analysis, a procedure for determining orbits geometrically using the graph of a function.

How does one use the graph of a function to display orbits? The procedure is quite easy (see Figure 1). Start with the graph of F and superimpose the diagonal line $y = x$. The orbit of a given point x_0 will be displayed along this diagonal. To begin, draw a vertical line from the diagonal to the graph, starting at (x_0, x_0) and ending at $(x_0, F(x_0))$. The second step is to draw a horizontal line from this point back to the diagonal, reaching the diagonal at $(F(x_0), F(x_0))$, which gives the second point on the orbit. Thus the process of computing F is given geometrically by going vertically from the diagonal to the graph, and then going horizontally from the graph to the diagonal. To compute further iterates of F the process is the same: move vertically from $(F(x_0), F(x_0))$ on the diagonal to the graph, and then horizontally back to the diagonal to reach the point $(F^2(x_0), F^2(x_0))$ on the diagonal. Thus we see displayed on the diagonal the first three points on the orbit of x_0 under F. Continuing in this fashion, we often see a "staircase" diagram which displays the orbit of x_0 along the diagonal. Figure 1 displays this procedure for $F(x) = \sqrt{x}$.

Note that graphical analysis immediately yields the fixed points of F: these are points of intersection of the graph with the diagonal, i.e., points

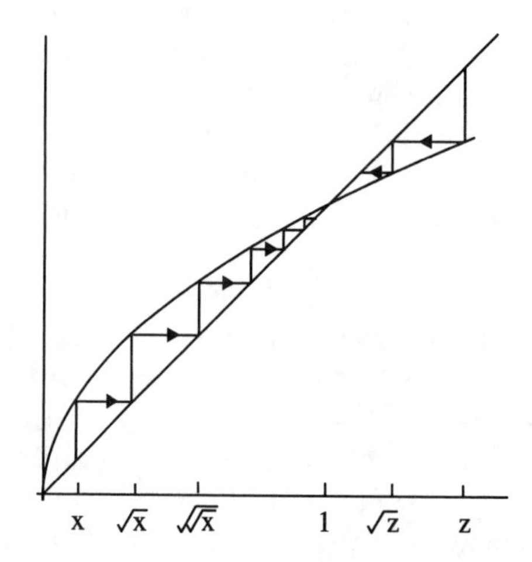

FIGURE 1. Graphical analysis of $F(x) = \sqrt{x}$

x_0 which satisfy $F(x_0) = x_0$. In the case of $\sqrt{x}$, the fixed points are 0 and 1. Note that, by graphical analysis, the orbit of any non-zero seed tends to the fixed point at 1. The staircase ascends toward 1 if the seed satisfies $0 < x_0 < 1$; it descends toward 1 if $x_0 > 1$. In this way, graphical analysis gives a powerful tool for visualizing orbits geometrically.

For simple functions such as $F(x) = x^2, x^3, \sqrt{x}$, and $1/x$, students can check easily the fate of any orbit using graphical analysis. This provides a good exercise for the student in that it reinforces the concept of the graph of a function (indeed it necessitates an accurate graph) as well as introducing a new concept simultaneously. In Figure 2 we display graphical analysis applied to $F(x) = \cos x$. Note that, as with the square root function, all orbits here also tend to a fixed point. This fixed point is impossible to determine algebraically, however, since we must solve the transcendental equation $\cos x = x$. Nevertheless, graphical analysis yields its existence together with the fact that all orbits tend to it.

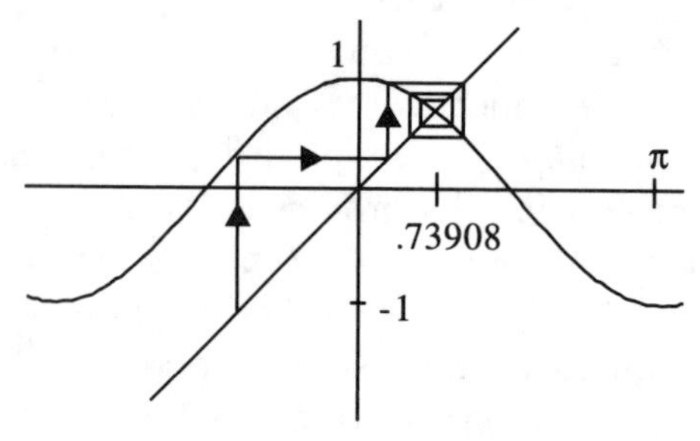

FIGURE 2. Graphical analysis of $F(x) = \cos(x)$

Not all iterations are as well behaved. In Figure 3 we display the result of iteration of the function $F(x) = 4x(1 - x)$. Note how complicated the

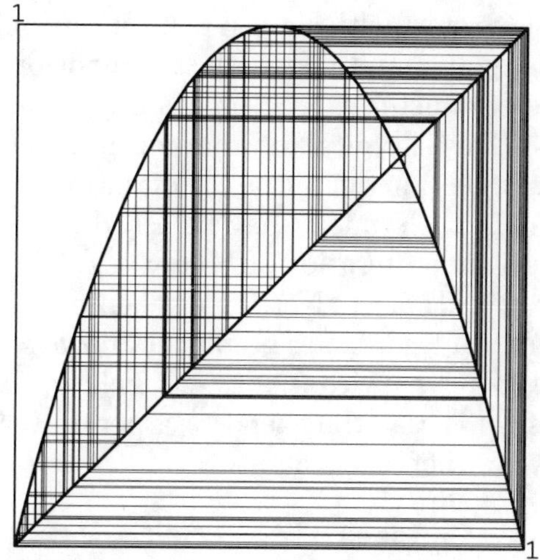

FIGURE 3. Graphical analysis of $F(x) = 4x(1-x)$

graphical analysis is: This is one view of what mathematicians now call chaos.

Incidentally, graphical analysis is a built in feature of the TI-82 graphing calculator, and it can be programmed on most other calculators.

3. Iteration via Computer

At this point, less than two weeks into the standard calculus course, the student may be exposed to a laboratory experiment that foreshadows much of the remaining topics in dynamics. If a computer lab is available, the student may be asked to report on the behavior of orbits of $F_c(x) = cx(1-x)$ for various values of the parameter c. As long as $c > 1$, it is a fact that all orbits of points x_0 with $x_0 < 0$ or $x_0 > 1$ tend to $-\infty$. This fact may be seen using graphical analysis. As an example, the orbit of $x_0 = 2$ under $F(x) = 2x(1-x)$ is given by $x_0 = 2, x_1 = -4, x_3 = -40, x_4 = -3280, \ldots,$ and the orbit of $x_0 = 1.1$ is

$$\begin{aligned} x_0 &= 1.1 \\ x_1 &= -0.22 \\ x_3 &= -0.5368 \\ x_4 &= -1.6499\ldots \\ x_5 &= -8.7442\ldots \\ x_6 &= -170.4109\ldots \\ &\vdots \end{aligned}$$

Therefore, all of the interesting dynamical behavior occurs for $0 \leq x_0 \leq 1$. For various values of c in the range $1 \leq c \leq 4$, the student may observe a wide variety of dynamical behaviors.

The experiment is to start with a given point in the unit interval (usually $x_0 = 1/2$, the critical point) and record the behavior of the orbit of x_0 for various values of the parameter c. The program to print this orbit is quite simple. See Figure 4. This is typical of programs in dynamical systems: the code is often quite simple but the output is often quite remarkable. Using the program ITERATE, the student may be asked to compute the orbit of 0.5 for various c values under iteration of $F_c(x) = cx(1-x)$. Several orbits of $x_0 = 0.5$ are displayed in Table 1. Note that when $c = 1.5$, the orbit of $1/2$ tends to a fixed point. Indeed, using graphical analysis, it is easy to check that the orbit of any point x_0 with $0 < x_0 < 1$ behaves in this manner. On the other hand, most of these orbits tend to a period 2 cycle when $c = 3.1$ and to a period 4 cycle when $c = 3.5$.

```
input x          \%\%\% (seed)
input c          \%\%\% (parameter)
input n          \%\%\% (max number of iterations)
i = 0
do while i <= n
    print i, x
    x = c*x*(1-x)
    i = i+1
end
```

FIGURE 4. The program ITERATE

One of the principal benefits that may be derived from this or similar experiments is that students see first-hand the notion of convergence and non-convergence. For values of $c < 3.6$, it appears that all orbits eventually tend somewhere (perhaps to a fixed point or a cycle). For $3.6 < c \leq 4$, many orbits seem to wander aimlessly about the unit interval, although there are c-values for which convergence may be observed. The c-values for which no convergence or cyclic behavior is observed is what we will describe below as chaotic behavior. In any event, the phenomenon of non-convergence arises naturally in the setting of iteration, unlike traditional calculus settings, where limits which fail to exist often seem like pathological cases to students.

We remark that iteration demands a different notion of convergence as convergence to cycles is clearly allowable. However, experiments such as the above make this concept easy to digest.

4. Attracting and Repelling Cycles

The experiment in the previous section highlights the fact that fixed and periodic points come in two quite distinct varieties. For all $c > 1$, F_c has two fixed points in the unit interval, at 0 and at $x_c = (c-1)/c$. When $c \leq 3$ we may easily find x_c because the orbit of any other point in $0 < x < 1$ tends

Iterate	$c = 1.5$	$c = 3.2$	$c = 3.5$
0	0.5	0.5	0.5
1	0.375	0.8	0.875
2	0.3515625	0.512	0.3828125
3	0.341949462	0.7995392	0.826934814
4	0.337530041	0.512884056	0.500897694
5	0.335405268	0.799468803	0.874997179
6	0.334362861	0.513018994	0.382819903
7	0.333846507	0.799457618	0.826940887
8	0.333589525	0.513040431	0.500883795
9	0.33346133	0.79945583	0.874997266
10	0.333397307	0.513043857	0.382819676
11	0.333365314	0.799455544	0.826940701
12	0.333349322	0.513044405	0.500884222
13	0.333341327	0.799455499	0.874997263
14	0.33333733	0.513044492	0.382819683
15	0.333335331	0.799455491	0.826940706
16	0.333334332	0.513044506	0.500884209
17	0.333333832	0.79945549	0.874997263
18	0.333333583	0.513044509	0.382819683
19	0.333333458	0.79945549	0.826940706
20	0.333333395	0.513044509	0.50088421
21	0.333333364	0.79945549	0.874997263
22	0.333333348	0.513044509	0.382819683
23	0.333333341	0.79945549	0.826940706
24	0.333333337	0.513044509	0.50088421
25	0.333333335	0.79945549	0.874997263
26	0.333333334	0.513044509	0.382819683
27	0.333333333	0.79945549	0.826940706
28	0.333333333	0.513044509	0.50088421
29	0.333333333	0.79945549	0.874997263
30	0.333333333	0.513044509	0.382819683
31	0.333333333	0.79945549	0.826940706
32	0.333333333	0.513044509	0.50088421
33	0.333333333	0.79945549	0.874997263
34	0.333333333	0.513044509	0.382819683
35	0.333333333	0.79945549	0.826940706
36	0.333333333	0.513044509	0.50088421
37	0.333333333	0.79945549	0.874997263
38	0.333333333	0.513044509	0.382819683
39	0.333333333	0.79945549	0.826940706

TABLE 1. The orbit of .5 for various c-values. This orbit is attracted to a fixed point when $c = 1.5$, to a cycle of period 2 when $c = 3.2$, and to a cycle of period 4 when $c = 3.5$.

to it. When $c > 3$, this ceases to be the case. In the former cases, x_c is called an attracting fixed point; in the latter, x_c is repelling. The computer experiment above shows that attracting fixed points are visible whereas repelling fixed points are invisible to the computer. To be precise, a fixed point x_0 for F is called *attracting* if there is an interval (a, b) containing x_0 and having the property that if $x \in (a, b)$ then $F(x) \in (a, b)$ and moreover, $F^n(x) \to x_0$ as $n \to \infty$. The fixed point is *repelling* if the orbit of any $x \in (a, b)$ (except $x = x_0$) eventually leaves (a, b). Attracting and repelling periodic cycles are defined analogously.

This is where calculus enters the picture. It is easy to see that a fixed point x_0 for F is attracting if $|F'(x_0)| < 1$. Similarly, x_0 is repelling if $|F'(x_0)| > 1$. See Figure 5.

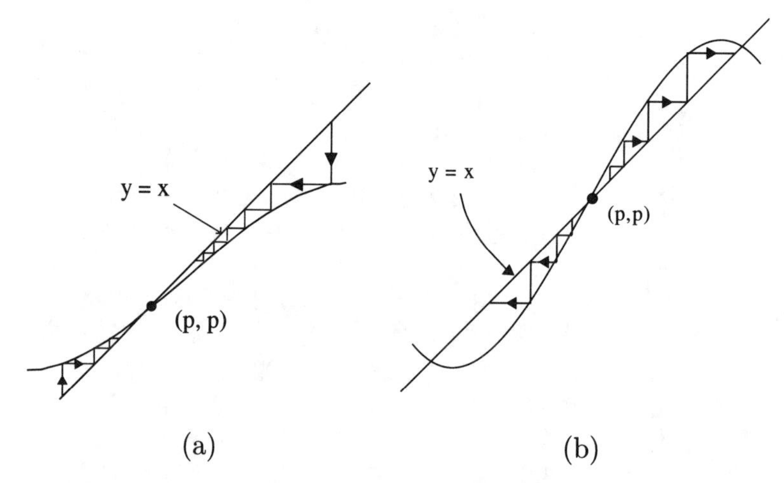

FIGURE 5. (a) The fixed point p is attracting: $|F'(p)| < 1$. (b)The fixed point p is repelling: $|F'(p)| > 1$.

As an example, we know that $F(x) = x^2$ has two fixed points, at 0 and at 1. We have $F'(0) = 0$, so 0 is an attracting fixed point. Also, $F'(1) = 2$, so 1 is repelling. This is illustrated nicely by graphical analysis. See Figure 6.

The determination of whether a cycle is attracting or repelling provides a nice application of the chain rule. Suppose x_0 lies on a cycle of period n for F. Then, arguing as above, x_0 is an attracting periodic point if $|(F^n)'(x_0)| < 1$. But, by the chain rule,

$$(F^n)'(x_0) = F'(F^{n-1}(x_0)) \cdot \ldots \cdot F'(F(x_0)) \cdot F'(x_0)$$

For example, the points 0 and -1 lie on a cycle of period 2 for $G(x) = x^2 - 1$. This cycle is attracting since

$$\begin{aligned}
(G^2)'(0) &= G'(G(0)) \cdot G'(0) \\
&= G'(-1) \cdot G'(0) \\
&= -2 \cdot 0 \\
&= 0
\end{aligned}$$

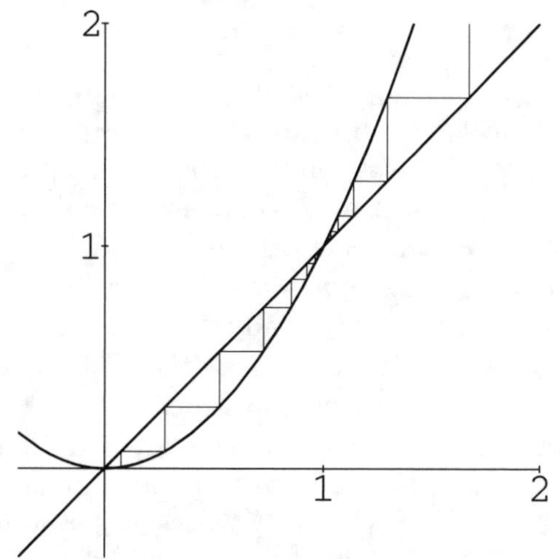

FIGURE 6. $F(x) = x^2$ has a repelling fixed point at 1 and an attracting fixed point at 0.

As another example, consider $F(x) = -x^3$. This function has a 2-cycle given by ± 1, since $F(1) = -1$ and $F(-1) = 1$. We have $F'(x) = -3x^2$ so $F'(1) = -3$ and $F'(-1) = -3$. Therefore, $(F^2)'(1) = F'(F(1)) \cdot F'(1) = -3 \cdot -3 = 9$, so this cycle is repelling. This is readily observed using graphical analysis. See Figure 7. Indeed, analyzing all orbits of a given dynamical system using both graphical analysis and the above techniques provides a valuable (and often challenging) exercise for students.

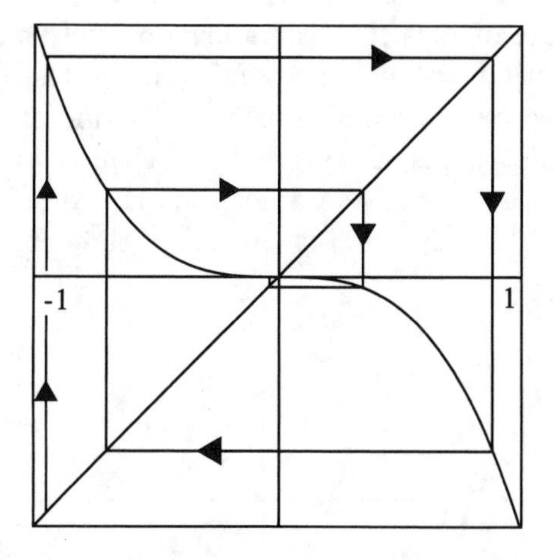

FIGURE 7. Graphical analysis of $F(x) = -x^3$.

5. Chaos

One of the most interesting mathematical discoveries of the past quarter century is the fact that very simple dynamical systems may behave "chaotically." Despite the fact that this phenomenon is a topic of much contemporary interest in mathematics, it may still be explained in elementary calculus courses. Moreover, as we will see when we discuss Newton's method, chaos occurs in a great many simple dynamical systems.

While there is not yet a universally accepted definition of chaos, one definite ingredient of any definition is *sensitive dependence on initial conditions.* An orbit is said to have sensitive dependence if nearby orbits behave in a vastly different manner. To be precise, the orbit of x_0 has sensitive dependence if there exists $K > 0$ such that, for any $\delta > 0$, there exists x_1 with $|x_1 - x_0| < \delta$ but $|F^n(x_1) - F^n(x_0)| > K$ for some integer n. That is, arbitrarily near x_0 there exist initial conditions x_1 whose orbit eventually separates from that of x_0 by at least K units. We also call such orbits *chaotic orbits.*

According to this definition, repelling fixed and periodic points are always chaotic. There are many other kinds of chaotic orbits. Here is an example of a dynamical system with many chaotic orbits. We will meet this function in a completely different setting in the next section.

Consider the doubling function

$$D(x) = \begin{cases} 2x, & 0 \leq x < \frac{1}{2} \\ 2x - 1, & \frac{1}{2} \leq x < 1 \end{cases}$$

Note that D is defined on the interval $[0, 1)$. Since the derivative of D is always 2, each iteration of D doubles the distance between corresponding points on different orbits, at least until these points appear on different sides of $1/2$. Hence all orbits in the unit interval are chaotic.

Another property of D is that there are infinitely many periodic points. Figure 8 indicates that the graph of D^n crosses the diagonal exactly 2^n times; each of these fixed points of D^n is a periodic point of D. Indeed, it can be shown that if p/q is rational with $(p, q) = 1$ and q odd, then p/q lies on a cycle for D. For example, the orbit of $1/7$ under doubling is $1/7, 2/7, 4/7, 1/7, \ldots$ which is a 3-cycle. Also, the orbit of $1/5$ is a 4-cycle: $1/5, 2/5, 4/5, 3/5, 1/5, \ldots$.

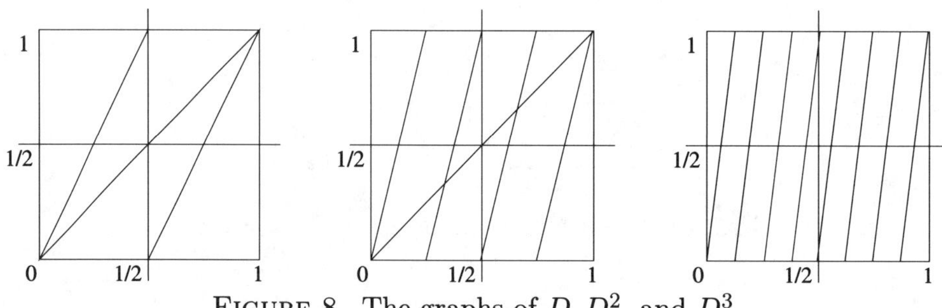

FIGURE 8. The graphs of D, D^2, and D^3.

A third property of D that is typical of chaotic systems is the fact that most orbits are not periodic. Indeed, the orbits of most points tend to fill out the interval $[0,1)$ in a dense fashion. This can be proved rigorously in an elementary setting. Let $x_0 \in [0,1)$ and consider the orbit of x_0. Assign an infinite sequence $S(x_0)$ of 0's and 1's to x_0 according to the following prescription.

$$S(x_0) = s_0 s_1 s_2 \ldots$$

where $s_j = 0$ if $D^j(x_0) \in [0,1/2)$, $s_j = 1$ if $D^j(x_0) \in [1/2,1)$. The sequence $S(x_0)$ gives the *itinerary* of the orbit of x_0 in the sense that $S(x_0)$ tells whether the jth iterate of x_0 lies in the left or right half of the unit interval. But $S(x_0)$ has another interpretation: $S(x_0)$ is just the binary expansion of x_0. Thus there are clearly many, many orbits whose itineraries are non-periodic.

This interpretation of the orbits of D also shows why all orbits are chaotic. For a given point x_0, we can typically "know" its binary representation with only finite precision, i.e., up to, say, l binary digits. This means, then, that after l iterations of T we will have obliterated any knowledge whatsoever of the location of points on the orbit of x_0.

One other issue that sometimes surfaces regarding the doubling function is the behavior of computed orbits. On machines that use binary arithmetic, students always observe that each computed orbit always ends up eventually fixed at 0. Of course, this stems from the fact that the typical x_0-value is represented in finite binary form. Each successive application of D effectively removes one of the digits in this binary representation, thus leading to the above behavior. This is an excellent lesson for students to learn: the computer may lie!

For example, the binary representation of $1/3$ is $.010101\ldots$ since

$$\begin{aligned}
\frac{1}{3} &= \frac{0}{2} + \frac{1}{2^2} + \frac{0}{2^3} + \frac{1}{2^4} + \frac{0}{2^5} + \cdots \\
&= \frac{1}{4} + \frac{1}{4^2} + \frac{1}{4^3} + \cdots \\
&= \frac{\frac{1}{4}}{1 - \frac{1}{4}} = \frac{1}{3}
\end{aligned}$$

However, the computer stores only a finite number of these binary digits, and these are removed one by one as we iterate the doubling function. For more details on the dynamics of the doubling function, we refer to [**2**].

6. Newton's Method

With iteration, graphical analysis, attracting fixed points, and chaos as concepts in hand, the introduction of Newton's method in the calculus course becomes a central topic in the course, rather than a peripheral curiosity. Indeed, Newton's method utilizes all of the preceding topics in an essential fashion.

As is well known, if $P(x)$ is a polynomial, the associated Newton iteration function

$$N(x) = x - \frac{P(x)}{P'(x)}$$

has the property that x_0 is a root of P if and only if x_0 is an attracting fixed point of N. Indeed, $N'(x_0) = 0$ if x_0 is a simple root of P. Thus, to find a root of P, all we need do is select some random initial condition x_0 and compute the orbit of x_0 under N. Hopefully, this orbit will converge to one of the attracting fixed points of N, i.e., to one of the roots of P. This, of course, need not be the case, since the orbit of x_0 may be attracted to an attracting cycle or, even worse, may behave chaotically.

Newton's method together with graphical analysis provide a natural and ideal place in the curriculum for students to manipulate and comprehend the graph of "complicated" functions. For example, the Newton iteration function corresponding to $P(x) = x^2 - 1$ is given by

$$N(x) = \frac{1}{2}(x + \frac{1}{x})$$

and the corresponding graphical analysis is depicted in Figure 9.

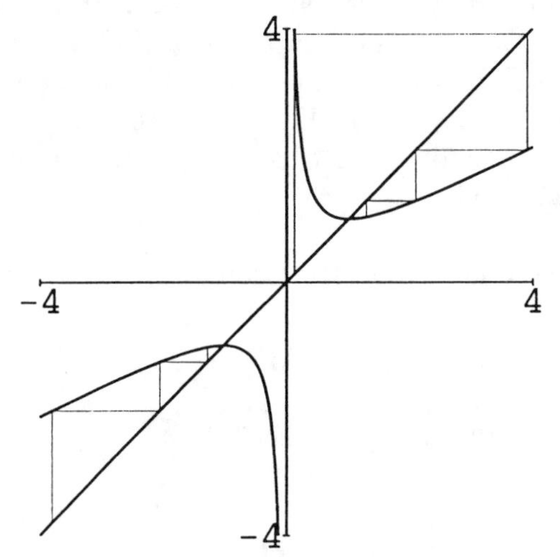

FIGURE 9. Newton iteration for $P(x) = x^2 - 1$.

For the cubic function $Q(x) = x(x^2 - 5)$, the corresponding Newton function is more complicated

$$N(x) = \frac{2x^3}{3x^2 - 5}.$$

Nonetheless, students can be expected to understand the graphs of such functions and perform the graphical analysis, as in Figure 10.

This last example gives one simple reason why Newton's method sometimes fails to converge: note that the points $+1$ and -1 lie on a cycle of

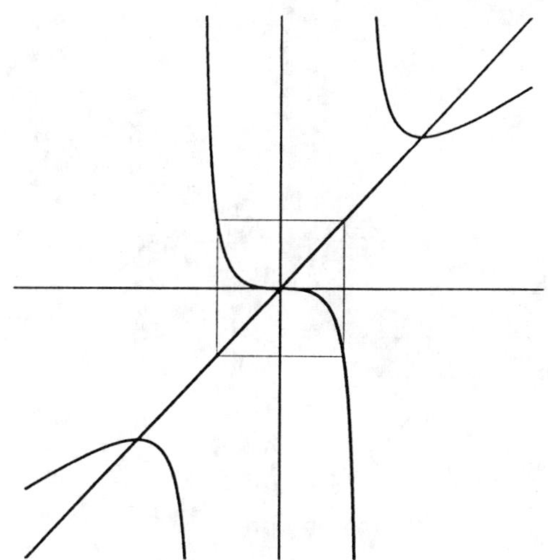

FIGURE 10. Newton iteration for $Q(x) = x(x^2 - 5)$.

period 2. Thus the perfectly natural initial guess of 1 leads to no convergence of Newton's method.

For polynomials of degree four or more, the set of points at which Newton's method fails on the real line is more complicated. If the polynomial has real, distinct roots, it turns out that the set of initial points at which Newton's method fails is a Cantor set. This is discussed in an award winning paper in the American Mathematical Monthly by Saari and Urenko [**7**]. Here we see a simple fractal appearing as the "interesting" set of points in a dynamical system, a fact that occurs over and over again in dynamics.

Another manner in which Newton's method fails to converge occurs when the original function is not differentiable at a root. For example, the Newton iteration corresponding to $F(x) = x^{1/3}$ is $N(x) = -2x$. Using graphical analysis we see that 0 is a repelling fixed point for N and all orbits tend away from 0.

Finally, it is interesting to ask what happens when we apply Newton's method to the polynomial $P(x) = x^2 + 1$. The Newton iteration function in this case is

$$N(x) = \frac{1}{2}(x - \frac{1}{x}).$$

Graphical analysis shows that orbits of this function tend to behave quite chaotically. See Figure 11. In fact, this function has dynamics that are exactly the same as the doubling function introduced in the previous section.

Recall that

$$D(x) = \begin{cases} 2x, & 0 \le x < \frac{1}{2} \\ 2x - 1, & \frac{1}{2} \le x < 1 \end{cases}$$

Consider the function

$$C : [0, 1) - \{1/2\} \to \mathbf{R}$$

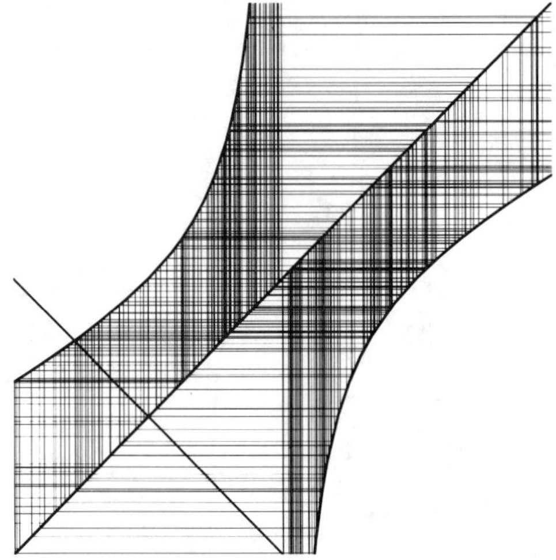

FIGURE 11. Newton iteration for $N(x) = \frac{1}{2}(x - \frac{1}{x})$

given by $C(x) = \cot \pi x$. We have

$$\begin{aligned} C \circ D(x) &= \cot(\pi \cdot D(x)) \\ &= \cot(2\pi x) \\ &= \frac{\cos^2(\pi x) - \sin^2(\pi x)}{2\sin(\pi x)\cos(\pi x)} \\ &= \frac{1}{2}\left(\cot(\pi x) - \frac{1}{\cot(\pi x)}\right) \\ &= N \circ C(x). \end{aligned}$$

This fact means that C carries orbits of D onto orbits of N, because

$$N^n \circ C(x) = C \circ D^n(x).$$

So, if x lies on a cycle of period n for D, then $C(x)$ lies on a similar cycle for N! Moreover, since all orbits of D are chaotic, the same is true for all orbits of N! Thus we see that chaos occurs in even the most unexpected places: Newton's method for a quadratic polynomial!

These ideas are by no means new: they go back to Cayley in the nineteenth century. Indeed, Cayley showed that, for a complex quadratic polynomial, Newton's method failed to converge only on the perpendicular bisector of the roots. On this line, Newton's iteration behaved similarly to the doubling function. Cayley even went so far as to announce that he planned a similar solution for cubic polynomials. But this paper never appeared, for reasons that are only nowadays becoming clear. The chaotic set for Newton's method for a complex cubic polynomial is hardly a line! It is often a very complicated fractal—the Julia set of the Newton iteration function. While this topic is too advanced for a standard calculus course, it nevertheless makes a wonderful subject for computer experimentation in a lab. We have found that, even if students do not have access to a lab in which to perform such experiments themselves, they love to see the computer-generated images of the Julia sets for Newton's method. They readily understand what these pictures mean and are most intrigued by the fact that these pictures were first seen a mere ten years ago. I often conclude my lecture on Newton's method by showing students a few slides of Newton's method in the plane. I have found that this gives students more than just a peek at pretty pictures. Indeed, these images give students a glimpse of what goes on in research mathematics. Many students find this quite enticing: they often tell me that they never imagined how beautiful mathematics could be.

References

[1] Blaine, L., "Theory vs. Computation in Some Very Simple Dynamical Systems," *Coll. Math. J.* **22** (1991), 42-44.

[2] Devaney, R. L., *A First Course in Chaotic Dynamical Systems,* Addison-Wesley Co., Reading MA, 1992.

[3] ———, *Chaos, Fractals, and Dynamics: Computer Experiments in Mathematics,* Addison-Wesley Co., Menlo Park CA, 1989.

[4] ———, "Putting Chaos into the Classroom," *Discrete Mathematics across the Curriculum K–12,* 1991 NCTM Yearbook (Margaret J. Kenney and Christian R. Hirsch, eds.), NCTM, Reston VA, 1991, pp. 184-194.

[5] Parris, R., "The Root Finding Route to Chaos," *Coll. Math. J.* **22** (1991), 48-55.

[6] Strang, G., "A Chaotic Search for i," *Coll. Math. J.* **22** (1991), 3-12.

[7] Saari, D. G., and J. Urenko, "Newton's Method, Circle Maps, and Chaotic Motion," *Amer. Math. Monthly.* **91** (1984), 3-17.

DEPARTMENT OF MATHEMATICS, BOSTON UNIVERSITY, BOSTON, MA 02215
E-mail address: `bob@math.bu.edu`

DIMACS Series in Discrete Mathematics
and Theoretical Computer Science
Volume **36**, 1997

Making a Difference with Difference Equations

John A. Dossey

The use of difference equations in the modeling of processes and in the description of change is a topic usually studied by advanced students in mathematics. However, many examples of arithmetical growth are found in elementary mathematics, and the topic of difference equations used to model such growth already enters the present school curriculum in many hidden ways.

At the college level, difference equations have been reserved for a lead-in to differential equations or for students studying actuarial mathematics. In recent years, study of the topic has moved somewhat earlier in the college curriculum and toward a broader spectrum of mathematics students. The inclusion of difference equations in introductory courses on discrete mathematics for undergraduates in the first two years of the curriculum has heightened the visibility of difference equations, their applications, and their power. The next decade should see the increased recognition of difference equations as an important topic for the K-12 curriculum.

Discrete mathematics, and difference equations in particular, has received a great deal of interest in recent years [**5, 10, 12, 13, 14**]. However, it is time to recognize that many of the topics related to difference equations have application much earlier in the school mathematics curriculum, especially if students are to come to see mathematics as a discipline having connections to patterns they observe in their everyday worlds. Informal introduction to difference equations as ways of counting, as ways of relating successive items in a pattern, as ways of thinking about processes, should enter the school mathematics curriculum in the upper elementary grades.

Such recommendations were made in 1963 in the report of the Cambridge Conference on School Mathematics [**3**], but, unfortunately, these recommendations never took hold. With the current interest in reform in school mathematics and supporting recommendations from the NCTM Standards [**12**], difference equations may once again have an opportunity to make a difference in students' mathematical experiences. But before we explore that

1991 *Mathematics Subject Classification.* Primary 39A10, 00A05, 00A35.

possibility, it is best to develop the topic of difference equations and their applications, and their relevance to school mathematics.

1. Difference equations

Difference equations have played an important role in mathematics across time. They appear informally in the work of the Greeks as they considered the values of terms in sequences that represented convergent processes [**2**], in the work of Fibonacci [**7, 11, 16**] in the 13th century as he modeled the growth of rabbit populations in his text *Liber Abaci*, and in the work of 16th and 17th century analysts as they struggled to place calculus on a firm footing [**2**]. In each of these, and many other settings, mathematicians have turned to examining the values of successive terms in some sequence $a_0, a_1, a_2, a_3, a_4, \ldots$, and to understanding the process of change as one moves from one value in the sequence to subsequent values. This study of change is usually captured in an expression similar to that which describes the growth of Fibonacci's population of rabbit pairs:

$$\begin{aligned} r_n &= r_{n-1} + r_{n-2}, \quad \text{for } n = 2, 3, 4, \ldots \\ r_0 &= r_1 = 1. \end{aligned}$$

This famous difference equation sets the initial conditions for the pairs of rabbits in years 0 and 1 and then describes, via the difference equation $r_n = r_{n-1} + r_{n-2}$, the way in which that population from year 2 forward grows.[1] The process of iterating the difference equation gives:

	n	r_n or Population
Initial time	0	1
End Year 1	1	1
End Year 2	2	1 + 1 or 2
End Year 3	3	2 + 1 or 3
End Year 4	4	3 + 2 or 5....

While many in mathematics are familiar with the Fibonacci sequence, 1, 1, 2, 3, 5, 8, 13, 21, ..., and its many famous properties, they do not know the central role that other difference equations have played in the mathematics curriculum, even within the last two centuries. However, students of G. Chrystal's 1886 work, *Textbook of Algebra* [**4**], and Hall and Knight's 1887 classic *Higher Algebra* [**9**] were well aware of difference equations and their power in illuminating the nature of change in finite processes.

[1] Editors' note: See Kowalczyk's article in this volume for an elementary approach to deriving this relationship.

2. School Mathematics and Difference Equations

Difference equations first enter the present school curriculum in a hidden way as students come to study arithmetic and geometric sequences in upper middle school and in secondary school mathematics. These important concepts capture mathematically the essence of two of the most powerful forms of change. Unfortunately, these formats for modeling change are not exploited to their maximum potential. Instead of developing arithmetic and geometric sequences as symbolic number patterns, curricula should focus on the types of change they describe.

Many examples of arithmetic growth are found in elementary mathematics, ranging from the multiples of 3 (3, 6, 9, 12, ...); to the effect of simple interest at 5% on a $100 investment over years ($100, $105, $110, ...); to the growth in values of a linear function $f(x) = 3x + 5$ evaluated at nonnegative integer values $(5, 8, 11, 14, \dots)$. Formally, this arithmetic growth can be described by the difference equation

$$\begin{aligned} A_n &= A_{n-1} + d \ \text{ for } \ n = 1, 2, 3, \dots, \\ A_0 &= a. \end{aligned}$$

From the initial value of a, each successive term in an arithmetic sequence is found by adding the common difference of $d : a, a + d, a + 2d, a + 3d, \dots$. The central property signaling the underlying difference equation model in these sequences is the fact that the subtractive comparison $A_n - A_{n-1}$ between successive terms is a constant difference d.

Many examples of geometric growth are also found in elementary mathematics, from the study of the powers of 5 (5, 25, 125, 625, ...); to the effect of compound interest at 5% on a $100 investment over years ($100, $105.00, $110.25, $115.76, ...); to the growth in values of the exponential function $g(x) = e^x$ evaluated at nonnegative integer values $(1, e, e^2, e^3, \dots)$. Geometric growth is formally described by the difference equation

$$\begin{aligned} A_n &= rA_{n-1} \text{ for } n = 1, 2, 3, \dots, \\ A_0 &= a. \end{aligned}$$

From the initial value of a, each successive term in a geometric sequence is found by multiplying the preceding term by the common factor of r: $a, ar, ar^2, ar^3, \dots$. The central property signaling the underlying difference equation model in these sequences is the fact that the comparison ratio A_n/A_{n-1} between successive terms is the constant r. These examples reflect the occurrence of difference equations, either formally or implicitly, in a variety of levels and contexts within the K-12 curriculum. However, too often, students only see the beauty and power of difference equations in the second-year algebra or precalculus curriculum in their study of sequences and series. Even in these cases, the actual difference equation structure is often bypassed to move quickly to establishing formulas for the nth terms in arithmetic and geometric sequences, and sums for the first n terms in progressions involving terms from either type of sequence. In doing so, students are rushed pass the central notions of change, rate of change, and the powerful insight that difference equations and their study provide for both modeling common situations and preparing a base for further study in mathematics.

Both of the difference equations, $A_n = A_{n-1} + d$ and $A_n = rA_{n-1}$, are special cases of the broader class of difference equations $f_n = Af_{n-1} + B$, known as first-order linear difference equations. Linear because of the general form, and first-order because the value of the nth case is dependent on only the preceding value f_{n-1} in the sequence. The Fibonacci difference equation observed earlier is a second-order linear difference equation, since the value of any term depends on the values of two preceding terms.

3. Modeling with First-Order Linear Difference Equations

The study of difference equations provides valuable opportunities to introduce students to the richness of applying mathematics. Consider the following examples reflecting the use of difference equations to explain the long-term effects of quantitative decisions in real-world settings.

Forestry. The Clear Lake Pine Company owns a timber stand with 7000 pine trees [**6**]. Each year the company harvests 12% of its trees and plants 600 seedlings. They are particularly interested in the pine tree population in this timber stand in 10 years and in the long-range future. Examining this situation for the nature of change from one year to another, one can build the first-order linear difference equation model for the pine population p_n in year n as follows:

$$\begin{aligned} p_n &= 0.88p_{n-1} + 600 \\ p_0 &= 7000. \end{aligned}$$

One can iterate this model on a TI-81 with the following keystrokes: `7000`, `ENTER`, `0.88 * 2nd ANS + 600`, and then `ENTER` to get the value for p_1, `ENTER` for the value of p_2, and `ENTER` for successive values of p_i, as the value of i increases. The first 26 values for the sequence are shown on the calculator screens in Figure 1.

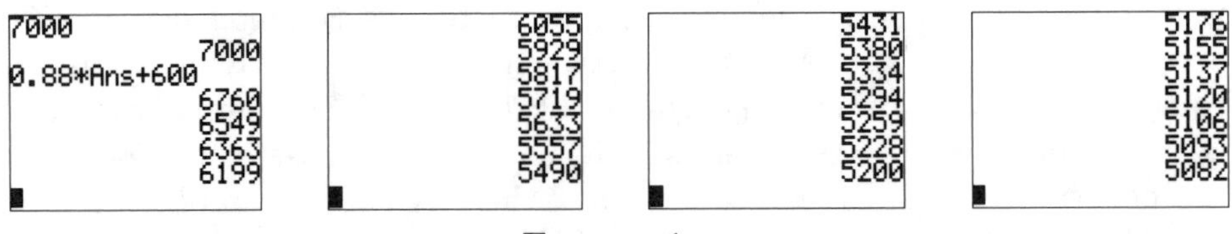

FIGURE 1.

Continuing to iterate the difference equation forward, one gets a pattern indicating that the pine tree population tends to stabilize at 5000 over the long haul. An examination of the data above shows that after ten years the population would be 5557. In the next section we will present and discuss the general solution for a first-order linear difference equation $f_n = Af_{n-1} + B$; this will reveal that, except in the case of arithmetic growth when A is 1, the solution always models exponential growth. Employing the solution form for first-order, linear difference equations (see Section 4), one arrives at the general value for the nth year following the initiation of the process:

$$p_n = 2000(0.88)^n + 5000.$$

Examining this as a function describing the pine tree population at each year following the first count of 7000 trees, it is easy to see that this model predicts an eventual steady state population of 5000 trees in the stand. When viewed graphically in Figure 2, using a continuous graph to make the trend more visible, one sees this developing over the first 100 years; each horizontal interval in Figure 2 represents 10 years and each vertical interval represents 1000 trees.

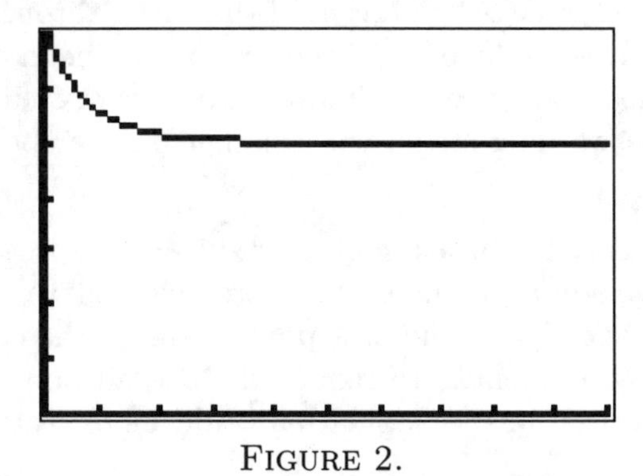

FIGURE 2.

This limiting process can be viewed in a step-wise fashion using the cobweb approach discussed in [**15**] and [**1**]. This approach allows for an investigation of the relationship between the difference equation, the initial value, and the existence of limiting values for a sequence. For example, the graph of the line $y = 0.88x + 600$ considered together with the graph of the auxiliary line $y = x$, allows one to establish geometrically the long-term behavior of the process described by the difference equation.

In Figure 3 we have entered the y-value of 7000 for an initial population on the line $y = x$; we then drop vertically to the line $y = 0.88x+600$ defining the difference equation to enter this as an x value for determining the next y value, then horizontally to the $y = x$ line to get the next x value, then vertically to the difference equation line for the next y value, transforming back and forth from present value to next value as the $y = x$ line transfers the output value at one stage to the input value at the next stage. The resulting pattern provides a graphical picture of the convergence of the tree population to the limiting value of 5000 pine trees. Figure 3 shows the TI-81 cobweb graph of this transformation of values.

Comparisons between the various possible values of the "slope" of a first-order linear difference equation quickly shows that the process converges for "slopes" where the absolute value of A is less than 1.

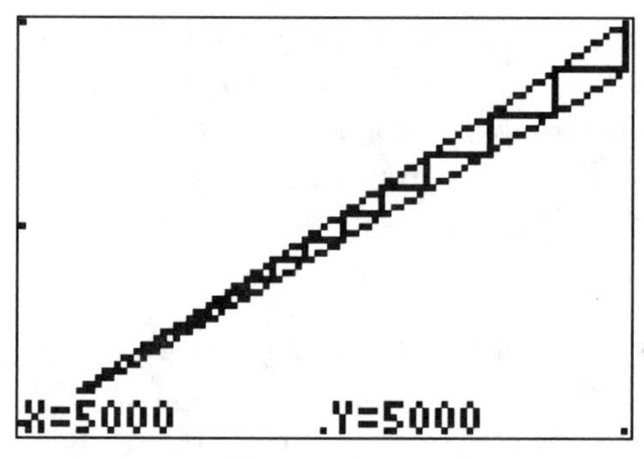

FIGURE 3.

Medicine. Suppose that a person takes a dose of medicine containing 16 units of a particular drug every four hours. Further, suppose that, on average, the person eliminates 25 percent of that drug through body functions every four hours. How many units of the drug would be in the person's body after the fourth dose? after the 10th dose? If a level of 80 units is considered to be an overdose warning stage, will this person ever be in danger? (See [**15**].)

Insect Control. In an attempt to reduce the spread of a pesticide-resistant fruit fly, sterilized male flies were released to mate with fertile females to cut the growth of the pest population. If the effect of this effort in a controlled environment is to reduce the overall population by 3% per month, what reduction in the population could be expected in half-a-year? in one year?

Investments. Which of the following is a better investment scheme to employ over a thirty year period? Invest $500 per month over the entire period at 5% interest compounded monthly or invest $1000 per month over the last 15 years at 6% interest compounded monthly.

4. The Structure of Difference Equations

First-order linear difference equations $f_n = Af_{n-1}+B$ have a rich theory that is carefully explicated in a number of sources [**8, 15**]. The search for a solution for the equation leads to a rich understanding of what "solution" means. Consider the difference equation $f_n = 2f_{n-1} + 1$, where $f_0 = 1$. Working backward from $f_n = 2(2f_{n-2}+1)+1$ with continued resubstitution and consolidation, one arrives at the formula

$$f_n = 2^{n+1} - 1.$$

This "solution" for the difference equation is a function of n that satisfies the original difference equation and the initial condition $f_0 = 1$. If the initial value f_0 is different from 1, then the situation is somewhat different. In this setting, $f_n = 2^n f_0 + 2n - 1$ satisfies the difference equation for any f_0. Hence, as in indefinite integration, where there is a family of solutions in the absence of a fixed initial condition, here too there is a family of functions that satisfy the difference equation; and when an initial condition is specified, there is in fact a unique solution [**8**]. In general, a first-order, linear difference equation of the form

$$f_n = Af_{n-1} + B,$$

with a given initial value of f_0, has the unique solution:

$$f_n = \begin{cases} A^n(f_0 + \frac{B}{A-1}) - \frac{B}{A-1} & \text{if } A \neq 1 \\ f_0 + nB & \text{if } A = 1. \end{cases}$$

Examination of this solution shows that in the case that $A = 1$ and $B = d$, the situation described by the difference equation is the general arithmetic sequence. If its initial term f_0 is a, its $(n+1)$st term f_n is given by $a + nd$. If, on the other hand, $A = r$ and $B = 0$, then we have the general geometric sequence, with the exception that when $A = r = 1$ the situation represents the constant arithmetic sequence with each term equal to r. When $A = r \neq 1$ and $f_0 = a$, then the $(n+1)$st term is given by ar^n. Thus, the above solution includes general forms for the nth terms of arithmetic and geometric sequences of $a+(n-1)d$ and $ar^n - 1$ as we normally see them in school mathematics.

The instructor who reviews the theory of difference equations [**8, 15**] will find that the subject provides a host of opportunities for innovative and explorative approaches in courses at a more advanced level. Spreadsheets can be used to calculate the first 15 values, for example, of a homogeneous first-order difference equation, that is, one where every term of the equation involves the recursive variable x_i for some value of i. Students can be asked to conjecture a general form for the general solution, as well as particular solutions given specific initial values, for such equations. Using the method of undetermined coefficients [**8**], forms can be generalized to describe general and specific solutions to non-homogeneous, first-order difference equations. Again, calculating the first 15 values of the non-homogeneous difference equations, students can use their solutions for their homogeneous portions and the "residuals" observed between these values and the calculated values, term-by-term, for the first 15 terms of the non-homogeneous equations. Fitting models to these "residuals," students can develop forms for the general and particular solutions to non-homogeneous, linear difference equations.

These procedures involve pattern recognition, modeling, and function knowledge, and exemplify a connected approach to doing mathematics. Students come to see the development of the theory as a unified process, rather than as the development of a series of specific formulas for solution forms for difference equations. The process can be further extended by examining the solution of the second order difference equations, using the characteristic equation and systems of first-order equations with elementary linear algebra, employing eigenvalues and eigenvectors. Each step up the ladder of difference equations is only a slight generalization of the process for modeling and solving problems at the lower level. Students get a chance to see how mathematics is developed, with the mathematics at one level supporting problem solving at the next level; thus, difference equations provide a strong model for illustrating a number of the issues central to a mathematics program based on the NCTM Standards.

5. Conclusion

The study of first-order, linear difference equations provides a strong thread through the curriculum, pulling many diverse topics together into a more cohesive whole. The nature of arithmetic and geometric patterns, viewed as sequences, can be analyzed in terms of the type of change taking place in each. A constant difference d signals an underlying arithmetic model of the form $y_n = y_{n-1}+d$. When the quotient of consecutive terms y_n/y_{n-1} is a constant ratio r, the model signaled is geometric of the form $y_n = ry_{n-1}$. In other settings, the determination of the type of model to describe the relationship between successive states of the situation requires considerably more effort and insight on the part of the modeler. The examination of the terms of each sequence through the iteration of the difference equations defining them allows the consideration of the relationships between these two dominant patterns. In both cases students have an opportunity to look at situations where time enters as a variable in discrete settings where it is generally easy both to compute and interpret the values.

The consideration of the rate of change in the middle grades, along with data and its graphic representation, provides an early introduction to the concepts which later will generalize into slope in linear equations in algebra and to the derivative in the study of curves in calculus. The movement to finding closed form functional representations for f_n as a function of n, starting with a difference equation, gives a new meaning to the nature of a "solution" for most students. This is especially powerful, as students note the difference between specific, or particular, and general solutions. This sets a solid base for later work with indefinite integrals and differential equations in the calculus. Finally, the cobweb graph approach to studying successive values resulting from difference equations provides strong insight into functions, their representations, and the long-term behavior of discrete processes.

The study of first-order linear difference equations can be used to connect the study of arithmetic and geometric sequences, and to set the stage for the concept of general and specific solutions to both difference and differential equations. But more than this, this class of difference equations provides a workplace for examining change and the nature of change.

References

[1] Bannard, D. N., "Making Connections Through Iteration", in [**10**].
[2] Boyer, C. B., *A History of Mathematics*, J. Wiley & Sons, New York NY, 1968.

[3] Cambridge Conference on School Mathematics. *Goals for School Mathematics: the Report of the Cambridge Conference on School Mathematics*, Houghton Mifflin, Boston MA, 1963.
[4] Chrystal, G., *Algebra: An Elementary Textbook for the Higher Classes of Secondary Schools and for Colleges*, 7th ed., 2 vols., Chelsea, New York NY, 1964 (original work published, 1886).
[5] Dossey, J. A., "Discrete Mathematics: The Math for our Time", in [**10**].
[6] ———, A. D. Otto, L. E. Spence, and C. Vanden Eynden, *Discrete Mathematics*, (2nd ed.). Harper Collins, New York NY, 1993.
[7] Garland, T. H., *Fascinating Fibonaccis*, Dale Seymour Publications, Palo Alto CA, 1987.
[8] Goldberg, S., *Introduction to Difference Equations*, J. Wiley & Sons, New York NY, 1958.
[9] Hall, H. S., and S. R. Knight, *Higher Algebra: A Sequel to Elementary Algebra for Schools*, Macmillan & Company, New York NY, 1964 (original work published, 1887).
[10] Hirsch, Christian R., and Margaret J. Kenney, eds. *Discrete Mathematics Across the Curriculum, K-12*, Yearbook of the National Council of Teachers of Mathematics, Reston VA, 1991.
[11] Hoggatt, V. E., Jr., *Fibonacci and Lucas Numbers*, Houghton Mifflin, Boston MA, 1969.
[12] National Council of Teachers of Mathematics, *Curriculum and Evaluation Standards for School Mathematics*, NCTM, Reston VA, 1989.
[13] ———, *Discrete Mathematics and the Secondary Mathematics Curriculum*, NCTM, Reston VA, 1990.
[14] Sandefur, J. T., Jr., "Discrete Mathematics: The Mathematics we all Need," In C. Hirsch and M. Zweng, eds., *The Secondary School Mathematics Curriculum*, NCTM, Reston VA, 1985.
[15] ———, *Discrete Dynamical Systems: Theory and Applications*, Oxford University Press, New York NY, 1990.
[16] Vorob'ev, N. N., *Fibonacci Numbers*, Blaisdel, New York NY, 1951.

Illinois State University, Normal, IL 61790
E-mail address: `jdossey@math.ilstu.edu`

DIMACS Series in Discrete Mathematics
and Theoretical Computer Science
Volume **36**, 1997

Discrete Mathematical Modeling In The Secondary Curriculum: Rationale and Examples from The Core-Plus Mathematics Project

Eric W. Hart

Discrete Mathematical Modeling in the Secondary Curriculum

Discrete mathematics is an important branch of mathematics that has been widely recommended for inclusion in the secondary curriculum [**13, 2, 6, 11**]. But which parts of discrete mathematics should be included, and how should they be incorporated? This article attempts to answer these two questions. The answers proposed have been applied as guidelines for weaving discrete mathematics into a new integrated high school mathematics curriculum: The Core-Plus Mathematics Project curriculum.

What Is The Core-Plus Mathematics Project?

The Core-Plus Mathematics Project (CPMP), funded by several grants from the National Science Foundation, is developing, field testing, and evaluating an integrated four-year high school mathematics curriculum. The first three years are designed to fulfill the mathematical needs of both college-bound and employment-bound students, while the fourth-year course focuses on the transition to college mathematics. The program features mathematical modeling, student investigation, integrated content, and appropriate use of the numeric, graphic, programming, and link capabilities of modern calculators.

Interwoven strands of algebra and functions, geometry and trigonometry, probability and statistics, and discrete mathematics are developed each year in the context of realistic applications. Connected 5-week units are comprised of several multi-day lessons centered on core mathematical ideas. Units are connected by common themes of data, shape, change, chance, and representation; by common topics such as matrices, symmetry, and

1991 *Mathematics Subject Classification.* Primary 00A05, 00A35.

curve-fitting; and by mathematical modes of thought such as reasoning, visualization, and recursive thinking. Guided student investigations lead to construction of important mathematics that makes sense to students and in turn enables them to make sense of new situations and problems.

Each course is developed through a 4-year cycle of writing, field testing, and revising.[1] Field testing is carried out in about 50 schools nationwide. As of Fall 1997, Courses 1 and 2 are completed and published by Everyday Learning Corporation, under the title *Contemporary Mathematics in Context.* Courses 3 and 4 will be completed in subsequent years.

Extensive evaluation research is being conducted in 36 schools around the country. Results from the first two years of the three-year national field test show that the curriculum is well-received and successful, as measured by surveys, case studies, student interviews, teacher reports, local tests, and national normed tests. For example, data show that CPMP students' growth on the mathematics portion of the Iowa Test of Educational Development (ITED) is significantly greater than that of comparable students in traditional curricula [**21**].

What is Discrete Mathematics?

Before deciding which parts of discrete mathematics to include in any curriculum, we need to know what discrete mathematics is.[2] There have been a variety of definitions given over the years: "Discrete mathematics is the mathematics of decision making for finite settings" ([**2**], p.1); "Discrete mathematics describes processes that consist of a sequence of individual steps" ([**4**], p. *xv*); "Discrete mathematics potentially involves the study of objects and ideas that can be divided up into 'separate' or 'discontinuous' parts" ([**3**], p. 1); "A good short answer contrasts 'discrete' topics with those that are 'continuous.' " ([**1**], p. *ix*).

The working definition of discrete mathematics in the CPMP curriculum is the following: Discrete mathematics consists of concepts and techniques for modeling and solving problems involving finite processes and discrete phenomena. More specifically, discrete mathematics deals with problems that involve enumeration, decision-making in finite settings, relationships among a finite number of elements, and sequential change. Central themes of discrete mathematics in CPMP are existence (Does a solution exist?), algorithmic problem solving (Can you efficiently construct a solution?), and optimization (Which solution is best?).

---

[1]Development team members are Christian Hirsch, director, Western Michigan University; Gail Burrill, University of Wisconsin, Madison; Arthur Coxford, University of Michigan; James Fey, University of Maryland; Eric Hart, Western Michigan University; Harold Schoen, University of Iowa; and Ann Watkins, California State University, Northridge. The Core-Plus Mathematics Project is supported by NSF grant no. MDR-9255257.

[2]Editors' note: See also the articles by Maurer and Rosenstein in Section 3 of this volume, which address the issue of defining discrete mathematics.

Which Discrete Mathematics Topics Should Be Included in the Secondary Curriculum?

The list of topics and areas that have been listed under the title "discrete mathematics" is quite long indeed, including graph theory, game theory, difference equations (also called recurrence relations), combinatorics, operations research, management science, logic, algorithms, matrices, applied modern algebra, finite probability, coding theory, linear programming, and so on. What from this list should be included in the high school mathematics curriculum? This question clearly has many possible answers, but there is some consensus.

The NCTM Curriculum Standards [**13**] recommends the following topics: graph theory, matrices, sequences, recurrence relations, algorithms, systematic counting, finite probability, and linear programming. A number of NSF discrete-mathematics teacher-enhancement projects of the 1990's — for example, those directed by Hart [**8**], Kenney [**10**], Rosenstein [**17**], and Sandefur [**18**], have had a content focus on graph theory, iteration and recursion, social choice theory, matrices, and combinatorics. These same five major topics make up most of the first published high school discrete mathematics text [**1**].

Roughly the same topics are developed in the Core-Plus Mathematics Project curriculum. In particular, the CPMP curriculum includes:

- **a:** graph theory — using vertex-edge models to study relationships among a finite number of elements, as in a transportation network or a predator-prey food web;
- **b:** social choice theory — such as the mathematics of voting, fair division, apportionment, and cooperation and competition;
- **c:** combinatorics — systematic counting;
- **d:** recursion — the method of describing sequential change by indicating how the next stage of a process is determined from previous stages; and
- **e:** matrices — used to represent and solve problems from a variety of real-world settings while connecting important ideas from different strands of mathematics.

Before continuing the discussion of what discrete mathematics to include and how to include it, we should consider a more fundamental question.

Why Should Discrete Mathematics Be Part of the High School Curriculum?

The high school mathematics curriculum is already quite full. Why should we make room for discrete mathematics? There are at least three compelling reasons: it's good mathematics; it's useful mathematics; and it's pedagogically powerful mathematics.

Discrete mathematics is good mathematics. The five major topics mentioned above are thriving, active research areas in mathematics. The mathematics is pretty and profound. Studying discrete mathematics will give students a broader view of the whole field of mathematics.

Discrete mathematics is useful mathematics. Discrete mathematics provides rich and powerful mathematical models that are invaluable for making sense of the world we live in. Applications of these five topics come from a variety of settings, including project management, communication networks, scheduling, routing, manufacturing, lotteries, voting, fair division, finance, population growth, inventory control, wildlife management, and social relations, to name just a few. As Dossey states, "Discrete mathematics is the math of our time" ([**3**], p. 1).

Discrete mathematics is pedagogically powerful mathematics. Discrete mathematics is a potent vehicle for achieving many goals of mathematics education. In the course of investigating important concepts in discrete mathematics, students learn and apply powerful habits of mind, like mathematical modeling, algorithmic problem solving, and recursive thinking. Many students' old beliefs about mathematics are challenged and changed. They see that mathematics is active and alive, as they come face-to-face with some of the current frontiers of mathematics in, for example, graph theory or modeling with recursion. They see that mathematics is useful and modern, as they study the ubiquitous contemporary applications of discrete mathematics. Even the beliefs students may have about their own ability to learn mathematics can be dramatically changed when they find that many discrete mathematics topics are new and accessible and do not have a plethora of technical prerequisites.

How Should Discrete Mathematics be Incorporated Into the Secondary Curriculum?

Two complementary answers are proposed here. Discrete mathematics should be woven into an overall integrated mathematics curriculum, and the emphasis should be on discrete mathematical modeling. Both approaches are implemented in the CPMP curriculum, as now discussed.

Specific units focusing on discrete mathematics are woven into each of the integrated CPMP courses. For example, there are units entitled *Graph Models, Matrix Models, Network Optimization, Discrete Models of Change, Modeling Public Opinion*, and *Counting Models*, all of which are connected to each other as well as to the other CPMP units. In addition, topics and themes of discrete mathematics, such as matrices, recursive thinking, optimization, and algorithmic problem solving, permeate the entire curriculum. Finally, the discrete mathematics strand of the CPMP curriculum is connected to the other strands of algebra and functions, geometry and trigonometry, and statistics and probability by the common themes of data, shape, representation, and change.

The emphasis throughout the CPMP discrete mathematics strand is on discrete mathematical modeling. Students are engaged, both in groups and individually, in making sense of realistic situations by constructing, operating on, analyzing, and interpreting discrete mathematical models. The rest of this article is devoted to two examples of discrete mathematical modeling in the CPMP curriculum.

Discrete Mathematical Modeling in CPMP

Two examples are presented here. One example involves modeling with recursion, taken from a unit in Course 3 entitled *Discrete Models of Change*, and the other uses vertex-edge graph models, from the *Graph Models* unit in Course 1. Both are pulled out of context from the complete units in which they appear, but they give a flavor of discrete mathematical modeling in the CPMP curriculum.

Each example also illustrates the style of active learning and teaching that is characteristic of the CPMP curriculum. Lessons are launched with a brief whole-class brainstorming session related to a given real-world situation. This is built into the curriculum materials, as seen below, in boxes entitled "Think About This Situation." The goal is to motivate the lesson and get some important questions on the table. Next, students work in teams on guided investigations related to the launching situation. Through these investigations students explore and apply important mathematical concepts and methods. The investigations are followed by a "Checkpoint" section. The checkpoint consists of a few questions that summarize the lesson so far. Teachers typcially lead a whole-class discussion of the checkpoint questions. The goal is to provide a class-generated summary and to ensure that students have indeed learned the targeted concepts and methods. Finally, to make sure that students can apply what they have learned on their own, there is an additional brief investigation entitled "On Your Own." These features of the CPMP curriculum are illustrated in the following sample student investigations (addressed to a student reader).

EXAMPLE 1: Modeling Sequential Change Using Recursion

We live in a changing world. Mathematics can be used to help describe and understand patterns of change. Examples you have already studied include using equations, tables, and graphs to investigate linear and exponential patterns of change, using coordinates and matrices to study geometric transformations, and using trigonometry to study periodic change. Another important pattern of change is sequential change, for example, change from year to year. You have already used equations involving the words NOW and NEXT to study this type of change. In this unit you will continue the study of sequential change.

Think About This Situation

Wildlife management has become an increasingly important issue as modern civilization puts greater demands on wildlife habitat. As an example, consider a fishing pond that is stocked from a nearby hatchery. Suppose you are in charge of managing the fish population in the pond.

a: What are some factors to consider in managing the fish population in the pond? List as many factors as you can.

b: How could you figure out the current size of the fish population?

c: Why would it be useful to be able to predict the year-to-year changes in the fish population? Why would knowledge of long-term population changes be useful?

INVESTIGATION 1.1: Modeling Population Change. In this lesson you will build and use a mathematical model to help you predict the changing fish population.

1. So far you have very little information about the fishing pond that you are supposed to manage. What additional information do you need in order to predict changes in the size of the fish population over time? Make a list.
2. A typical first step in mathematical modeling is simplifying the problem and deciding on some reasonable assumptions. Three pieces of information that you may have listed above are: the initial size of the fish population in the pond, the annual growth rate of the population, and the annual restocking amount, that is, the number of fish that are added to the pond each year. For the rest of this investigation, use just the following three assumptions:
   - There are 3000 fish currently in the pond.
   - Regardless of restocking, the population decreases by 20% each year due to the combined effect of all causes, including natural births and deaths and fish being caught.
   - 1000 fish are added at the end of each year.

   Using these assumptions, you can build a mathematical model to analyze the population growth in the pond.
   (a) Estimate the fish population in the pond after 1 year. After 2 years.
   (b) What is the population after 3 years? Explain how you figured it out.
   (c) Write an equation using the words NOW and NEXT that models this situation.
   (d) Use the equation from part (c) and the ANS key on your calculator to find the population after 7 years. Explain how the

keystrokes you use on the calculator correspond to the words NOW and NEXT in the equation.

3. Think about the long-term population of fish in the pond.
   (a) Do you think the population will grow without bound? Level off? Die out? Make a conjecture about the long-term population.
   (b) Compute the long-term population. Was your conjecture correct? Explain, in terms of the fishing pond ecology, why the long-term population you have computed is reasonable.
4. Does the fish population change faster around year 5 or around year 25? How can you tell?

INVESTIGATION 1.2: What if ...? Think about what happens to the long-term population if certain conditions change.

1. What are the three key conditions in this problem?
2. What do you think will happen to the long-term population if the initial population is different but all other conditions remain the same? Make an educated guess, and then check your guess by finding the long-term population when the initial population is 0, 2000, 4000, and 10,000. Describe the pattern of change in long-term population as initial population varies.
3. What happens to the long-term population if the annual re-stocking amount is 500, and all other conditions are as in the original assumptions? How about if the annual re-stocking amount is 2000? 4000? Describe the relationship between long-term population and re-stocking amount.
4. What happens to the long-term population if the annual decrease rate is 10%, and all other conditions are the same as in the original assumptions? How about if the annual decrease rate is 40%? 60%? Describe any patterns that you see.
5. Now consider the case where conditions are such that the fish population shows an annual rate of *increase*.
   a: What do you think will happen to the long-term population if the population *increases* at an annual rate? Make a conjecture and then test it by trying at least two different annual increase rates.
   b: Write equations using NOW and NEXT that represent your two test cases.
   c: Do you think it is reasonable to model the population of a fish in a pond with an annual rate of increase? Why or why not?

√ Checkpoint

Consider this equation: NEXT = 0.6 NOW + 1500.

a: Describe a fish population problem that could be modeled by this equation.

b: What additional information is needed to be able to use this equation to predict long-term population?

c: What additional information is needed to be able to use this equation to predict the population in 5 years?

d: For a fish population situation modeled by an equation like the one above:

- If the initial population doubles, what will happen to the long-term population?
- If the annual re-stocking amount doubles, what will happen to the long-term population?
- If the annual population decrease rate doubles, what will happen to the long-term population?

e: How would you modify the equation above if it is to represent a situation where the fish population increases annually at a rate of 15%? What effect does such an increase rate have on the long-term population?

Be prepared to share your group's thinking with the entire class.

The fish population problem you have investigated involves sequential change, since the change takes place sequentially or step-by-step. In this case the step-by-step change in population is recorded year-by-year. You have used the terms NOW and NEXT to describe the sequential change. This method of describing the next step in terms of previous steps is called **recursion**. Situations involving sequential change are sometimes called **discrete dynamical systems**. A discrete dynamical system is a situation (system) involving change (dynamical), where the nature of the change is step-by-step (discrete). An important part of analyzing discrete dynamical systems is determining long-term behavior, like what you did when you found the long-term population of fish.

On Your Own. A hospital patient is taking an antibiotic to treat an infection. He initially takes a 30mg dose, and then takes another 10mg at the end of every six hour period thereafter. Through natural body metabolism 20% of the antibiotic is eliminated from his system every six hours.

a: Estimate the amount of the antibiotic in his system after the first six hours. After the second six hours.

b: Write an equation using the words NOW and NEXT that models this situation.

c: Find the amount of antibiotic in his system after two weeks.

d: His doctor decides to modify the prescription so that the long-term amount of antibiotic in his system will be about 25 mg. How should the prescription be modified?

Note to the Teacher: Some subsequent investigations in this unit include the following ideas and topics[3]:

- Continued development and use of subscript notation;
- A discrete (recursive) view of linear, exponential, and power functions, through investigation of arithmetic sequences, geometric sequences, and finite differences;
- Function iteration, including some analysis of fixed points and cycles;
- Graphical analysis (including "cobweb" diagrams);
- Much of the present work with recursion is summarized and formalized in terms of affine recurrence relations, $A_n = rA_{n-1} + b$.

EXAMPLE 2: Managing Conflicts with Vertex-Edge Graphs

Have you ever noticed how many different radio channels there are? Each radio station has its own transmitter which broadcasts on a particular channel, or frequency. The Federal Communications Commission (FCC) makes sure that the broadcast from one radio station does not interfere with the broadcast from any other radio station. This is done by assigning an appropriate frequency to each station. The FCC requires that if two stations are within transmitting range of each other, they must use different frequencies. Otherwise, you might tune into "ROCK 101.7" and get Mozart instead!

Think About This Situation

Seven new radio stations are planning to start broadcasting in the same region of the country. The FCC wants to assign a frequency to each station so that no two stations interfere with each other. The FCC also wants to assign the fewest possible number of new frequencies.

a: What factors need to be considered before the frequencies can be assigned?

b: What method can the FCC use to assign the frequencies?

INVESTIGATION 2.1: Building a Model. Suppose that because of geographic conditions and the strength of each station's transmitter, the FCC determines that stations within 500 miles of each other must be assigned different frequencies, otherwise their broadcasts will interfere with each other. The location of the seven stations is shown on the grid in Figure 1. A side of each small square on the grid represents 100 miles.

[3]Editors' note: See the article by Dossey, in this volume, for a further discussion of sequences, iteration, and recurrence relations.

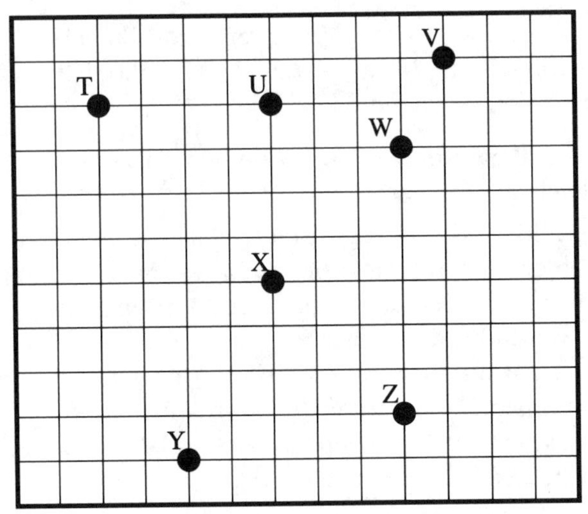

FIGURE 1. Scale: grid lines are 100 miles apart.

1. Working on your own, figure out how many different frequencies are needed for the seven radio stations. Remember that stations 500 miles or *less* apart must have different frequencies, but stations more than 500 miles apart can use the same frequency. *Try to use as few frequencies as possible.*
2. Compare your answer with others in your group.
   (a) Did everyone use the same number of frequencies? Reach agreement in your group about the minimum number of frequencies needed for the seven radio stations.
   (b) Given two particular stations, is it possible that one person assigns them the same frequency and another person assigns them different frequencies, and yet both assignments are acceptable? Explain.

   It is possible in this case to find the minimum number of frequencies by trial and error. However, a more systematic method is needed for more complicated situations, such as when there are many more radio stations. One way to solve this problem more systematically is to model the problem with a graph similar to those in the previous lesson. Remember, to model a problem with a graph, you must first decide what the vertices and edges represent.
3. Working on your own, begin modeling this problem with a graph.
   (a) What should the vertices represent?
   (b) How will you decide whether or not to connect two vertices with an edge? Complete this statement:

   Two vertices are connected by an edge if . . .

   (c) Now that you have specified the vertices and edges, draw a graph for this problem.

4. Compare your graph with others in your group.
   (a) Did everyone in your group define the vertices and edges in the same way? Discuss any differences.
   (b) For a given situation, suppose that two people define the vertices and edges in two different ways. Is it possible that both ways accurately represent the situation? Explain your reasoning.
   (c) For a given situation, suppose that two people define the vertices and edges in the same way. Is it possible that their graphs could look different but both be correct? Explain your reasoning.
5. A common choice is to let vertices represent the radio stations. Edges might be thought of in two ways, as described in parts (a) and (b) below.
   (a) You might connect two vertices by an edge whenever the stations they represent are 500 miles or *less* apart. Did anyone in your group do this? If not, draw a graph where two vertices are connected by an edge whenever the stations they represent are 500 miles or *less* apart.
   (b) You might connect two vertices by an edge whenever the stations they represent are *more* than 500 miles apart. Did anyone in your group do this? If not, draw a graph where two vertices are connected by an edge whenever the stations they represent are *more* than 500 miles apart.
   (c) Compare the graphs from parts (a) and (b).
      - Are both graphs accurate ways of representing the situation?
      - Which graph do you think will be more useful and easier to use as a mathematical model for this situation? Why?
6. For the rest of this investigation, you will use the graph where edges connect vertices that are 500 miles or *less* apart. Make sure you have a neat copy of this graph.
   (a) Are vertices (stations) X and W connected by an edge? Are they 500 miles or less apart? Will their broadcasts interfere with each other?
   (b) Are vertices (stations) Y and Z connected by an edge? Will their broadcasts interfere with each other?
   (c) Compare your graph to the graph in Figure 2.
      - Does this graph also represent the radio-station problem?
      - What criteria can you use to decide if two graphs both represent the same situation?
7. So far you have a model that shows all the radio stations and which stations are within 500 miles of each other. The goal is to assign frequencies so that there will be no interference between stations. You still need to build the frequencies into the model. So, as the last

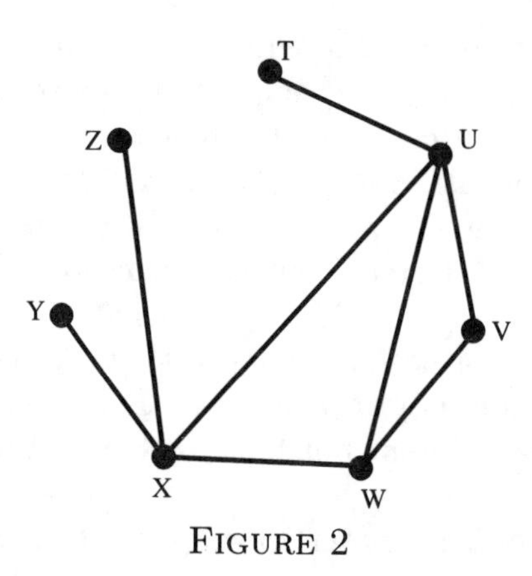

FIGURE 2

step in building the graph model, represent the frequencies as **colors**. To **color a graph** means to assign colors to the vertices so that two vertices connected by an edge have different colors.

You can now think about the problem in terms of *coloring the vertices of a graph.* The table below contains statements about stations and frequencies in the left-hand column and corresponding statements about vertices and colors in the right-hand column. Write statements to complete the right-hand column of the table.

Statements about stations and frequencies	Statements about vertices and colors
Two stations have different frequencies.	Two vertices have different colors.
Find a way to assign frequencies so that stations within 500 miles of each other get different frequencies.	
Use the smallest number of new frequencies	

8. Now color your graph for the radio-station problem. That is, assign a color to each vertex so that any two vertices that are connected by an edge have different colors. Try to use as few colors as possible. You can use colored pencils or just the names of some colors to do the coloring. Color or write a color name next to each vertex.
9. Compare your coloring with that of another group.
   (a) Are both colorings legitimate? That is, do they satisfy the condition that vertices connected by an edge must have different colors?
   (b) Do both colorings use the same number of colors to color the vertices of the graph?
   (c) Reach agreement about the minimum number of colors needed. Explain, in writing, why the graph cannot be colored with fewer colors.

(d) Given two particular vertices, is it possible that one group assigns them the same color and another group assigns them different colors, and yet both assignments are acceptable? Why or why not?

(e) What is the connection between graph coloring and assigning frequencies to radio stations? As you answer this question, compare the results of this activity to those in Activity 2.

10. Think about the strategy you used in Activity 8 to color the radio-station graph with as few colors as possible.

(a) Write down a step-by-step description of your coloring strategy. Write the description so that your strategy can be applied to graphs other than just the radio-station graph.

(b) Use the description of your strategy to color a copy of the graph in Figure 3.

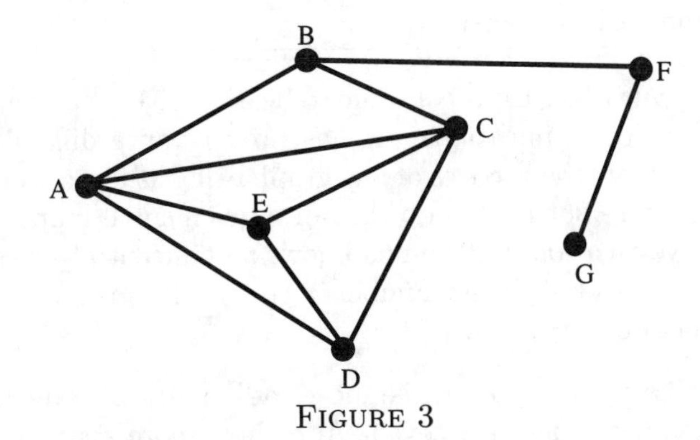

FIGURE 3

(c) Refine the directions for your coloring strategy so that any one of your classmates could follow the directions.

11. Exchange your written coloring directions with another group. Then do the following:

(a) Use the other group's directions to color a second copy of the graph in Figure 3. The other group will be doing the same thing with your directions.

(b) Compare your colorings with the other group's colorings.

(i) Are they the same?

(ii) Are they each legitimate colorings?

(iii) Do they each use the least number of colors possible? Reach agreement with the other group about the minimum number of colors needed to color the graph.

(c) Discuss any problems that came up with either group's coloring directions. If necessary, rewrite your directions so that they work better and are easier to follow.

As you saw in the previous lesson, a careful list of directions for carrying out a procedure is called an *algorithm*. Designing and applying algorithms is an important method for solving problems. There are many possible algorithms for coloring the vertices of any graph, including the ones you developed.

√ Checkpoint

a: What do the vertices, edges, and colors represent in the graph model that you have been using for the radio-station problem?
b: How does "coloring a graph" help solve the radio-station problem?
c: In what ways can two graphs differ and yet still both accurately represent a given situation?
d: What are some strengths and weaknesses of the graph-coloring algorithm created by your group?

Be prepared to share your group's thinking and coloring algorithm with the entire class. Decide as a class on algorithms that seem most efficient and easily understood.

Graph-coloring algorithms continue to be an active area of mathematical research with many applications. It has proven quite difficult to find an algorithm that colors the vertices of any graph using as few colors as possible. You can often figure out how to do this for a given small graph, as you have done in this investigation, but no one knows an efficient algorithm that will color *any* graph with the *fewest* number of colors. This is a famous unsolved problem in mathematics.

On Your Own: Suppose three more radio stations want to move into the same region with the other seven. Add three more stations to a copy of the grid on page 30 so that at least two of them are within 500 miles of one of the existing seven stations. Then use graph coloring to assign frequencies to all ten stations so that their broadcasts do not interfere with each other.

Note to the Teacher: After this initial vertex-coloring investigation, students then apply vertex coloring to solve a variety of other problems, such as scheduling meetings without conflicts, setting up efficient tour schedules, coloring maps, and setting up an optimal emergency evacuation plan for a hospital.[4] Several other graph theory topics are studied in the CPMP curriculum. These topics, along with a sample application, are listed below.

- Euler paths (find efficient snow plow routes)
- Critical paths and PERT charts (schedule large projects)
- Hamiltonian paths (rank players in a a tournament)
- Shortest paths (measure degree of influence within a social group)
- Minimal spanning trees (set up efficient computer networks)
- Traveling salesperson type problems (manufacture integrated circuit boards)

[4]Editors' note: For another application of graph coloring, see the article by Picker in this volume.

Conclusion

The idea of including discrete mathematics in the secondary curriculum has been discussed for many years. The basic questions of what, when, and how are still tough questions that need to be carefully considered. This article has proposed one set of answers to these questions. The answers have been put into practice by weaving discrete mathematics into a new high school curriculum, the Core-Plus Mathematics Project curriculum. Discrete mathematics can and should be in the same league with our old friends algebra, geometry, trigonometry, statistics, and probability.

References

[1] Crisler, Nancy, Patience Fisher, and Gary Froelich, *Discrete Mathematics Through Applications*, W. H. Freeman and Company, New York, 1994. See also "A Discrete Mathematics Textbook for High Schools," by the same authors, in this volume.

[2] Dossey, John, *Discrete Mathematics and the Secondary Mathematics Curriculum*, NCTM, Reston VA, 1990.

[3] ———, "Discrete Mathematics: The Math for Our Time." in *Discrete Mathematics Across the Curriculum, K-12, 1991 Yearbook of the National Council of Teachers of Mathematics*, Christian R. Hirsch and Margaret J. Kenney, eds. NCTM, Reston VA, 1991, pp. 1-9.

[4] Epp, Susanna S., *Discrete Mathematics with Applications*, Wadsworth Publishing Company, Belmont CA, 1990.

[5] Garfunkel, Solomon, et al., *For All Practical Purposes: Introduction to Contemporary Mathematics*, 3rd edition., W. H. Freeman and Company, New York, 1994.

[6] Hart, Eric W., "Discrete Mathematics: An Exciting and Necessary Addition to the Secondary School Curriculum", in [**11**], pp. 67-77.

[7] ———, James Maltas, and Beverly Rich, "Implementing the NCTM Standards: Discrete Mathematics," *The Mathematics Teacher*, May, 1991, pp. 362-7.

[8] ———, and Harold Schoen, Directors of the NSF Teacher Enhancement Project: "Teachers As Leaders: Launching Mathematics Education Into the Nineties," 1989-93.

[9] Hirsch, Christian R., Arthur F. Coxford, James T. Fey, and Harold L. Schoen, "Core-plus mathematics: Teaching sensible mathematics in sense-making ways," *The Mathematics Teacher*, Nov. 1995, pp. 694-700.

[10] Kenney, Margaret, Director of the NSF Teacher Enhancement Project: "Implementing the NCTM Standard in Discrete Mathematics," 1992-97.

[11] ———, and Hirsch, Christian R., eds., *Discrete Mathematics Across the Curriculum, K-12*, Yearbook of the National Council of Teachers of Mathematics, NCTM, Reston VA, 1991.

[12] Malkevitch, Joseph and Walter Meyer, *Graphs, Models, and Finite Mathematics*, Prentice Hall, Englewood Cliffs NJ, 1974.

[13] National Council of Teachers of Mathematics, *Curriculum and Evaluation Standards for School Mathematics*, NCTM, Reston VA, 1989.

[14] ———, *Professional Standards for Teaching Mathematics*, NCTM, Reston VA, 1991.

[15] ———, *Assessment Standards for School Mathematics*, NCTM, Reston VA, 1995.

[16] Roberts, Fred, *Applied Combinatorics*, Prentice Hall, Englewood Cliffs NJ, 1984.

[17] Rosenstein, Joseph G., Director of the NSF Teacher Enhancement Project: "Leadership Program in Discrete Mathematics," 1989-97. See also "The Leadership Program in Discrete Mathematics," by Joseph Rosenstein and Valerie DeBellis, in this volume.

[18] Sandefur, James T., Director of the NSF Teacher Enhancement Project: "Leadership Training Institute in Dynamical Modeling." 1990-94.

[19] ______ "Drugs and Pollution in the Algebra Class," *The Mathematics Teacher*, February, 1992, pp. 139-145.

[20] ______ *Discrete Dynamical Modeling*, Oxford University Press, New York, 1993.

[21] Schoen, H. L., and S. W. Ziebarth, "High School Mathematics Curriculum Reform: Rationale, Research, and Recent Developments." in Hiebowitsh, Peter S., and William G. Wraga (eds.) *Annual Review of Research for School Leaders*, Macmillan, New York, (in press).

[22] Tannenbaum, Peter and Robert Arnold, *Excursions in Modern Mathematics*, 2nd ed., Prentice Hall, Englewood Cliffs NJ, 1995.

[23] Tobt, B. and T. Jensen, *Graph Coloring Problems*, Wiley, New York, 1995.

[24] Wilson, Robin J. and John J. Watkins, *Graphs: An Introductory Approach*, Wiley, New York, 1990.

WESTERN MICHIGAN UNIVERSITY, DEPARTMENT OF MATHEMATICS AND STATISTICS
Current address: 613 S. 2nd St., Fairfield, IA 52556
E-mail address: `ehart@mum.edu`

DIMACS Series in Discrete Mathematics
and Theoretical Computer Science
Volume **36**, 1997

A Discrete Mathematics Experience with General Mathematics Students

Bret Hoyer

1. The Need

- "When are we ever gonna use this?"
- "This is dumb; why do we have to do this?"
- "Math is hard!"
- "I hate math!"
- "Math!? Yuck!!"

These are some of the many, many comments I have heard students and students' parents make with regard to mathematics. Our mathematics department has struggled each year to get students excited enough to take three or four years of mathematics. Sustaining student interest is a terrific challenge. Students aspiring to attend college would take at least three years of mathematics — up to Algebra II, which would be taken typically in the Junior year. After this point, many students would choose either not to take a mathematics course during their Senior year, or to take Senior Advanced Math, only to drop after the first semester. We wanted to address this problem, as well as another problem: what to do with students who don't feel comfortable with Algebra II or Geometry? These students don't want to take Consumer/Applied Math because they don't feel challenged, and they fear the stereotype that usually accompanies that course. We were searching for ideas and/or materials to address both of these issues.

2. Goals and Objectives

We took a step towards solving this problem after we learned of the textbook *For All Practical Purposes: An Introduction to Contemporary Mathematics* [**1**]. I was introduced to this material in the University of Iowa's

---

1991 *Mathematics Subject Classification.* Primary 00A05, 00A35, 05C45.

"Teachers as Leaders" project during the summer of 1990.[1] My project group focused on the *For All Practical Purposes* materials. My group and the members of our mathematics department were very impressed with the twenty-six half-hour videos[2] that complement the text. Our students thought the idea of watching videos in a mathematics class was bizarre, but they didn't object. We felt that the material allowed for a wide range of student ability levels. We also felt that the material would be accessible to students who had some experience with high school geometry. I began by teaching one of the units, "Street Networks" (Euler circuits/paths), to my Algebra II and Advanced Math classes. My goals were as follows.

1. Engage the students in problem-solving activities that they would find "fun".
2. Introduce a discrete mathematics topic and its application to the real world.
3. Discuss the concept of an optimal solution when multiple solutions are possible.
4. Incorporate cooperative learning activities into a mathematics class.
5. Give students experience with non-routine problem-solving within a structured unit.
6. Engage those students who are starting to slide into the "I just want to get through Algebra II so I don't have to take any more math" mode.
7. Take a break from the traditional textbook during mid-February, when school days in Iowa become very long.

The students thoroughly enjoyed the material and, to my surprise, found it very easy. Students were actually doing mathematics during other classes! One particular student had covered his English notes with graphs such as that in Figure 1, in his attempts to find an Euler circuit.

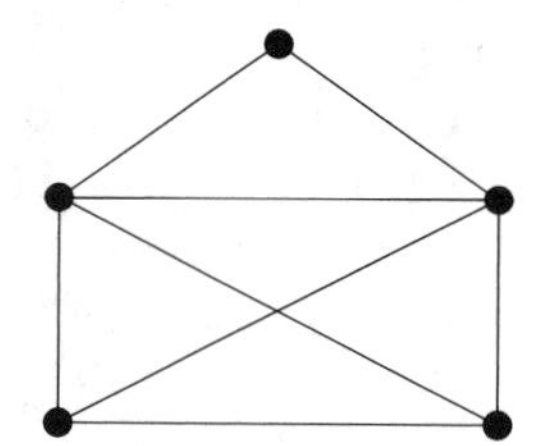

FIGURE 1. A graph without an Euler circuit

I shared my classroom experiences with my colleagues. Our Consumer/ Applied Math teacher was frustrated with her materials because her students were "bored". I suggested that she give these materials a try. She enjoyed

[1] Editors' note: This was an NSF Teacher Enhancement Program directed by Eric W. Hart and Harold Schoen, "Teachers as Leaders: Launching Math Education into the Nineties", 1989–93.

[2] The videos were put together by COMAP with funding from the Annenberg/CPB Project and the Carnegie Corporation of New York.

tremendous success. The students were excited! They did their homework and participated in class! They learned the material because it was fun and they saw the connection between the theory and the real world. The videos played a major role in helping the students see the connection. Each chapter has numerous exercises, and the teacher's manual gives additional exercises that are accessible to students with low ability levels. The text also includes exercises (and proofs) that will challenge the brightest Calculus students. We convinced our administration to offer a new course (we called it Contemporary Mathematics) during the 1992-93 school year. We crossed our fingers, hoping there would be enough interest to sustain the course. The Consumer/Applied Math teacher mentioned earlier has thoroughly enjoyed teaching the material. I have had the opportunity to weave the unit on "Street Networks" into my Algebra I, Geometry, and General Math courses. I have also enjoyed a great deal of success; many parents at our open house last fall shared with me comments about their children's excitement.

3. Teaching Strategies

The only real difficulty we have faced is the reading level of the material. The textbook was designed for a course that would meet a mathematics requirement in a liberal arts college. In this section, I present cooperative learning activities which help address this issue. To facilitate these cooperative learning activities, I have round tables in my classroom. Here are some of the techniques I use.

Jigsawing. I use jigsawing to give all the students an opportunity to practice teaching. I first divide the students into base groups of three or four students, and assign different exercises or activities to the members of each base group. Then, I form new solution groups consisting of students who have the same exercise. Once the solution groups have completed their exercises, the students all return to their base groups, where they teach the other members of their base group what they have learned.

KWL. In the KWL strategy, the "K" stands for "know", the "W" stands for "want", and the "L" stands for "learned". Each base group identifies what they "know" about an assignment or topic. After a few minutes, they identify what they "want" to know about the homework or topic. After a couple more minutes two people from each base group are selected to travel to another group, for comparison of "know" and "want" lists. After comparing notes, most of the "want" list questions should be answered. Students then return to their base groups to summarize what they have "learned".

Think-pair-share. In this strategy, the students work individually on an exercise (think), then partner up (pair) to share their solutions.

4. A Sample Lesson—Street Networks

I decided that the best way to approach the street networks material with my freshman General Math students was to forget about the textbook. I also approached it a little bit differently than the text in that I started with a puzzle. I chose not to start with a real-world problem and model it with a graph because these particular students respond very positively to puzzles and games. Also, these students found it difficult (initially) to model real problems as graphs. I wanted to ease them into the mathematics so that their curiosity and excitement remained intact.

I started out by challenging my students to trace the edges of a graph like the one in Figure 1 without lifting their pencils or retracing any edges, but returning to their starting point. I told them that such a tracing was called an *Euler circuit.* Many of my students thought they had a solution to the puzzle, but discovered that they didn't when they tried to display it on the board. After a few attempts, a few of them announced that there was no solution. After about five minutes, I told them that there was in fact no solution to this puzzle (no Euler circuit for the graph). Next, we tried to find Euler circuits in the graphs in Figure 2. I have found, to a surprising degree, that the students develop excellent informal definitions and conjectures in their groups. All I have to do is give the students "enough" concrete examples.

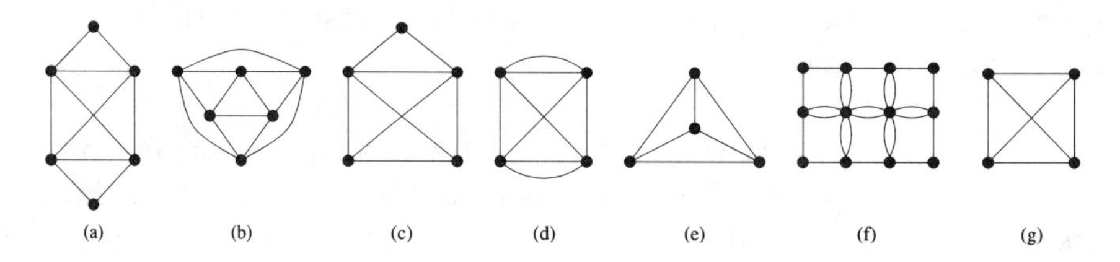

FIGURE 2. Some graphs for Euler circuit exercises: only graphs (a), (b), (d), and (f) have Euler circuits.

I showed them what the terms edge, vertex (vertices), and degree meant. We listed the degrees of the vertices on each graph and classified the graphs as either having or not having an Euler circuit. I asked the students to independently write down a rule for determining whether a graph has an Euler circuit. Many came up with valid ideas, but did not state a complete rule, for example, "If it is even, it will work," or "If the degree is an odd # it is not an Euler circuit." Students then discussed all of the individual rules in groups of four or five. Each group then came up with one rule. Different groups came up with one of the following two rules:

- All of the vertices must have an even degree for the graph to have an Euler circuit.
- If the graph has any vertices with odd degree, then it doesn't have an Euler circuit.

I explained that the two rules were equivalent.

I used the same method to present Euler *paths* (tracings which cover each edge exactly once but may not return to their starting points). We listed the same graphs, and found that only one had an Euler path which was not also a circuit. Eventually they discovered a rule for a graph with no Euler circuit to have an Euler path—exactly two vertices of odd degree. I then offered the graphs in Figure 3 and pointed out that none of them has an Euler path, although each has exactly two odd-degree vertices.

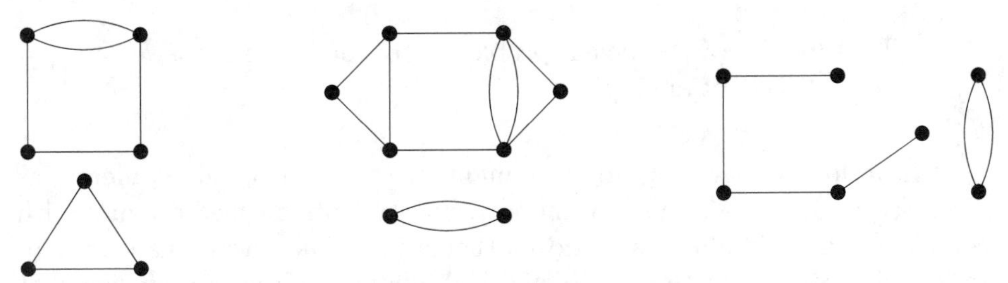

FIGURE 3. Three disconnected graphs

The students objected that these examples weren't graphs, because they weren't "connected"! I explained to them the concept of connectedness, and we revised our rules for the existence of Euler circuits and paths.

The next step in the unit was to apply Euler paths and circuits to a real-world situation. The following example is from pilot materials from the University of Iowa's Core Plus Mathematics Project [**2, 3**]. The diagram in Figure 4 represents a school floor plan. The black squares represent lockers.

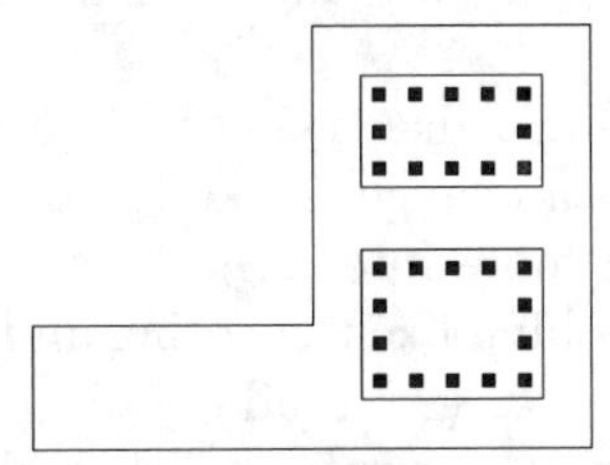

FIGURE 4. Floor plan of a school

The task is to decide whether you can paint all the lockers without retracing your steps with the heavy equipment. The challenge is to model the problem using a graph. The suggested answer was to treat portions of the hallways as vertices, and treat rows of lockers as edges (Figure 5b). Some of the students placed vertices at *each* hallway intersection, but still represented rows of lockers by edges (Figure 5c). Other students felt that the hallways, rather than the rows of lockers, should become the edges of the graph (Figure 5d), perhaps assuming that one could paint lockers on both sides of a hall at once.

We decided that determining what the edges represent is vital in modeling a real world situation with a graph.

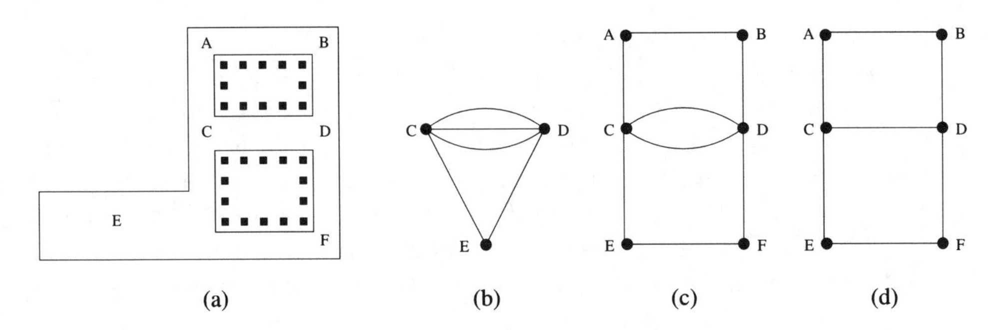

FIGURE 5. (a) Labeled problem, (b) suggested answer, (c)-(d) student answers

I have found that this unit met many of the needs of my students, as discussed in Sections 1 and 2. Generally, my students finished the unit with confidence in their abilities to do mathematics. They were challenged to reason and communicate in mathematical language. They recognized that what they were doing was important and of value.

I asked students to compare the units they completed on street networks and voting theory to previous work they had done in General Math. Here are a few of their responses.

- I like voting theory more than a lot of things that I have done in the past. I think it was fun. Street networks is fun too. I like working on these types of things because they make me think. I like these units more than a lot of the others. These units were easier for me than the others, they went a lot slower so I could keep up easier. (9th grade boy)
- I like the graph unit. I like trying to figure out if it is Euler circuit or not. This unit is a change. I like the graph unit better the voting because it is much more interesting and much fun and challenging. Pre Algebra was nothing like the voting and street networks. I'm glad I found something new. (9th grade boy)
- I do enjoy doing the theorys better than doing our books. It seems to help me more with my math. I am also learning more so I would encourage people to learn from the theorys other than just the mathematics book. (10th grade girl)

5. Assessment Procedures

Some of the procedures I use for assessment include paragraph responses, daily homework, individual projects, group projects and presentations, individual exams and quizzes in class, take home exams and quizzes, and group quizzes[3]. In the street-network unit, I used the paragraph response given

[3]In a group quiz, students work on the quiz independently for 10-15 minutes, then in groups, with each group turning in one paper. The students don't know when a group quiz

above in the previous section, take-home quizzes and exams, in-class exams, and daily homework activities.

6. Conclusion

This summary of the development and implementation of the unit on street networks illustrates some of the things that I try in my classroom. I feel it has been successful in meeting the goals I set and the needs of my students.

References

[1] Garfunkel, Solomon, et al. *For All Practical Purposes: Introduction to Contemporary Mathematics*, 3rd edition., W. H. Freeman and Company, New York, 1994.
[2] Hart, Eric W., "Discrete Mathematical Modeling In The Secondary Curriculum: Rationale and Examples from The Core-Plus Mathematics Project," this volume.
[3] Hirsch, Christian R., Arthur F. Coxford, James T. Fey, and Harold L. Schoen, "Core-plus mathematics: Teaching sensible mathematics in sense-making ways," *The Mathematics Teacher*, Nov. 1995, pp. 694–700.

JOHN F. KENNEDY HIGH SCHOOL, 4545 WENIG RD. NE, CEDAR RAPIDS, IOWA 52402
E-mail address: bhoyer@cedar-rapids.k12.ia.us

is going to be given, so they must assume each quiz will be individual. This encourages maximum effort on all quizzes all of the time.

DIMACS Series in Discrete Mathematics
and Theoretical Computer Science
Volume **36**, 1997

Algorithms, Algebra, and the Computer Lab

Philip G. Lewis

One thing leads to another.

First I agreed to help teach the introductory programming course in our new high school computer department. That meant I had to learn some Logo. Meanwhile, back in the math department, I was teaching a little bit of linear algebra—vectors, vector-valued functions, linear transformations, matrices. I noticed then that a Logo[1] list and a vector are one and the same thing. Of course I needed programming practice, so, for a month I spent my free time getting Logo to know everything I knew about linear algebra. I had so much fun doing this, that it seemed natural to propose to teach a mathematics course meeting in our new computer lab, "a course that really uses a computer language to express and develop the mathematics." I did; the course was successful; it resulted in a book [**6**], *Approaching Precalculus Mathematics Discretely*; the book led to my participation at the DIMACS conference[2] and the conference led to this article.

So here are some questions I'll try to answer. What's discrete, different, and worth preserving about the course I taught and tried to encapsulate in the book? And are there general principles that might be applied to other courses?

Let's take note first of the role of discrete mathematics. The study of algorithms and algorithmics permeates discrete mathematics. Students learned most of the mathematics by writing algorithms and running them on the computer. I'll argue here that concentrating on the process of constructing and analyzing algorithms is a productive way to teach algebra, provided that certain conditions are satisfied.

What were the salient characteristics of the course?

1991 *Mathematics Subject Classification.* Primary 00A05, 00A35.

[1]Logo is a dialect of Lisp that avoids its insistence on parentheses. Logo's primary data objects are numbers, words (numbers are also words), and lists. Lists of numbers can, of course, represent vectors, lists of lists matrices.

[2]Editors' note: This conference, held in 1992, is described in the Preface and Introduction to this volume.

1. All classes, about twenty-two students to a class, met in the computer room. They were primarily college-bound juniors at an upper (but not honors) achievement level. Most of the time there was one computer to a student, although a few students worked closely enough together so that their efforts could be considered a team effort. Classes were scheduled four times a week, but the room was free for lab work on the fifth day.
2. On a day that required new material, I usually began by introducing the basic concept together with relevant Logo notation and then set a task that required students to construct the algorithms necessary to develop the concept. For example, an introduction of the concept of vector as ordered pair (trivial to these students) led them to write Logo programs first to establish an axis system given two scalar inputs (`AXES 20 20`), then to draw a given vector in the appropriate position to the scale of the axis system (`VECTDRAW [5 7]`).[3]

   The concept of a workspace saved to disk allowed students to build and save an increasingly complex algorithmic structure having the flavor, if not the logical structure, of an axiomatic mathematical system. For example, given definitions of vector addition (`VSUM`) and scalar multiplication (`SCALAR`), the student could define an algorithm to perform a linear combination of two vectors (`LC`) by invoking the two algorithms `VSUM` and `SCALAR`:[4]

```
TO LC :N1 :N2 :VECT1 :VECT2
  OUTPUT VSUM (SCALAR :N1 :VECT1)
              (SCALAR :N2 :VECT2)
END
```

3. I tried to set a sufficient number of tasks to occupy the day following a class. This let me spend at least half of the class periods walking around, making suggestions and debugging programs. At regular intervals I'd have an interactive class—a codification and catch-up period during which students would make sure that everyone had an updated set of procedures. I regularly asked students who came up with versions of algorithms that were noteworthy to present them for class discussion and possible inclusion in the "official" algorithm structure that constituted a record of the mathematics that the class had accomplished. This set of current procedures was rather like a class notebook, except that the notes were a compilation of the best of the class-generated algorithms. As the course progressed, some of these would get reworked to reflect the increasing sophistication of

[3]Logo is a prefix language—that is, the name of the procedure precedes its inputs. Notice that `AXES` takes two numerical inputs while `VECTDRAW` takes one numerical list input.

[4]For a procedure to provide an input for another procedure, Logo requires the programmer to declare the output. Notice that the output is the vector sum (`VSUM`) of two vectors, each obtained from a scalar product of a number and a vector.

the students and the mathematics. Students kept their own up-to-date version of the current record on disk. Often, however, these disks would differ significantly from the "official" record because students tended to develop a strong fondness for including their own working procedures rather than the "official" version.

4. The goal from the start was always to develop the mathematics that could be applied to computer graphics, so during the most successful part of the course, the need to "get somewhere" kept the class together as a class. This cohesiveness diminished somewhat toward the end of the course as some students, after completing the computer graphics work, moved on to other algebraic topics (e.g. polynomial operations from a vector point of view, polar graphing) while their colleagues completed the computer graphics section.

Both the students and I had very positive feelings about the course. Some of us talked then about why it worked and I have done a lot of thinking about it since. Here, in no particular order, are the results of some of these reflections.

First, it must be said that the computer is a wonderful critic. It is ruthlessly impartial and relentlessly logical—like a force of nature. When a student writes a program that doesn't work, *it* doesn't work like a machine doesn't work, and it can be made to work by a combination of techniques ranging from logical analysis to random tinkering. Furthermore, a student who fixed a program by random tinkering was almost never completely satisfied. Eventually he or she would come back to the question "Now, why the hell did that work?"

Some of the characteristics of a class conducted in a computer lab are unique. The teacher tends to be moved out of the feedback loop, taking on more the role of mentor as he or she analyzes and debugs programs. It turns out to be much more satisfying to answer the question "Why doesn't this work?" than to evaluate the solution to test problems. If the goal of a class is to have its members construct original programs, students come to class in the middle of ongoing work. They don't want to be interrupted by instruction, so a teacher finds it much more difficult to teach. It amounts to a bad-news good-news joke: they won't listen so you can't teach; on the other hand, they're learning so you don't *have* to teach.

If the task is to write a program to teach a computer how to do something, the process of learning the mathematics is different. First you must comprehend the idea; then you must write an algorithm to express it. While it is possible to comprehend a concept without understanding it and still design the algorithm to express it, once the algorithm works, at least you feel better about understanding the concept on which the algorithm is based. While this feeling may have been illusory for some students, the psychological impact seemed strong enough. After all, you own something that you construct yourself. And even though most students don't get a chance to own original concepts, they seem to be able to acquire ownership by constructing algorithms to implement the concepts. The effect of so doing makes a big difference.

We talk a lot about the heuristics of problem solving without teaching them. The act of solving the problems inherent in constructing a hierarchical system of algorithms naturally focuses on these heuristics. I mentioned the problem of defining LC to output a linear combination of two vectors and two scalars. Many students initially solve the task by rebuilding VSUM and SCALAR inside LC, producing procedures like this:

```
TO LC :N1 :N2 :VECT1 :VECT2
  OUTPUT LIST SUM (PRODUCT :N1 FIRST :VECT1)
                  (PRODUCT :N2 FIRST :VECT2)
              SUM (PRODUCT :N1 LAST :VECT1)
                  (PRODUCT :N2 LAST :VECT2)
END
```

instead of:

```
TO LC :N1 :N2 :VECT1 :VECT2
  OUTPUT VSUM (SCALAR :N1 :VECT1)
              (SCALAR :N2 :VECT2)
END
```

After seeing a few examples of this sort, they not only see the power of building complex procedures out of simpler ones, but tend to use the principle by dividing a complex problem into simpler parts, solving these, and then incorporating the solutions into the solution of the larger task. The resulting structure is analogous to the axiomatic structure of a mathematical system: the algorithm for LC depends completely on the status of the two algorithms for vector addition and scalar multiplication. It is possible for students to construct algorithms from the "top down," saying, in effect, "I don't know how to add vectors and I don't know how to multiply by scalars, but if I did, then here's the correct algorithm for LC. In such cases, the definitions of VSUM and SCALAR have a status similar to bypassed lemmas. LC can be defined and Logo will accept the definition, but it won't work until the two algorithms on which it depends are themselves successfully defined.

Once you have solved a lot of algorithmic problems, you have in effect developed a set of tools for getting the computer to do mathematics. This provides a strong impetus on the one hand to extend the mathematical concepts into new territories (we can handle matrices in two space, why not matrices in n-space?) and on the other to see one thing as like another—adding polynomials is just like adding vectors! Consequently the student's investment in constructing an algorithmic system has great potential mathematical payoffs. There is no intellectual satisfaction comparable to discovering that one's solution to one problem can be made to apply to another problem in a totally different environment.

The choice of computer language is important. Sudents who follow an algorithmic approach have a sense of the dependent structure of mathematics—provided the language is modular like Logo. They have a sense of function in terms of input and output—provided that the language takes a functional form like Logo. And they will be comfortable with recursive definitions and the foundations of mathematical induction—provided that the language is as dependent on recursion as Logo. The latter point is both positive and negative: you can't do anything powerful without recursion in Logo, but because it is so dependent on recursion, you can't easily look at the world iteratively. Therefore my students were good at formulating recursive definitons, but not very good at formulating or analyzing iterative definitions. On the other hand, being able to look at a definition like SCALAR and state that it *must* work on any n-dimensional vector requires that you know what mathematical induction is all about, and that is a big plus in the high school.[5]

```
TO SCALAR :NUM :VECT
  IF EMPTYP :VECT [OUTPUT :VECT]
  OUTPUT FPUT (PRODUCT :NUM FIRST :VECT)
              SCALAR :NUM BUTFIRST :VECT
END
```

Many (I'd like to think most) students would analyze the definition this way, saying something like "Well, if it works for a vector of dimension k, then it will work for one of dimension $k + 1$. So let's look at the first case. It outputs the empty list when :VECT is empty, so it *has* to work."

Teaching students to construct algorithms is hard; getting them to analyze the structure of those algorithms is harder. If, however, the algorithms are implemented on a computer, analysis is a natural consequence of creation. Some algorithms clearly run more slowly than others. Some are clearly more elegantly expressed than others. In no other course have I found students to be so aware of efficiency and elegance as desirable characteristics. Here is one example. A student wrote a program to invert a two-by-two matrix (in Logo this is a list of two lists—e.g. [[1 2][3 4]]) by first writing a procedure DET to obtain the determinant of the matrix:

```
TO MATINV :MAT
  OUTPUT MATRIX VECT (LAST LASTVECT :MAT)/DET :MAT
                     (MINUS LAST FIRSTVECT :MAT)/DET :MAT
                VECT (MINUS FIRST LASTVECT :MAT)/DET :MAT
                     (FIRST FIRSTVECT :MAT)/DET :MAT
END
```

where the procedure VECT makes a vector out of two numbers, the procedures FIRSTVECT and LASTVECT retrieve the appropriate vectors from a matrix, and MATRIX makes a matrix out of two vectors. I was particularly pleased that the student had renamed familiar procedures like FIRST to make the algorithm mathematically readable, and complimented him, but was taken

[5]SCALAR provides a good indication of Logo's recursive power. It depends on several list operating procedures: FPUT inserts an object into the front of a list. FIRST outputs the first element in a list and BUTFIRST outputs a list comprised of all but the first of a list. The procedure SCALAR tests its input vector to see if it is empty (EMPTYP). If it is, it outputs the empty list. Otherwise, it multiplies the number (:NUM) by the first element of the list and inserts this into the list that results from calling SCALAR recursively on the BUTFIRST of the list. When the list is empty, SCALAR outputs the list comprised of all of the products of :NUM and the elements of :VECT.

aback by another student's comment, "That's really inefficient!" On being pressed, she observed devastatingly "Look, `DET` is being called four times when once will do." It is hard to imagine that kind of algorithmic analysis occurring in the absence of the computer.

There is one other component of the course that should be mentioned. A student with a clever solution to a problem invariably had the opportunity to present it to the class. Because any program was one that classmates could use, the audience tended to be enthusiastic in their appreciation. Having a potential audience for their work led some students into new intellectual territory.

I have made a case for some of the virtues of taking an algorithmic approach to vector algebra. It is reasonable to ask if the approach makes sense in more elementary contexts. Two of us have taught a first-year algebra course that takes an algorithmic approach to teaching functions. We have observed some of the same sorts of instructional dividends I have mentioned and, as a result of working with younger and more mathematically unsophisticated students, are developing the theory that students initially come to algebra with a well developed view of function as algorithm. Unfortunately, our curriculum forces them to a set-theoretic view at the start and only returns to algorithms when they are so thoroughly brainwashed that thinking algorithmically has become difficult. Wouldn't it be logical to reverse the curriculum structure and begin with an algorithmic view of function? This argument has led two of us to produce a proposal for a computer-based beginning algebra course in which the algorithmic component is significant. I hope to be reporting on that some day.

References

NOTE: Following this note is a list of books for those interested in using Logo to teach mathematics in a computer lab setting. Following are two appropriate versions of Logo:

A: The most accessible vanilla (i.e., standard, no-frills) Logo for the IBM PC as well as the Macintosh was developed by Brian Harvey and is free for the downloading. The URL is `http://http.cs.berkeley.edu/~bh`. The materials were originally developed on the BBC Acorn Computer, which had a resident version of Logo.

B: Paradigm Software has a powerful object-oriented version of the language for the Macintosh. This comes in a student version or a full-fledged development version. Either version of Object Logo is capable of being run in a "vanilla" mode. The address is Paradigm Software P.O. Box 2995, Cambridge, Massachusetts 02238.

[1] Abelson, Harold, and Andrea H. deSesa, *Turtle Geometry*, MIT Press, 1980.

[2] Abelson, Harold, and Amanda Abelson, *Logo for the Macintosh, and Introduction Through Object Logo*, Paradigm Software, Cambridge MA, 1992.

[3] Cuocco, Albert, *Investigations in Algebra*, MIT Press, Cambridge MA, 1989.

[4] Clayson, James, *Visual Modeling with Logo*, MIT Press, Cambridge MA, 1988.

[5] Harvey, Brian, *Computer Science Logo Style, Volume 1, Intermediate Programming*, MIT Press, Cambridge MA, 1986.

[6] Lewis, Philip G., *Approaching Precalc. Math. Discretely*, MIT Press, 1990.

LINCOLN-SUDBURY (MA) REGIONAL HIGH SCHOOL (RETIRED)

E-mail address: `pgl@world.std.com`

DIMACS Series in Discrete Mathematics
and Theoretical Computer Science
Volume **36**, 1997

Discrete Mathematics is Already in the Classroom — But It's Hiding

Joan Reinthaler

The last several years have presented high school teachers with an almost irresistible buffet of mathematical goodies to add to the curriculum—real-world applications, recursion, fractals and chaos, data analysis, and applications of matrices, to name just a few—and of course, a working competency with the technology that has made all this possible. Discrete mathematics is the umbrella that covers a lot of these topics and, despite the excitement these ideas have engendered in many teachers and their students, when the reality of closely prescribed curricula rears its ugly head, the inevitable question we ask is "How can I find room to fit this in?" This is not a naive question and it deserves more attention than it generally gets from a mathematics community of scholars and university teachers whose curricula are not similarly mandated.

As a first step in answering this question, I suggest that aspects of discrete mathematics are already in the curriculum but that we as teachers, and some of the textbooks we use, tend to ignore them. Many introductory algebra books gloss over the difference between a continuous domain, as in the relationship between distance and time, and a discrete domain, as in the relationship between the price of a carton of milk and the amount of milk the carton holds, or in the relationship between frequency of cricket chirps and the temperature. These are wonderful problems, but we lose an opportunity to make distinctions between discrete and continuous domains and ranges when we treat them all identically, or when, automatically, we connect the points on their graphs. We need instead to recognize problems that involve discrete domains (problems involving money, for instance, or numbers of things like pencils or people or raffle tickets), indeed to be on the lookout for them, to rejoice in them and to use them as a jumping-off place for creative and productive investigations.

Here is an example of the sort of material found in many introductory algebra courses and frequently treated as if the domain were continuous.

1991 *Mathematics Subject Classification.* Primary 00A35, 00A05.

A shoemaker makes moccasins and boots and needs 2 square feet of leather for each moccasin and 3 square feet of leather for each boot. 20 square feet of leather are available.

At this point students are used to being asked a question, the obvious one being something about how many boots and moccasins are made, or they may be asked to draw a graph that represents the information given, or to write an inequality that models the information. The usual responses to these requests are

$$2M + 3B \leq 20,$$

or a graph that looks something like that in Figure 1(a). The assumption is made that $M \geq 0$ and $B \geq 0$ and that both range and domain are continuous (and usually the student is unaware that he or she has made such assumptions).

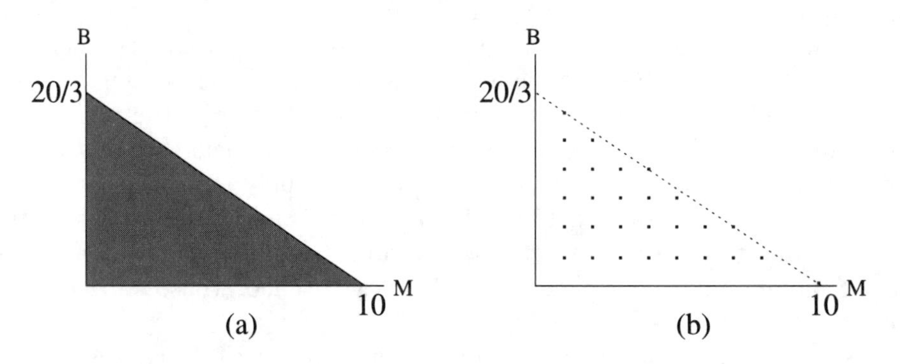

FIGURE 1. (a) Graph of $2M+3B \leq 20$; (b) Graph of integer lattice points satisfying $2M + 3B \leq 20$.

However, the situation described above doesn't conclude with a question. My experience has been that this often leads students to creative investigations and more widely ranging discussions. Students may well ask the very questions listed above and may provide the same answers, but every time I have used this approach, some students have begun to question the validity of these conventional answers when they begin to ask other questions as well. How many different ways can a shoemaker use her resources? Can you make a fraction of a boot? Do you have to manufacture boots and moccasins in pairs? Are the relations and graphs shown above appropriate models of the situation? Are there better models?

At this point, we can begin to explore the difference between the graph of a *region* and that of a *lattice*, as shown in Figure 1(b).) We can ask about the number of solutions. We can discuss the nature of the boundary and look at the number of solutions that lie on the boundary. All of these investigations lie entirely within the scope of material that is included in even the most conservative algebra curriculum. This sort of discussion can begin to give students insight into what is meant by "Discrete Mathematics".

Here is another example of the kind of material that is already at our fingertips just waiting to be investigated in the context of a discrete domain. It is a problem from a popular precalculus textbook.

> *Sally's office has a system to let people know when the department will have a meeting. Sally calls three people. Then, those three people each call three other people, and so on, until the whole department is notified. If it takes ten minutes for a person to call three people, and all calls are completed within 30 minutes, how many people will be called in the last round?*

The assumption usually made is that the appropriate model that will be discovered for this problem is the exponential function $y = 3^x$, with no mention of the fact that in this case x (the number of rounds) and y (the number of calls made in a round) are discrete quantities. The answer given in the book, $3^3 = 27$, arises from simply substituting into the equation. The problem can also be investigated by examining a tree diagram as in Figure 2, and any teacher who wishes to find ways to make use of the standard tools of discrete math should take such an opportunity to do this. However, if the problem is discussed explicitly as one involving discrete mathematics, a very different solution can be found and, along the way, students can have further experience with the behavior of discrete systems.

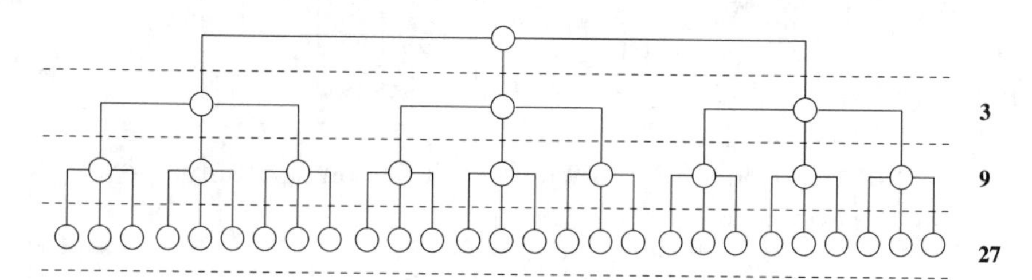

FIGURE 2. Standard phone tree: each person calls three people, waiting until the previous round is complete before beginning their own calls.

The answer, 27, arises only if it is assumed that the second generation of calls are not begun until *all three* of the first calls are made, and in general, that the $n+1$st generation of calls are not made until the nth generation is completed. Suppose, instead, that we assume that each call takes one time slot ($3\frac{1}{3}$ minutes in this problem as originally stated) and that a person called in time slot t places her first call in time slot $t+1$ instead of waiting for the next generation. Then, as explained below, the sequence of numbers of people called in each generation is a term in a Fibonacci-like sequence. (The number of calls still grows exponentially.) If each person makes only *two* calls, and G_t is the number of calls made in time slot t, the terms follow the standard Fibonacci pattern,

$$G_{t+2} = G_{t+1} + G_t \quad (\text{for } t \geq 1),$$

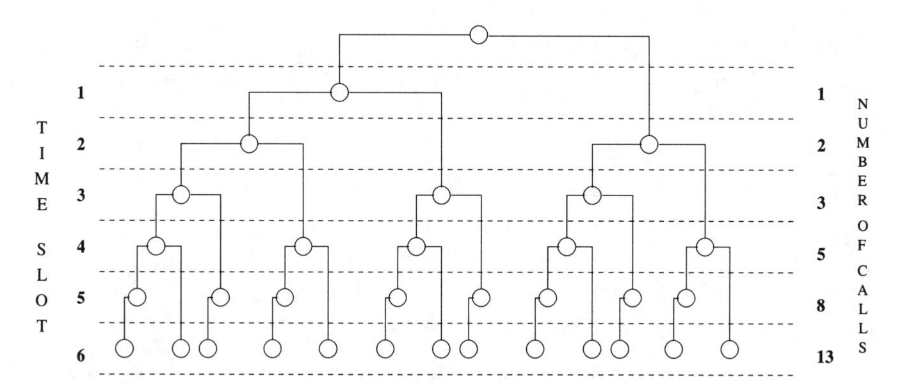

FIGURE 3. Phone tree in which each person calls two people, but each person called can start their first call in the next time slot.

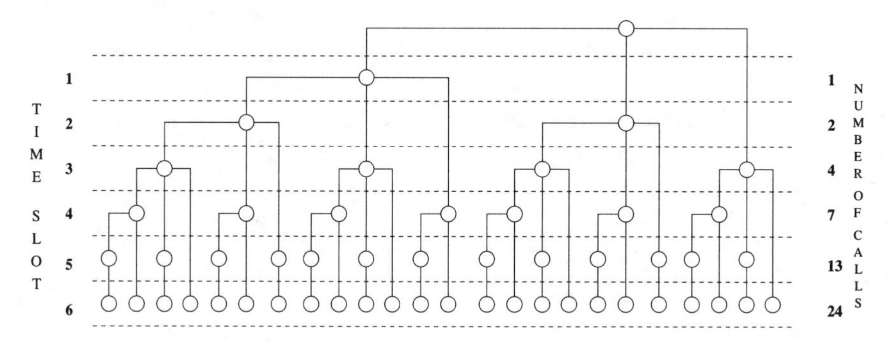

FIGURE 4. Phone tree in which each person calls three people in succession.

where $G_1 = 1$ and $G_2 = 2$. (See Figure 3.) If each person makes three calls, as in the original problem, then the terms grow according to the recursive relation

$$G_{t+3} = G_{t+2} + G_{t+1} + G_t, \quad (\text{for } t \geq 1),$$

where $G_1 = 1$, $G_2 = 2$, and $G_3 = 4$. (See Figure 4.) In this case, the last round occurs at $t = 9$, and G_9, the number of calls made, is 149 instead of 27!

Given this model, questions can be asked about the most efficient calling system for a given number of people. Or, if each person calls m other people ($m = 2, 3, 4, \dots$), what happens to the number of calls in the final time slot?

One last example of the opportunities offered by standard problems is an extension of this one, found in an intermediate algebra book:

> *The length and the width of a rectangle are in the ratio of 3:2. If each dimension is increased by 4 inches, the new length and width are in the ratio of 7:5. Find the original dimensions.*

Since the domain for lengths is continuous, the solution of this problem is easily found by standard algebraic methods. However here is a problem that, at first glance, looks very similar:

> *As Michael Jordan steps to the foul line for two shots, the announcer reports that his foul-shooting percentage for this year is 78%. Jordan makes one and misses one. The next time he steps to the line, the announcer reports that he is shooting at 76%. How many foul shots has he attempted this year?*

In this problem the domain, the number of shots, is discrete, and the percents reported are the ratios of shots made to shots attempted, rounded to the nearest hundredth.[1] This means that, if m stands for the number of shots made and a stands for the number of shots attempted, then

$$.775 \leq \frac{m}{a} < .785 \text{ and } .755 \leq \frac{m+1}{a+2} < .765.$$

There is more than one possible answer to this problem (for example, $a = 36$, $m = 28$ or $a = 40$, $m = 31$ are solutions) and finding them all is not easy. (You can start with a diagram like that used in Figure 1(b).) However, the analysis that leads to the system of inequalities is, in itself, a useful exercise and one that arises only because of the discrete nature of the situation.

Investigations such as the ones outlined above do not represent digressions from the standard curriculum. They arise from problems that abound in traditional texts and provide teachers an opportunity to bring discrete mathematics into the classroom naturally and within the constraints of an existing curriculum. What it takes to do this are teachers who can recognize the problems in their texts that are good jumping-off points for discrete investigations, and who are interested in pursuing the mathematics that reveals itself when they and their students jump.

THE SIDWELL FRIENDS SCHOOL, 3825 WISCONSIN AVENUE, N. W., WASHINGTON, D. C. 20016

E-mail address: `joanr@umd5.umd.edu`

[1]Editors' comment: See also Pollak's article in this volume, Section 3, example (c), for further discussion of this problem.

DIMACS Series in Discrete Mathematics
and Theoretical Computer Science
Volume **36**, 1997

Integrating Discrete Mathematics into the Curriculum: An Example

James T. Sandefur

Many teachers believe that they are being asked to teach discrete mathematics in addition to the mathematics that is already part of their curriculum. Since there isn't adequate time to cover the existing curriculum, it seems impossible to add discrete mathematics except on enrichment days, such as the day before Christmas break. But one can effectively cover a lot of mathematics by spreading an appropriate structured activity over several classes. This approach integrates discrete mathematics into the existing curriculum, results in deeper student understanding, and can be accomplished in about the same amount of time as is presently devoted to existing topics. This approach, which involves using algebraic, geometric, and discrete topics to study complex problems, is the approach recommended by the *NCTM Curriculum and Evaluation Standards* [**3**].

Let me relate an example that illustrates this point. At first glance this example is a simple combinatorics problem. But further reflection shows that it involves the standard algebra concepts of (1) functions, (2) domain and range, (3) the formula for the sum of the first n integers, and (4) parabolas, as well as the discrete topics of recursion and graph theory. Two high school teachers that I have worked with, Linda Mencarini and Nancy Wheeler,[1] introduced me to the "handshake problem" by describing how they use it in their algebra classes. When entering class on the first day of school, they ask all their students to shake hands with each other. The essence of the problem is for the students to compute how many handshakes took place.

While I learned this approach from these teachers, I will relate how I adapted this example for my precalculus class. As did Ms. Wheeler and Ms. Mencarini, on the first day of class I asked my students to shake hands with everyone else in the class. There were 31 people in the class, including myself. After a few moments of disbelief and several minutes of chaos,

1991 *Mathematics Subject Classification.* Primary 00A35, 00A05.

[1]They both teach at Rockville High School, 2100 Baltimore Rd., Rockville, MD 20851.

everyone settled back into their seats. I then asked them how they knew that they had shaken everyone else's hand. The general consensus was that no one did shake everyone else's hand. We then discussed how we could be sure that everyone had shaken everyone else's hand. One student suggested that everyone wear name tags using the numbers from 1 to 31 instead of their names. Everyone would also have a sheet with the numbers from 1 to 31 on it. When you shake someone's hand, you cross off their number from your sheet. That way, everyone could be sure that they shook everyone else's hand.

Another student suggested that each person, one at a time, could get up and go around the room shaking everyone else's hand, while everyone else remained seated. The class decided that this method would be too time consuming. After trying this method on a group of size 6, it was seen that this method results in everyone shaking everyone else's hand twice. In particular, Sue would shake Sam's hand when Sue was going around and Sam was seated, and Sam would shake Sue's hand when Sam went around and Sue was seated. A variation on this plan was developed in which the first person would go around and shake everyone's hand. The second person would shake everyone's hand except the first person, whose hand had already been shaken. In general, each person would get up and shake the hands of those that had not already gotten up.

This discussion is, in my mind, what it means to do mathematics. Such discussions occur far too infrequently in mathematics class, or any class for that matter.

I now asked how many handshakes had taken place, assuming everyone shook everyone else's hand. To help the students answer this question, I divided them into groups ranging from size 3 to size 6. I then suggested that everyone in each group shake hands with everyone else and compute how many handshakes take place. We recorded the number of handshakes for each size group. Several groups got the wrong answer, but after general class discussion, they figured out where they went wrong. For example, one group of 5 computed 20 handshakes, while a group of 6 counted 15 handshakes. Obviously one of them was wrong. I did not tell the class which group was wrong, but let them figure it out.

The data was recorded in a list of n, number of people, versus h, number of handshakes for n people. I served as a group of 1, with no handshakes. The other results were that for groups of size 2, 3, 4, 5, and 6, the number of handshakes was 1, 3, 6, 10, and 15, respectively.

Here I pointed out that we had a function. The input or independent variable is the number in the group, n. The output or dependent variable, $h(n)$, is the number of handshakes that take place if everyone in the group of n shakes hands once with everyone else. In particular, we now know that

$$\begin{aligned} h(1) &= 0, \\ h(2) &= 1, \\ h(3) &= 3, \\ h(4) &= 6, \\ h(5) &= 10, \\ h(6) &= 15. \end{aligned}$$

In an elementary school class, instead of discussing functions, the teacher could have the students plot the points (1,0), (2,1), (3,3), ...; that is, the size of the group on the horizontal axis and the number of handshakes on the vertical axis.

We then discussed how we could use this data to compute the number of handshakes that would take place in a class of 31. Again, I let the groups work on this problem for a while. Then the groups presented their approaches to the rest of the class.

Several groups observed that to compute the number of handshakes for each group, you added one less than the group size to the number of handshakes of the previous size group. For example, the number of handshakes for a group of size 6 (15 handshakes) is the number of handshakes for a group of size 5 (10 handshakes) plus one less than 6 (5 handshakes). They then predicted that the number of handshakes for a group of size 7 is the 15 handshakes of the size 6 group plus 6 more for a total of 21 handshakes. They used this method and their graphing calculators to compute that $h(31) = 465$.

I pointed out that, using function notation, their method could be summarized as

$$h(n) = h(n-1) + n - 1;$$

that is, $h(6) = h(5) + 5 = 10 + 5 = 15$, $h(7) = h(6) + 6 = 21$, and so forth. Many students did not like this method because it is time consuming to use. They asked what the formula was. I assured them that this is a perfectly legitimate formula, since it is easy to program a calculator or computer to use this formula to compute $h(n)$ for any given n. We did not actually program our calculators in this class.

Some groups noticed that $h(n)$ was the sum of the first $n-1$ integers. For example,

$$h(6) = 1 + 2 + 3 + 4 + 5 = 15.$$

So they knew that

$$h(31) = 1 + 2 + \ldots + 30.$$

Many of them knew there was a formula for this. Some remembered the right formula, $(30 \times 31)/2$, but others were off slightly, for example, forgetting to divide by 2. Nobody could explain why the formula worked.

I then reminded the class of their first approach to making sure everyone shook hands; that is, having one person at a time go around and shake hands with everyone else. The class quickly commented that each of 31 people would then shake hands with 30 others for a total of (31×30) handshakes. But they also realized that everyone shook hands with everyone else twice, so the correct answer should be $(31 \times 30)/2$. We quickly concluded that

$$h(n) = \frac{n(n-1)}{2}.$$

We had now 'proven' that

$$1 + 2 + \ldots + (n-1) = \frac{n(n-1)}{2}.$$

Several students commented that they remembered the formula as $n(n+1)/2$, so we also discussed the fact that

$$1 + 2 + \ldots + n = \frac{n(n+1)}{2}$$

was in reality the same formula. This was not a trivial observation and some time was spent on it. In an effort to help students understand that these two formulas describe the same rule, we wrote

$$1 + 2 + 3 + 4 + 5 = (6 \times 5)/2.$$

We then let $n = 5$, so $n + 1 = 6$ and our formula becomes $1 + 2 + \ldots + n = n(n+1)/2$. We then let $n = 6$, which means that $n - 1 = 5$. Substitution now gives $1 + 2 + \ldots + n - 1 = n(n-1)/2$. Even with this discussion, this concept was still fuzzy to some students.

Then they graphed the function

$$y_1 = \frac{x(x-1)}{2}$$

on their graphing calculators. They observed that the graph was a parabola. Using the 'trace' feature of the calculator, they could approximate the number of handshakes y_1 for any size group x. Using the 'zoom box' feature and the knowledge that the answer y_1 has to be an integer, they could get the exact answer this way. I also showed them that they could now store any value they wanted in x, say 31, and have the calculator give the value of y_1.

One of the goals of this precalculus class is to help students understand the concepts of roots and extrema of functions. Thus, we digressed from the handshake problem and used this opportunity to begin studying these topics and also to learn more about the graphing calculators. We did this by using the trace feature and the root feature to approximate the roots, $x = 0$ and $x = 1$, of this parabola. On reflection these roots are obvious, both because the function is already factored and because groups of size 0 and 1 will have no handshakes. The class also used the trace feature and minimum feature to find that the minimum of this parabola occurs at $x = \frac{1}{2}$ and $y = \frac{1}{8}$.

In addition, this problem generated a discussion of domain and range. For the handshake problem the domain is the set of nonnegative integers. But the function $f(x) = x(x-1)/2$ has the set of real numbers as its domain and numbers greater than or equal to $\frac{1}{8}$ as its range. Thus, domain and range of functions may depend on the context in which the function arises.

I assigned the following as homework. Draw a regular n-gon. A line connecting 2 vertices that goes through the interior of the figure is defined to be a diagonal. How many diagonals does a regular n-gon have and how many diagonals does a regular 25-gon have? For example, a triangle has 0 diagonals, a square has 2 diagonals, a pentagon has 5 diagonals, and a hexagon has 9 diagonals. (See Figure 1.) I suggested that students draw each of these figures to check the numbers. They had some difficulty developing a systematic method for drawing and counting the diagonals.

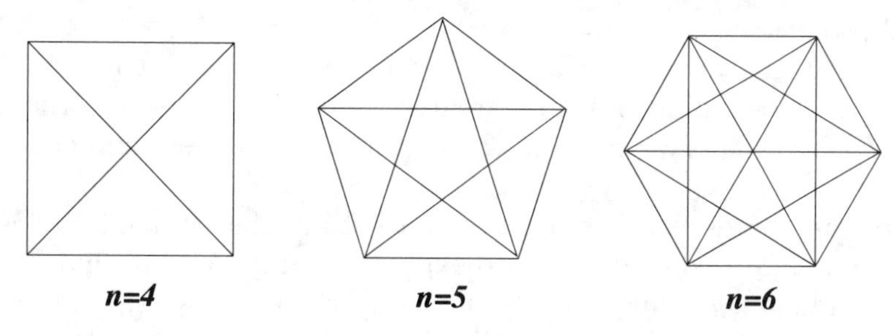

FIGURE 1.

Let $d(n)$ be the number of diagonals of an n-gon. Some students noticed that $d(4) = 2$, $d(5) = 2 + 3$, and $d(6) = 2 + 3 + 4$. They conjectured that

$$d(n) = 2 + 3 + 4 + \ldots + (n-2).$$

They simplified this to

$$d(n) = 1 + 2 + 3 + \ldots + (n-2) - 1 = \frac{(n-1)(n-2)}{2} - 1$$

using the formula for the sum of the first $n-2$ integers.

Other students saw that the total number of lines in an n-gon, including diagonals and sides, was the same as the number of handshakes in a group of size n, which was $n(n-1)/2$. Subtracting away the number of sides gives

$$d(n) = \frac{n(n-1)}{2} - n.$$

Some other students noticed that you can draw a diagonal from a point by drawing a line to any other point except the starting point itself and the points immediately on each side of it. Thus, you draw lines to $n-3$ other points. Since you are doing this for each of the n points, this give $n(n-3)$ lines. But, as in the handshake problem, you are drawing two lines between every pair of points, so the number of diagonals is half that amount:

$$d(n) = \frac{n(n-3)}{2}.$$

When the homework was due, I had students present their approaches. Which of the three above solutions is correct? I had the students simplify each of these expressions. Then they saw that they were all equal to

$$d(n) = \frac{n^2 - 3n}{2}.$$

This demonstrated that there are often many different, and correct, approaches to solving a problem, three in this case. In addition, the students received useful practice in manipulating expressions. Because many of the students were uncomfortable simplifying these expressions, I used this moment to hand out a worksheet to give them further practice at simplifying polynomials. Most of the students actually enjoyed this break from our more involved problem.

Note that the domain of $d(n) = n(n-3)/2$, in the context of this problem, is the set of all integers greater than or equal to 3, although in another context, $f(x) = x(x-3)/2$ has the set of all real numbers as its domain.

As before, we used this opportunity to study the parabola defined by this function and used calculators to explore its graph and to find its roots and the minimum y-value, which is $y = -\frac{9}{8}$ when $x = \frac{3}{2}$. (Note that this minimum has no significance for the diagonal problem.) In an abstract setting, this parabola has all real numbers greater than or equal to $-\frac{9}{8}$ as its range. Again we see that domain and range are dependent on the context.

The handshake problem and the diagonal problem can also be related to the discrete mathematical topic of graph theory. In graph theory, a graph is a set of vertices and a set of edges connecting pairs of vertices. When doing the handshake problem, you can have students plot a point (or vertex) on a paper for each student in the class. Two points are connected with an edge if those two students shake hands. When all students have shaken hands, then all pairs of vertices are connected with edges. This is called the complete graph on n vertices. The discussion of the handshake problem shows that the complete graph on n vertices has $n(n-1)/2$ edges. The diagonal problem is related in that the diagonals are what remain if you delete the edges around the 'outside' of the complete graph on n vertices. Thus, $n(n-.1)/2 - n$ edges remain. The edges around the 'outside' form a cycle that goes through each vertex once, arriving back where it started. Such a cycle is called a Hamilton cycle.

The handshake problem and diagonal problem can be used to motivate the study of graph theory. Conversely, a teacher studying graph theory can introduce the previous functions in the context of counting edges in a complete graph and a complete graph minus a Hamilton cycle. The handshake problem could then be introduced as an application of graph theory. There is no one correct method for introducing this material, and the mathematical topics introduced will vary depending on the classroom discussion.

The point of this extended discussion is to demonstrate how problems arising from discrete mathematics can lead to discussions that cut across the algebra curriculum. The previous discussion took several days to complete, including reviews of homework problems, the diagonal problem, simplification of polynomial expressions, and some factoring (with the approximation of roots on the calculator aiding in factoring). If a teacher's approach is to teach all the algebra first and then do this type of problem, then there will not be enough time in the school year to include discrete mathematics topics. Alternatively, if a teacher incorporates discussion of this problem into the teaching of parabolas, functions, domain and range, and summation formulas, then this discussion replaces a number of lectures that would have occurred at different points throughout the school year. My own experience, and that of many teachers that I have worked with, is that students gain understanding when using this approach, without suffering any loss in manipulative skills.

To effectively use this approach, teachers should look to the literature for additional problems. Many excellent problems appear in *Mathematics Teacher*, such as in Chopin [**1**] and Sandefur [**4, 5**]. Another excellent source is the NCTM 1991 Yearbook, *Discrete Mathematics across the Curriculum* [**2**]. *Finite Differences* [**6**] includes many examples of recursively defined functions, similar to the one developed from the handshake problem, which could be adapted to the classroom. Instead of following the specific instructions for teaching a problem, the teacher should be willing to let the classroom discussion broaden, depending on the questions and interests of the students. Also important is that teachers reflect on these problems, constantly asking themselves how the mathematics in the problem is connected to the mathematics in their curriculum, and how they can help students see these connections.

References

[1] Chopin, Jeffrey M. "Spiral through Recursion", *Mathematics Teacher* 87 (October 1994), pp. 504-8.

[2] Hirsch, Christian R., and Margaret J. Kenney, eds. *Discrete Mathematics Across the Curriculum, K-12*, Yearbook of the National Council of Teachers of Mathematics, Reston VA, 1991.

[3] National Council of Teachers of Mathematics, *Curriculum and Evaluation Standards for School Mathematics*, Reston VA, NCTM, 1989.

[4] Sandefur, James T. "Drugs and Pollution in the Algebra Classroom", *Mathematics Teacher* 85 (February 1992), pp. 139-45.

[5] ——— "Technology, Linear Equations, and Buying a Car", *Mathematics Teacher* 85 (October 1992), pp. 562-7.

[6] Seymour, Dale and Margaret Shedd, *Finite Differences*, Dale Seymour Publications, Palo Alto CA, 1973.

Department of Mathematics, Georgetown University, Washington, D. C. 20057-0001

E-mail address: `sandefur@guvax.georgetown.edu`

Section 6

High School Courses on Discrete Mathematics

The Status of Discrete Mathematics in the High Schools
HAROLD F. BAILEY
Page 311

Discrete Mathematics: A Fresh Start for Secondary Students
L. CHARLES BIEHL
Page 317

A Discrete Mathematics Textbook for High Schools
NANCY CRISLER, PATIENCE FISHER, AND GARY FROELICH
Page 323

DIMACS Series in Discrete Mathematics
and Theoretical Computer Science
Volume **36**, 1997

The Status of Discrete Mathematics in the High Schools

Harold F. Bailey

1. Introduction

The basis of this article was the data collected for my doctoral dissertation, *Discrete Mathematics in Undergraduate and High School Curricula* [**2**]. The dissertation was done under the supervision of Henry Pollak, Bruce Vogeli, and Gail Young at Teachers College, Columbia University.

One questionnaire was sent to 50 randomly selected American and foreign high schools and a second questionnaire was sent to 100 American and Canadian colleges. Thirty-four private and public high schools representing mathematics departments in 17 different states (including Alaska) responded. No foreign high schools responded. Fifty-eight American colleges responded. Both questionnaires were sent in the spring of 1993.

The high school questionnaire attempted to ascertain the following:

1. Are the high schools teaching discrete mathematics?
2. What topics are being taught?
3. What are the goals of discrete mathematics in the schools?
4. How will college placement be affected by a student having taken discrete mathematics in high school?

Similar questions were asked in the college questionnaire.

2. A Basic Assumption

Bogart ([**4**] p. 94), Ralston ([**3**] p. 80-1), and the Mathematical Association of America (MAA) ([**5**], p. 85) have made distinctions between finite mathematics and discrete mathematics.[1] The high school responses to the survey indicate that these distinctions are not universally accepted by the

1991 *Mathematics Subject Classification.* Primary 00A05, 00A35.

[1]The MAA places finite mathematics in the pre-calculus category and places discrete mathematics in the same category as calculus. Ralston and Bogart regard discrete mathematics as a necessary course for mathematics majors.

TABLE 1. Common Discrete Mathematics Topics

Combinatorics	Graph Theory
Probability	Matrices
Logic	Number Theory
Set Theory	Computability and Formal Languages
Trees	Finite State Machines
Functions & Relations	Algorithm Analysis
Recurrence Equations	Boolean Algebra
Abstract Algebra	
Computer Applications	

high schools.[2] In this report, discrete mathematics in the high schools is defined as a course which includes most of the topics contained in Table 1 and is called "discrete mathematics" by the respondent. A distinction between discrete mathematics and finite mathematics will be made only when respondents' replies make it necessary.

3. Summary of the Survey

3.1. The Data. Only 23.5% (8 out of 34)[3] of the high schools reported teaching discrete mathematics. Discrete mathematics is taken by juniors and seniors. There is some overlap with the AP calculus students, and the data indicate the calculus students are stronger mathematically than the discrete mathematics students.

Two of the eight respondents said that computers were used in their discrete mathematics course and all six who responded to the calculator question used calculators in discrete mathematics.

The topics that are taught in high school are similar to the topics taught in college. Seven out of the eight high school respondents teach combinatorics and eight out of eight teach set theory. Table 2 lists the topics and the extent to which they are included in college and high school discrete mathematics courses.

Six different textbooks are used in seven of the schools and one of the schools used "in-house materials".

3.2. Personal Comments. The data summarized above and in Table 2 give a picture of the status of discrete mathematics in high schools. The comments made by the respondents are also instructive. For example, respondents to the high school questionnaire presented problems which must be resolved before a discrete mathematics class can be offered. The prestige

[2]Several of the colleges in the survey also make no distinction between discrete mathematics and finite mathematics.

[3]One of these high schools did distinguish between finite mathematics and discrete mathematics. This school reported that finite mathematics is taught.

of calculus, the emotional and physical burden on the students, administrative and budgetary concerns and an insufficient number of students are all mentioned as reasons against an AP exam in discrete mathematics. These same reasons militate against a non-AP discrete mathematics course being taught.

The comments of respondents to the college questionnaire indicate strong, conflicting opinions concerning a high school discrete mathematics course. One respondent said, "[We] don't care where or how they learn the material." Another felt discrete mathematics was more appropriate as a high school course than calculus.

On the other side of the issue, one college respondent doubted that a high school discrete mathematics course could be "sufficiently sophisticated and challenging" to qualify as a college course. Another warned that discrete mathematics would become a review course where students would review material that they should have learned in previous courses.

One answer to the question concerning an AP exam in discrete mathematics was "multiple choice tests are no indication of a degree of mastery

TABLE 2. Discrete Mathematics Topics Included in High School and College Courses

Topic	High School Actual Number MAX = 8	High School Topic Frequency Rank	College Actual Number Reported MAX = 51	College Topic Frequency Rank
Set Theory	8	1	30	7
Combinatorics	7	2	47	1
Graph Theory	6	3	45	2
Functions & Relations	6	3	39	3
Trees	5	5	38	5
Probability	5	5	23	11
Logic	4	7	39	3
Recursion	4	7	37	6
Abstract Algebra	3	8	11	13
Boolean Algebra	1	10	26	8
Computer Applications	1	10	24	9
Algorithm Analysis	1	10	24	9
Finite State Machines	1	10	18	12
Computability and Formal Languages	1	10	10	14

Note: 58 colleges responded and 51 reported teaching discrete mathematics. 34 high schools responded and 8 reported teaching discrete mathematics.

of calculus." Clearly, the writer was opposed to an AP exam in discrete mathematics also.

3.3. Goals. The goals of ten high school respondents are listed below:

1. "To provide an outlet (or another course) for
   (a) Juniors that have completed AP Calculus
   (b) Students that want more math, but not necessarily calculus."
   (This respondent also expressed a desire to see an AP test offered for discrete mathematics.)
2. "An excellent math experience for brighter students."
3. "To introduce students to a higher level of abstract mathematics before calculus. Lay foundation for college level discrete mathematics work."
4. "To develop the student's mathematics maturity through the study of discrete mathematics and its applications."
5. "To provide another 4th year mathematics course for students."
6. "To familiarize the students with the basics of topics covered and introduce concepts beyond linear programming." (This respondent listed combinatorics, set theory, functions and relations, graph theory and matrix algebra as the topics taught in this course.)
7. "Finite mathematics is taught to our less able students as an applications course. After 4 years (8-11) of algebra, geometry, and trig, they will finally see some use for mathematics and gain a bit of knowledge in probability and statistics which they can probably take in college."
8. "Our goal is to cover some of the discrete topics as recommended in the NCTM curriculum standards." (This respondent indicated that discrete mathematics is not now offered, but it is being considered.)
9. "We have no room in our curriculum for discrete mathematics. We attempt to offer some of the discrete topics in other classes."
10. "Prepare kids for Babson, Bentley, Roger Williams, etc. Business oriented career."

Responses 1 through 6 indicate that discrete mathematics was thought of as an advanced course for mathematically talented students. Responses 7 and 10 indicate a course that the MAA would label finite mathematics.

3.4. Attitudes Toward an AP Exam in Discrete Mathematics. There is no overwhelming support for an AP exam from either the high school or college respondents. Seven out of sixteen high school respondents reported favoring an AP exam in discrete mathematics, while over 60% of the college respondents reported that they did not care whether an AP exam was given.[4]

[4]Similar results were obtained in *A Survey of the Feasibility of Offering a New Advanced Placement Course in Mathematical Science*, a College Board publication ([**1**] pp. 5,11).

4. Conclusions

4.1. Interest in Discrete Mathematics. The rate of response (68%) to the questionnaire and the fact that 77% of the respondents wanted a copy of the final report indicate the interest in discrete mathematics. Some of the reasons against offering a discrete mathematics course were recorded above. The comment that I found most interesting is that one high school noted that they could not teach discrete mathematics because it was not recognized by the state as a legitimate mathematics course.

Few (only 23.5%) of the high school respondents teach discrete mathematics. Several others are in the planning stage and may teach discrete mathematics in a few years. My own personal reading of the data and conversations with high school chairs indicate that many would like to teach such a course but would have great difficulty getting it started.

4.2. Allowing for Change. Those colleges attempting to meet the MAA's recommendation that discrete mathematics be taught in the first two years along with calculus are having difficulty squeezing it in. This is especially true if the colleges are attempting to meet the recommendation of a full year of discrete mathematics. In some cases this has resulted in students taking five mathematics courses in the first two years. A high school discrete mathematics course could relieve some of the pressure. Introducing a new course means something has to change. In the colleges, shortening the calculus course or taking multivariate calculus in the junior year are options—not universally accepted options.

Stephen Maurer's response to the arguments against introducing discrete mathematics is appropriate. He recognizes the validity of the arguments against introducing discrete mathematics, especially those arguments which decry the time discrete mathematics will take away from calculus, but he states,

> It would be best if everyone knew everything. Since this will never happen, we have to pick and choose what will be taught and we can at least insist that future mathematicians know "a lot of analysis" ([**6**] p. 5).

Maurer's remark was directed at undergraduate mathematics, and recognized that the time devoted to discrete mathematics will probably lessen the time devoted to calculus. He argues that discrete mathematics is important and students can still learn "a lot of analysis" even if discrete mathematics does take some time away from calculus. A similar argument could be made for high schools. If we accept that discrete mathematics is important for a high school mathematics curriculum and should be taught, then some parts of the present mathematics curriculum may have to be given less attention or even given up.

In high school, can discrete mathematics ever hope to match the prestige of calculus? Is discrete mathematics worthwhile for high school students? The answer to the first question is "maybe", but not for some time. The answer to the second question seems to be "yes", based on the interest in the survey and the comments of most of the respondents.

References

[1] Armstrong, James S. and Chancey O. Jones, *A Survey of the Feasibility of Offering a New Advanced Placement Course in the Mathematical Sciences*, Advanced Placement Program, The College Board, Educational Testing Service, Princeton NJ, 1987.

[2] Bailey, Harold F., *Discrete Matematics in Undergraduate and High School Curricula*, Doctoral Dissertation, Columbia University, 1995.

[3] Hirsch, Christian R., and Margaret J. Kenney, eds. *Discrete Mathematics Across the Curriculum, K-12*, Yearbook of the National Council of Teachers of Mathematics, Reston VA, 1991.

[4] Mathematical Association of America, *Committee Report on Discrete Mathematics in the First Two Years*, MAA Notes no. 15, Washington, D.C., 1989.

[5] Mathematical Association of America, *Statistical Abstract of Undergraduate Programs in the Mathematical Sciences and Computer Science in the United States, 1990-1991, CBMS Survey, First Two Years*, MAA Notes no. 23, Washington D. C., 1992.

[6] Maurer, Stephen B. "The Lessons of Williamstown," presented at the Sloan Foundation *Conference on New Directions in Two Year College Mathematics*, Atherton, CA, July 1994.

Department of Mathematics and Computer Science, College of Mount Saint Vincent, Riverdale, NY 10463

E-mail address: `hbailey@manhattan.edu`

DIMACS Series in Discrete Mathematics
and Theoretical Computer Science
Volume **36**, 1997

Discrete Mathematics: A Fresh Start for Secondary Students

L. Charles Biehl

Many students enrolled in a college preparatory curriculum in high school perform at or below acceptable levels in their mathematics courses. Following two years of algebra and a year of geometry, they face junior and senior level mathematics courses which ultimately do not serve their needs and goals. These are students of average ability, who will have college majors in arts, language, or social sciences, for whom a course such as precalculus will be their exit-level experience with mathematics. Given that a substantial number of these students struggle with algebra and other basic skills, it is questionable whether precalculus delivers the appropriate message to these students regarding the power of mathematics in their lives.

As an alternative, we have run a full-year course on applied mathematics and problem solving at McKean High School for three years, and have introduced a similar course at the Academy of Mathematics and Science at Wilmington High School. The course, called Mathematical Analysis, embraces the spirit of the National Council of Teachers of Mathematics *Curriculum and Evaluation Standards for School Mathematics*, and emphasizes discrete mathematics and probability and statistics. (A syllabus is included at the end of this article.) The course features many topics and activities designed to attract and hold the interest of the students, and gives them a sample of mathematics that is much more likely to be useful in later life than what is taught in precalculus or advanced algebra. Similar courses are currently under development or are being introduced at other schools, primarily by teachers who have had some experience and training in teaching discrete mathematics for the high school level.

We have analyzed student surveys and examples of student work to assess the impact of this course on the students' performance and perceptions. At the outset, the students demonstrate little or no knowledge of mathematics despite having passed "rigorous" courses in elementary algebra and geometry prior to entering this course. These are the same students who,

---

1991 *Mathematics Subject Classification.* Primary 00A05, 00A35.

when placed in a sequentially-structured course like advanced algebra or precalculus, tend to fall behind early and face an ever-increasing struggle to pass, let alone succeed. In the new course, the students show an improvement in grades, higher completion rate of work assigned outside of class, and a demonstrably more positive perception of the place of mathematics in their worlds. With the wide variety of topics from discrete mathematics and statistics, the course continually offers students a fresh start, reinforcing both the desire and ability to succeed.

The course uses mathematical modeling and problems from the outside world rather than a textbook, and emphasizes student projects. The topic *graphs and their applications* includes problems such as optimizing the pick-up routes for Goodwill Industries [**1**] or the rounds of the custodial staff at the school. As projects, students designed an optimum tour of the major attractions at a large amusement park nearby, and attempted to redesign garbage collection routes in Wilmington, Delaware. They designed and tested minimum cost networks for communications among Automatic Teller Machines (ATM's) and long-distance telephone systems. Their investigations have included analysis of various heuristics for their relative efficiency and complexity. Such investigations allow the teacher to introduce current research into the classroom, permitting the students to experience mathematics as a living and evolving science, rather than a static and closed field.

The *mathematics of social choice* allows students to investigate (for example) the inherent strengths and weaknesses of the processes used in the election of officials or the apportionment of representatives in the Federal government. Students apply their knowledge of American history in a mathematical context, only to discover that the historically important political question of apportionment has its underpinnings in mathematics. They also learn that the present resolution of the issue was brought about not by politicians but mathematicians, and that the question has not yet been fully resolved. Students also study the "fair" allocation of estates among heirs, or land among several prospective developers. In one class, students investigated the work of Steven Brams and Alan Taylor in the domain of "envy-free" fair division of continuous objects such as land or cake [**3**], research which was only months old. It was inspiring to have the students make the effort to digest this new work and communicate their findings directly to one of the authors, then receive his reply, including his gratitude for their willingness to perform such an investigation, as well as his comments on their work.

The unit on *iteration and recursion* included interdisciplinary projects using mathematics and art. By way of introduction, students used iteration as a modeling tool to investigate functions and their applications. They followed with the arithmetic of complex numbers and the iteration of functions containing complex numbers. This led into the study of the Mandelbrot and

Julia sets, followed by the iteration of functions containing affine transformations, in order to model natural objects such as clouds and plants. Students then applied these ideas to art projects. In one project, students took "zooms" of the Mandelbrot set and rendered them in clay, then fired and painted them to produce original sculpture. Some students created their own fractal designs using algebraic and geometric methods. One generated replacement algorithms using line segments and simple polygons rather than numbers. Another used cellular automata and oriented percolation models [**6**] to produce representations of oil spills and forest fires, then measured their fractal complexities in order to draw conclusions about their properties. One student even used an iterative structure to produce a "fractal" song for his band!

The unit on *probability and statistics* is naturally rich in applications. Student projects involve posing questions, collecting appropriate data, and using mathematical tools to analyze and interpret the data. Students have used probability to model natural phenomena such as forest fires, oil spills, and the spread of disease. One student used statistics in analyzing the design of paper airplanes. One analyzed the use of statistics in the media and the associated flaws, and another studied the use of statistics in the creation and introduction of new products on the market. Parents who use statistics in their professions have spoken with the class. Students have also made use of video and print materials for related projects, which were obtained from such sources as InFORMS [**5**] and textbooks such as *For All Practical Purposes* [**2**] and *Excursions in Modern Mathematics* [**7**].

Mathematical Analysis gives the students the opportunity to use much of their prior background, even if it is weak, in the investigation of new problem situations. It gives them knowledge of the breadth, and to some degree the depth, of contemporary mathematics. In our extensive surveys of the students, and interviews with both the students and teachers, there is evidence to suggest that the course not only increases students' awareness and appreciation of mathematics, but improves their own perceptions of their success as thinkers and problem solvers.

The student population in this course usually falls into three distinct groups (with occasional overlapping). First, there are those who have had minimal success in mathematics and simply wish to complete their studies with something "easier" than more algebra or precalculus. Second, there are those students who, while they have had moderate success in an algebra/geometry sequence, cannot demonstrate competence in topics essential to success in the precalculus/calculus curriculum. Finally, there are those who have been highly successful, are intrinsically motivated (perhaps taking precalculus or even calculus concurrently), and who wish to learn as much as possible while in high school. The heterogeneity of the class actually causes synergistic results, since those students who have formerly been reluctant to be actively engaged in the study of mathematics suddenly find themselves in the position of having just as much mathematical power as the "smart" ones. Maintaining a good balance of personality, ability, gender, and ethnic groups in a cooperative learning environment is the foundation for this synergy, at least after the initial acclimation process for those who have never worked cooperatively before. (N. B. This is still the majority of students as of this writing.) All aspects of cooperative learning are employed, in pairs or in groups of three to four, and group identities are easily established. Groups are restructured every eight to ten weeks. Student surveys indicate that the benefits of this environment, in conjunction with the course material, cross both gender and ethnic lines, especially when the roles assumed by students of varying backgrounds allow them to work from their present strengths, rather than trying to recover from prior weaknesses. The inclusion in the course of two or three of those students who have been highly successful prior to this course has had no detrimental effect on the others in the class. In fact, since all of the students are essentially on equal footing with the material, even preconceived notions of who the "smart" ones are rapidly fade, and the students work together very well.

The data regarding student perceptions of this course has also been highly encouraging. Students who have had little or no success before find that earning higher grades is highly rewarding. For all students, the variety and scope of the material provides great appeal and maintains interest. Although some question whether this course directly prepares them for college, they are developing thinking and problem-solving skills and experience that will be beneficial regardless of their college plans.

Cooperative learning is generally well-received. The students welcome the opportunity to discuss solutions with each other, and collaboration in the development of solutions, especially when solutions are not unique, is felt to be highly profitable. In at least one class, however, the data suggested mild discomfort with cooperative learning among a small number of minority students. The general feeling among these students was positive, but there was an underlying resistance to working collaboratively. These sentiments improved during the course of the year, possibly because of the variety and frequency of roles and partners that the students experienced. More exposure to collaboration and cooperation in the lower grades may serve to eliminate such feelings.

Following is the syllabus for our course. A section on trigonometry appears for historical reasons[1] and can be eliminated, allowing additional time for either discrete mathematics or probability and statistics.

Mathematical Analysis: Course of Study

1. Discrete Mathematics (one to one-and-a-half semesters)
   (a) Graphs and their applications
      (i) Eulerian paths and circuits

[1]In this school district, trigonometry is not a part of the algebra curriculum, and is introduced formally only in precalculus. This unit was included here in order that students not taking precalculus will have the opportunity to solve problems using right triangle trigonometry and trigonometric functions.

(ii) Hamiltonian paths and circuits
(iii) Applications of graph coloring
(iv) Shortest path problems
(v) Critical path analysis and dynamic programming
(vi) Optimum cost networks and spanning trees
(vii) Vertex coloring with applications
(b) Mathematics of social choice
(i) Election methods and strategies
(ii) Weighted voting schemes and power measures
(iii) Continuous and discrete fair division
(iv) History, methods and issues of apportionment
(c) Linear programming with applications
(d) Iteration and Recursion (including fractal geometry)
(e) Extended applications of mathematical modeling
2. Trigonometry (half semester if included)
(a) Trigonometric functions
(i) Angles in the coordinate plane
(ii) Trigonometric functions and their values
(iii) Solving right triangles and applications
(b) Graphs of trigonometric functions
(i) Using a graphing calculator and/or software
(ii) Radian measure
(iii) Period, amplitude and phase shift
(c) Trigonometry and Triangles
(i) Laws of sines and cosines
(ii) Solving triangles with applications
(iii) Vectors
3. Probability and Statistics (half to one semester)
(a) Measures of central tendency and dispersion
(b) Data collection, organization and analysis
(c) Graphic representation of data
(d) Collection and analysis of multivariate data
(e) Regression and correlation analysis
(f) Counting rules and rules of probability
(g) Applications and modeling

References

[1] Biehl, L. Charles, "Goodwill Tours," in Froelich, Gary, ed., *Math of the States: Lessons, Strategies, and Ideas from the Presidential Awardees in Mathematics*, COMAP, 1996.

[2] Garfunkel, Solomon, et al., *For All Practical Purposes: Introduction to Contemporary Mathematics*, 3rd edition., W. H. Freeman and Company, New York, 1994.

[3] Hively, Will, "Dividing the Spoils," *Discover*, March 1995, pp. 49-57.

[4] ———, "Fair Shares for All," *New Scientist*, 17 June 1995, pp. 42-46.

[5] Institute for Operations Research and Management Sciences (InFORMS), 940A Elkridge Landing Road, Linthicum, MD 21090-0909.

[6] Kaye, Brian H., *A Random Walk through Fractal Dimensions*, VCH Publishers, New York, 1989.

[7] Tannenbaum and Arnold, *Excursions in Modern Mathematics*, Prentice Hall, New York, 1992.

Academy of Mathematics and Science, Wilmington, Delaware
E-mail address: `biehl@dimacs.rutgers.edu`

DIMACS Series in Discrete Mathematics
and Theoretical Computer Science
Volume **36**, 1997

A Discrete Mathematics Textbook for High Schools

Nancy Crisler, Patience Fisher, and Gary Froelich

1. Background

The movement to include discrete mathematics in the schools has an interesting history, complete with prophets, pioneers, and struggles. First came the visionaries, or prophets, who saw the lack of discrete topics in schools and spoke of the importance of their inclusion. They were followed by pioneers, often disciples of the prophets, who overcame obstacles to develop discrete mathematics courses in their own schools and help spread the word to others.

The prophets and the pioneers are now watching their efforts bear fruit as discrete mathematics becomes commonplace in school curricula. The authors of this article think of their discrete mathematics textbook for high schools as a pioneering effort—one that reflects the vision and determination of many prophets and pioneers. The purpose of this article is to discuss the book's background, the process of writing and revising it, and its content.

The authors' experiences with discrete mathematics began in the early 1980s. Crisler attended institutes at Illinois State University, where she worked with John Dossey, Larry Spence, Albert Otto, and Charles Vanden Eynden. Froelich was appointed by the NCTM as its representative on the editorial panel of COMAP's HiMAP project, where he worked with a number of discrete mathematicians, including Margaret Cozzens, Sol Garfunkel, and Joe Malkevitch, to develop discrete mathematics modules appropriate for high schools. Fisher studied discrete mathematics in the Nebraska Math Scholars Program under Don Miller. All three of the authors began integrating discrete topics in existing courses at their high schools. Two of them (Fisher and Froelich) established discrete mathematics courses at their schools.

The authors' paths converged when the Council of Presidential Awardees in Mathematics (CPAM) asked them to serve as staff members for a discrete

1991 *Mathematics Subject Classification.* Primary 00A05, 00A35.

mathematics institute sponsored by CPAM and funded by the National Science Foundation.

Upon completion of the CPAM project, the authors served on a discrete mathematics task force convened by the NCTM with funding from the Exxon Education Foundation and directed by former NCTM president John Dossey. The purpose of the task force was to amplify the discrete mathematics standard included in the NCTM's *Curriculum and Evaluation Standards for School Mathematics* [**3**], and to further encourage implementation of discrete topics in the schools. The 1990 task force report, *Discrete Mathematics and the Secondary Mathematics Curriculum* [**2**] (available from the NCTM, 1906 Association Drive, Reston VA 22091), recommends integration of discrete topics in existing secondary courses and outlines a separate discrete mathematics course for high schools.

Shortly after the report appeared, one of the authors (Froelich) approached Sol Garfunkel, the executive director of COMAP, about a widely recognized obstacle to implementation of discrete topics in schools–the lack of a suitable textbook for high school discrete mathematics courses. COMAP offered to fund writing of a textbook, and a team of teachers began work on a first draft.

The authors used their own students, particularly those in the discrete mathematics courses taught by two of them (Fisher and Froelich), to test and revise much of the material as it was written. A complete first draft was tested in several high schools in the 1990-1991 school year, and revisions were tested during the 1991-1992 and 1992-1993 school years. In the summer of 1992 the NCTM, with funding from the National Science Foundation, began a discrete mathematics teacher enhancement program under the direction of Margaret Kenney. The program conducted an institute at Boston College in the summer of 1992 and at six sites around the country in 1993. Drafts of the text were tested in these institutes, and the published version was used in the summer of 1994 and 1995.

The text was published by W. H. Freeman & Co. in early 1994 under the title *Discrete Mathematics Through Applications* [**1**] and is accompanied by an instructor manual and disk of software for the first two chapters. It is currently used primarily as a high school text, but is also used in community colleges and in preservice and inservice teacher education programs.

2. Organizing and Revising the Book

Writing and revision of the book were guided by the vision of those who influenced the authors, the authors' own experiences with the materials, and the experiences of those who tested the various drafts.

The authors used the NCTM report, *Discrete Mathematics and the Secondary Mathematics Curriculum* [**2**], as a guide for organization of the book's content. The report divides content into five broad areas: social decision making, graph theory, counting techniques, matrix models, and the mathematics of iteration. It also identifies six unifying themes that flow throughout the five content areas: modeling, use of technology, algorithmic thinking, recursive thinking, decision making, and mathematical induction. The influence of the report's recommended content can be seen by inspecting the book's table of contents. The themes, however, are woven throughout the book's lessons, and an appreciation of their presence requires a more careful inspection.

Identification of the intended audience was another important decision that preceded the first draft. The authors' experience working with their own students and with other teachers in the CPAM institutes indicated a need for a text suitable for courses aimed at college-intending students interested in business, social sciences, and law. The authors also hoped to provide worthwhile experiences for students interested in mathematics, engineering, and the physical and biological sciences. Attention to the needs and abilities of the primary audience, however, required that the treatment place greater emphasis on concepts and applications than on abstraction and symbol pushing.

The style of writing also received much discussion when the first draft was planned. In order to encourage the kind of learning espoused by the NCTM Standards, the authors chose to write lessons and exercises designed to promote student involvement, critical thinking, and discovery. The lessons were to be written to the student, avoid excessive length, and include group activities. Exercises were to provide opportunities for practice, but also experiences in which students could participate in the construction of ideas.

In establishing the order of topics, the authors decided to place social decision making first, partly because of their own success using these topics to capture student interest. Election theory, for example, appeals to students because of its obvious applications, but also because the teacher can easily develop lessons around data obtained from the students. Another reason for placing social decision making first is that these topics provide a good arena in which to begin development of the six unifying themes mentioned previously.

Planning the treatment of individual topics proved most difficult with counting/probability. One reason, of course, is that these topics are difficult for many students. However, a more important reason when developing a textbook for general use is that the amount of previous student exposure to counting/probability varies considerably. Students in one class have no prior experience; those in another have studied much of the content in an Algebra II or a statistics/probability course. Therefore, the authors decided to make the book somewhat less dependent on completion of the counting/probability material than on completion of other topics.

The first field test produced generally favorable results. Most teachers liked the sequencing of topics and approach, and revisions did not change either the order of topics or the informal, student-centered flavor of the early drafts. Some of the teachers who liked the approach admitted their enthusiasm was not immediate, but that it developed gradually as they progressed through the chapters. Materials designed for a student-centered environment seemed to make teachers and students who are used to a teacher-centered classroom feel uncomfortable at first. The discomfort tends to diminish as students acquire a taste for an active role in the learning process.

Perhaps as one might expect of a new initiative in times of great change, comments of field testers did not achieve a consensus. For example, a teacher who did not approve of the development of new ideas in the exercises said, "The exercises were appropriate, but teaching in the exercises (introducing new vocabulary) is confusing and inappropriate." Much more common, however, were sentiments like those of the teacher who wrote, "My judgment as to whether a text is poor, adequate, good, or great is based on the problem sets. Problems should vary in difficulty, vary in degree of open-endedness, and vary in format. Many texts ask one type of question (such as simplify or solve) and follow it with thirty similar problem types. You have opted not to do this and that is what makes your book so valuable to me."

With two exceptions, revisions after the field tests consisted of relatively minor clarifications. One of these involved the treatment of recursion. Of the book's major topics, recursion was the only one which proved difficult for a significant number of the students who used early versions of the material. Many teachers feel that the difficulty students have with recursion lies not as much in the concept itself as in the symbolism attached to it. Thus, revised versions of the text introduce recursive symbolism in the first lesson of the first chapter and continue to expose students to it until recursion is treated carefully in the final chapter. Field testers, including one of the authors (Froelich), noticed considerable improvement in student understanding of recursion with this approach. The gradual development of recursion means, of course, that deviation from the established order of topics could have an adverse effect on student mastery of recursion and should not be done casually.

The second exception will not surprise experienced mathematics teachers. The treatment of mathematical induction was also revised. Induction, perhaps unlike recursion, is conceptually difficult. Students who are good symbol jugglers often perform the steps properly while displaying little understanding of the process. The book strives to improve student understanding of induction by using an approach that reflects the way in which induction is actually used. Students first collect data, then examine the data and conjecture a formula before attempting an inductive verification of the formula's validity. However, because of the relative conceptual difficulty of induction, the text is not as dependent on mastery of it as of, say, recursion.

A brief summary of the book, *Discrete Mathematics Through Applications* [**1**], consisting of a short description and sample problem for each chapter, follows.[1]

3. Chapter by Chapter Descriptions

Chapter 1: Election Theory. The chapter opens with an activity in which students vote, then invent their own methods of determining a group ranking. Several commonly used group-ranking methods are examined and their flaws uncovered. Students consider the work of Kenneth Arrow, who proved that all group-ranking methods will occasionally violate at least one of several reasonable conditions. Approval voting and weighted voting are also discussed. Three computer programs enhance the study of group-ranking methods and voter power in weighted voting situations.

> *In the 1912 presidential election, 45% of the voters preferred Wilson to Roosevelt to Taft; 30% preferred Roosevelt to Taft to Wilson; 25% preferred Taft to Roosevelt to Wilson. Students are expected to note that the plurality winner was ranked last by a majority of voters, discuss how some voters might have changed the results by voting strategically, and suggest alternate methods of determining a winner.*

Chapter 2: Fair Division. The chapter's opening activity asks students to examine several division scenarios: one involving the division of a piece of cake between two children, another about the disposition of a house when there are two heirs, and a third in which the seats in a student council are to be divided among a school's classes. A discussion of fairness criteria results, and a search for division algorithms that satisfy important criteria begins. Students examine an estate division algorithm with an appealing paradox: each of the heirs gets more than he/she thinks he/she deserves. The examination of apportionment algorithms adds a bit of American history by examining methods named after Alexander Hamilton, Thomas Jefferson, and Daniel Webster. The method currently used to apportion the United States House of Representatives is discussed, as well as the work of Balinski and Young, who proved an apportionment result similar to Arrow's in election theory. Algorithms for the fair division of a cake among several individuals include one that is simulated by a computer program on the accompanying disk. The chapter closes with a section on mathematical induction. Slow pacing is recommended in the induction material, with students working in groups on two or three problems a day for about a week.

---

[1] The book itself, and the accompanying instructor manual (which includes numerous teaching tips based on the authors' own experiences and those of field testers) and software, are all available from W. H Freeman, 41 Madison Avenue, New York NY 10010, 1-800-347-9405.

States A, B and C have populations of 647, 247, and 106 respectively. There are 100 seats to be apportioned among them. Students are expected to compare the apportionment results from several common methods and to discuss paradoxes that result when populations change slightly.

Chapter 3: Matrix Operations and Applications. This first chapter on matrices is placed here because basic matrix techniques are useful for the study of graphs in the next two chapters. Addition and multiplication of matrices are introduced as common-sense counterparts of, and useful shortcuts for, everyday calculations. Students are encouraged to use calculators with matrix functions. The chapter closes with two sections on the Leslie matrix model for the growth of wildlife populations. Teachers whose students have previous experience with matrices sometimes treat the early parts of the chapter lightly.

An artist fashions plates and bowls from small pieces of wood. Plates require 100 pieces of ebony, 800 pieces of walnut, 600 pieces of rosewood, and 400 pieces of maple. A large bowl requires 200, 1200, 1000, and 800 pieces of the respective types of wood; a small bowl requires 50, 500, 450, and 400. Students are expected to organize the data into a matrix and to construct an application of matrix multiplication that is meaningful in this context.

Chapter 4: Graphs and Their Applications. The chapter begins with an activity in which students attempt to determine the time needed to complete a school yearbook from information about various tasks, their times, and precedence relationships (information about the tasks that must be completed before others can begin). The investigation leads to the introduction of a graph (sometimes called a network) as an organizational tool. Students also explore real-world situations that can be modeled by graphs in which visiting each vertex of the graph is essential, and others in which traversing each edge is necessary. A discussion of graph coloring problems closes the chapter.

Final examinations for a summer school program require that six different tests be given. From a table showing the people taking each final, students are expected to model the situation with a graph and determine the minimum number of time slots necessary to schedule the exams without conflicts.

Chapter 5: More Graphs, Subgraphs, and Trees. The chapter opens with a discussion of planar graphs (those that can be drawn without crossing edges) that demonstrates the relationship between these graphs and the graph coloring problems that closed the previous chapter. Students explore situations, such as the well-known traveling salesperson problem, in which weights (price of a plane ticket, mileage, etc.) are attached to the edges of a graph, and the object is to find a path through the graph that minimizes the sum of the related weights. Algorithms that solve such problems are examined. The chapter closes with an examination of a special type of graph called a tree.

From the position at which it is currently located, a robot is programmed to find the shortest path for a trip from one location to another. Students are given a graph showing several key locations and the distances between them. They are expected to discuss several routes that the robot could take and show how an algorithm programmed into a computer could identify the shortest such route.

Chapter 6: Counting and Probability. The chapter begins with a situation in which the members of an organization are planning a lottery-type fund raiser. Students examine several suggestions for the design of the contest and explore the question "How many?" as it relates to the number of winners the organization might expect. The text uses a variety of applications to develop standard counting techniques. A similar approach is used to develop basic probability concepts, including addition and multiplication principles, conditional probability, and binomial probability. Trees are used to organize conditional situations. Extended exercise sets accommodate the needs of students with little previous exposure to counting and probability.

Some Americans favor mandatory HIV screening for workers in health care and other professions. If a medical test is 98% accurate, then why do so many people who test positive for a disease not have it? Given additional information about the incidence of the disease, students are expected to discuss the likelihood that a person testing positive for the disease actually has it.

Chapter 7: Matrices Revisited. This chapter is a collection of several real-world situations in which matrix models are valuable. It is placed after the counting/probability chapter because some of the models are dependent on a knowledge of probability. The first model discussed is the Leontief input-output model, which is used to determine production levels needed to meet estimated internal and external demands for a company's products. This chapter also explores Markov chains (multi-step probability models with many applications) and closes with a discussion of basic game theory, the topic of the 1994 Nobel Prize for economics.

A manufacturing company has divisions in Massachusetts, Nebraska and California. Each division of the company uses products produced by itself and by the other two divisions. Given information about the amount of usage, students are expected to model the situation with a matrix and a graph.

> *From estimates of consumer demand for the company's products students are expected to determine the levels of production necessary to meet both external and internal demands.*

Chapter 8: Recursion. Because of the experience with patterns and recursive symbolism woven through the previous chapters, students begin this chapter already able to "speak the language." The first lesson extends existing skills to the realm of technology, including computer and calculator algorithms and computer spreadsheets. Situations modeled by a single operation (either addition/subtraction or multiplication/division) are examined first, followed by those that require two operations. Financial applications are given a prominent place throughout the chapter, which closes with a discussion of cobweb diagrams (an important graphical representation of recursive models). The closing material opens the door for projects related to fractals, a favorite topic of some teachers.

> *Joan has $5,000 in an account to which she adds $100 monthly. The account pays 6.4% interest compounded monthly. Students are asked to create a model for the growth of the account and use their model to investigate the value of the account after a period of several years. They are also expected to determine the time needed for the account to grow to a given amount. Students are encouraged to use a variety of methods to answer questions once they have created their model.*

References

[1] Crisler, Nancy, Patience Fisher, and Gary Froelich, *Discrete Mathematics Through Applications*, New York, W. H. Freeman and Company, 1994.

[2] Dossey, John, *Discrete Mathematics and the Secondary Mathematics Curriculum*, NCTM, Reston VA, 1990.

[3] National Council of Teachers of Mathematics, *Curriculum and Evaluation Standards for School Mathematics*, NCTM, Reston VA, 1989.

PATTONVILLE PUBLIC SCHOOLS, ST. LOUIS COUNTY, MO
E-mail address: `6162698@mcimail.com`

UNIVERSITY OF NEBRASKA, LINCOLN, NE
E-mail address: `pfisher@unlinfo.unl.edu`

CONSORTIUM FOR MATHEMATICS AND ITS APPLICATIONS, LEXINGTON, MA
E-mail address: `g.froelich@comap.com`

Section 7

Discrete Mathematics and Computer Science

Computer Science, Problem Solving, and Discrete Mathematics
PETER B. HENDERSON
Page 333

The Role of Computer Science and Discrete Mathematics
in the High School Curriculum
VIERA K. PROULX
Page 343

DIMACS Series in Discrete Mathematics
and Theoretical Computer Science
Volume **36**, 1997

Computer Science, Problem Solving, and Discrete Mathematics

Peter B. Henderson

1. Introduction

We are now living in the information age, and information is one key to success in today's world[1]. Indeed, without information it is difficult to even survive in our modern society. We are becoming dependent upon video services such as cable and VCRs, audio services such as cellular phones and 800 and 900 numbers, and network-based communication technology. The emerging multi-media and network technology will provide vast resources of information available at our fingertips from anywhere.

This has created an essential connection between discrete mathematics and computer science. First, computer technology is primarily responsible for the current boom in discrete mathematics. Without computers, most applications whose solution requires discrete mathematics would not be feasible, making such mathematics concepts far less relevant. Second, technology is changing the way we teach and students learn in both mathematics and computer science. Computer-based technology is being used for instruction and as a classroom tool for problem-solving. For example, graphing calculators are required for many college mathematics courses. Educational TV, multi-media, and interaction through computer networks are becoming important vehicles for instruction and learning. Third, for those planning to work in technical fields, particularly computing or computer science, discrete mathematics and mathematical problem-solving principles have become essential foundations. Accordingly, these concepts should be taught early in the curriculum, whether in high school computer science or mathematics, or at the university level. This important relationship between mathematics and computer science is rarely stressed in high school, or even in some college computer science curricula.

1991 *Mathematics Subject Classification.* Primary 00A05, 00A35, 68R99.

[1]It is estimated that the average person in America today processes in one day the same amount of information a person 100 years ago processed in a year.

Many introductory computer science courses do not require significant knowledge of mathematics. Students who have very little mathematical background can learn programming skills [**9**]. Computer science graduates often lack formal mathematical training, yet they are capable of developing usable applications software. Students learn the *how to* rather than the *why* of computer science; they are trained as "technicians", not "engineers". In most engineering curricula mathematical foundations are developed early and are then used as the foundations of engineering design. Computer science educators have done it backwards. As David Gries has pointed out, we do the mathematics after the programming [**12**]. This is a sign of the immaturity of the discipline of computer science.

Currently, more computer science students at the college level are now being exposed to discrete mathematics earlier in the curriculum. There are several mathematically-based approaches such as those espoused by Abelson and Sussman [**1**], Baxter, Dubinsky and Levine [**7**], Henderson [**16**], Gries and Schneider [**13**], and a host of others. These advocate teaching discrete mathematics and logical reasoning early; however, there are only a handful of colleges where such courses are pre-requisites for the first programming course. There are several reasons. Traditionally the first computer science course has been a programming course, and change is difficult. More important is that most of the students taking the first computer science course lack mathematical maturity — making programming courses easier to teach than other courses.

I believe that students should learn general problem-solving skills [**25, 29, 26, 2, 23, 20, 4, 18, 19**], and discrete mathematics concepts [**10, 13**] prior to learning formal computer programming [**16**]. Mathematics involves many of the most powerful and general problem-solving tools students can learn. Although one does not necessarily require mathematical skills to be a reasonably competent programmer, mathematics is essential for reasoning in domains outside the narrow range of programming. It not only provides a common language for expressing ideas, but it is an extremely powerful tool for thinking about and representing problems. For instance, without mathematics, it is impossible to demonstrate conclusively the correctness of a software system.

This paper provides a framework for a course which serves the needs of students planning to study computer science and which focuses on discrete mathematics and general problem-solving skills. Courses using this philosophy have been taught at the State University of New York at Stony Brook for the past seven years (under the title *Foundations of Computer Science*), at several other colleges and universities on Long Island, and at three high schools, as a precursor to the Advanced Placement Computer Science course. The main features of the course are described in the next section.

2. Course Framework for *Foundations of Computer Science*

The approach in *Foundations of Computer Science* integrates the teaching of general problem-solving skills with discrete mathematics and computer science concepts. Connections are made using laboratory exercises [**15**] which exploit available computing technology. For example, mathematical logic and logical problem-solving skills are emphasized in a computer science context using logic-based theorem provers and PROLOG (PROgramming with LOGic language); using Standard ML, a powerful functional language, students extend their understanding of mathematical functions and learn to use recursion as a problem-solving tool [**17**]. Some of the mathematically-based approaches reinforce concepts through associated computer exercises. For instance, exercises using the functional programming language Scheme, a derivative of LISP, are given in Abelson and Sussman [**1**], and exercises in ISETL, an interactive set manipulation language, are used by Baxter, Dubinsky and Levine; exercises in Gries and Schneider emphasize the use of pencil and paper rather than the computer. In addition, there are numerous computer-based tutorials for important discrete mathematics concepts such as sets, relations and graphs. Such practical experiences are an extremely important part of the learning activities of today's student.

Problem-solving and discrete mathematics, two important foundations of computer science are discussed below.

2.1. Problem Solving. In order for students to learn general problem-solving techniques they need to solve lots of problems. There are numerous sources of fun and challenging problems which also serve to convey underlying mathematical concepts. These include, among others, problem-solving and recreational mathematics books, game magazines, articles in newspapers, and even placemats at fast-food restaurants. Students should be encouraged to engage in solving problems at all levels, whenever and wherever they encounter them. Such engagement can be as basic as trying to find the shortest route for a trip, or discovering why an appliance does not work. At an advanced level students can work on research projects. However, throwing problems at students without any pedagogical goals is not productive.

It is important for students to learn to look for patterns in problems. Patterns are relevant to most problem-solving activities, but especially those that occur in computer science. Why? Because the development of algorithms involves the discovery of general patterns or basic structures in a problem. The process of developing an algorithm often includes identifying suitable data representations, looking for and implementing a logical decision structure, and finding and/or implementing iterative structures. The patterns involved are frequently recursive and/or inductive in nature. We will come back to this shortly.

A second important reason for students to solve a large number of varied problems is that they learn problem-decomposition strategies. These are strategies for reducing a problem into sub-problems which can more easily be solved. For example, consider the following problem [**29**].

> *Nine adults and two children want to cross a river, using a raft that will carry either one adult or the two children. How can they all get cross the river?*

A problem solver is provided with the initial state and the desired goal state. One solution strategy is to identify "safe" intermediate states. Finding such states is usually a trial-and-error process applying some heuristics. For a selected intermediate state, call it IS, the problem solver looks for ways to get from the initial state to state IS, and then get from IS to the final state. So a larger problem is reduced to two smaller or sub-problems. This is an important problem-solving technique: to identify and solve the relevant sub-problems and use these sub-problem solutions to solve the original problem. This is also a foundation of algorithmic problem-solving. In computer science it is usually called top-down development, stepwise refinement, object-oriented decomposition, etc.

Computer science students who first learn general problem-solving techniques can apply these techniques when developing algorithms. In cognitive science this is called "transfer" — using a technique developed to solve problems in one domain to solve problems in other domains. Unfortunately, many students are very poor "transfer-ers". However, knowing some general problem-solving strategies makes it easier to learn others that are application specific. A second advantage of emphasizing general problem-solving is that students begin to learn how to "transfer" better: they become more mature. Also, focusing on only one problem-solving technique, such as algorithmic problem-solving, encourages students to think very narrowly and they often fail to consider the whole scope of the problem.

One concept which students discover by solving numerous problems is the principle of backtracking, which is an organized approach to enumerating all possible solutions. Backtracking is a very important concept in computer science. It is used in many searching strategies and is the primary technique used in artificial intelligence applications (e.g., speech recognition and chess playing programs).

In our *Foundations of Computer Science* course at SUNY Stony Brook we have found that all problems, from simple word problems to complex logical problems, intrinsically require some problem-solving strategy that is used in computer science. However, some problems are richer than others. Accordingly problems are carefully selected to emphasize the important concepts we believe students should learn. The focus is on problems which require students to identify structure, and those whose solutions involve inductive and recursive patterns. Two such problems are given below.

Problem 1: Toothpick Boxes

Four toothpicks can be used to make a single square (Figure 1), seven to make a row of two squares (Figure 2), ten to make a row of three squares (Figure 3), and so on.

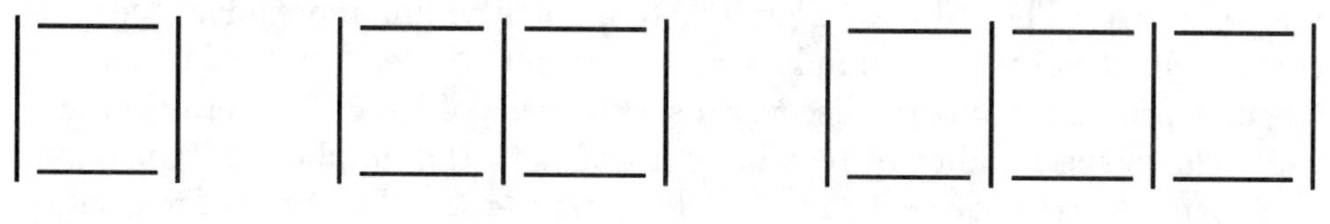

Figure 1 Figure 2 Figure 3

1. *How many toothpicks are required to make a row of 19 squares? How about 43? Now try 67?*
2. *In general, to make a row of $N > 1$ squares, how many toothpicks are required? Express your answer in terms of N.*
3. *We define a joint to be a point where two or more toothpicks touch. For example, there are 4 joints in a single square, and 6 joints in a row of two squares. How many joints are there in a row of 6 squares? How about 24? Try 56?*
4. *In general, how many joints are there in a row of $N > 1$ squares? Express your answer in terms of N.*
5. *A closed region is any region enclosed by toothpicks. For example, there is one closed region in a single square, and two in a row of two squares. In general, how many closed regions are there in a row of $N > 1$ squares?*
6. *Discover a equation relating the number of toothpicks, the number of joints and the number of closed regions. Give an argument justifying your answer.*

Problem 2: Jail Doors[2] [28]

In a certain prison there are 1000 jail cells in a row. A jailer, carrying out the terms of a partial amnesty, unlocked every jail cell in this row. Next, starting with the first he locked every second cell. Then, starting from the first, he turned the key in every third cell, locking those which were open and opening those which were locked. The jailer continued this pattern of locking and unlocking every nth cell on the nth trip until eventually he could not repeat this process. Now, those prisoners whose cells were unlocked were allowed to go free. In which cells were the lucky prisoners? How many times did the jailer have to walk down the row of cells performing this procedure? In general, if there were an arbitrary number of cells N, in

[2]Editors' note: An equivalent problem appears in Leibowitz' article in this volume.

terms of N how many times did the jailer have to walk down the row of cells? Which cells were unlocked?

These two problems require looking for and expressing inductively defined patterns. They also are a prelude to understanding more advanced concepts such as mathematical induction and graphs. Indeed, in the last part of the first problem students discover a special case of Euler's general formula that relates the number of regions of a graph to the number of its vertices and edges (`#vertices - #edges + #regions = 2`). The type of reasoning inherent in finding solutions to such problems is central to computer-based problem-solving. Since iterations in algorithms use inductively defined patterns, if students understand mathematical induction then they can reason more clearly about the algorithms they read and create. Furthermore, all important data structures in computer science can be understood inductively, and most can be concisely defined using recursive structures such as lists or trees.

2.2. Discrete Mathematics. Mathematically-based thinking, applications, and concepts are central to everything a computer scientist does. It is important for students to become comfortable with mathematics and to appreciate its importance in computer science. For this and many other reasons, computer science students should be exposed to mathematical ideas and concepts as early as possible.

Mathematics provides a language for thinking about a wide range of problems and for expressing solutions to such problems. Many of the current software applications and programming paradigms can be traced back to mathematical foundations. For instance, relational database technology is founded upon principles of mathematical relations; SETL [**8**] and ISETL [**7**] are programming languages based upon sets as the key concept; Prolog[**6**] is one of the best known programming languages derived from the principles of mathematical logic; and numerous other important programming languages have mathematical origins such as LISP, Standard ML (SML), Hope, Miranda, Haskel, and Gofer.

What discrete mathematics topics should be covered in a mathematically-based first course for students interested in computer science? Different educators would weight each topic differently depending upon their knowledge and goals. We use some guiding principles. Students should be exposed to different computer science-based paradigms for thinking and problem-solving, especially those based on mathematical concepts. Accordingly, logic, sets, relations and functions are important as noted in the previous paragraph. Graphs are an fundamental tool for problem abstraction and reasoning in general. The fundamental concepts of induction and recursion should be woven into the course from the first day, and they should be constantly reinforced, as they are difficult concepts for students to grasp. An introduction to and applications of general proof techniques are necessary in a first course [**10, 13**]. Of less importance initially are counting principles, combinatorics, probability, algebraic structures, boolean algebra, matrices, automata, and even algorithms. To provide a flavor for the way material could be presented, the areas of logical reasoning and recursive problem-solving using a functional language are briefly discussed.

2.2.1. *Mathematical Logic and Logical Reasoning.* Logical reasoning is an important part of mathematics, as is learning to use logic as a tool for problem-solving, design, and analysis in computer science [**22, 11**]. Games and puzzles provide a good vehicle for getting students to practice logical reasoning. For example, consider the logical problem below from Averbach and Chen [**4**]. This is a relatively straightforward problem which can be solved using techniques from propositional logic.

> *Either Lucretia is forceful or she is creative. If Lucretia is forceful, then she will be a good executive. It is not possible that Lucretia is both efficient and creative. If she is not efficient, then either she is forceful or she will be a good executive. Can you conclude that Lucretia will be a good executive?*

In our *Foundations of Computer Science* course the laboratory component includes working with simple theorem provers for both propositional and predicate logic. The above problem can easily be encoded and solved using such tools. Once students have learned the mathematical fundamentals of logic and have experimented with these two theorem provers, they can use this knowledge to understand how ProLog [**6**] works and to use ProLog to solve some simple problems. When students learn ProLog the traditional way they see only its goal-directed problem-solving strategy. However, with knowledge of logic and resolution they also understand how ProLog was derived from mathematical concepts and is founded on basic problem-solving strategies such as identifying and solving subgoals and backtracking. This demonstrates one significant relationship between mathematics, problem-solving, and computer science, and further reinforces the importance of each in the context of this course.

2.2.2. *Recursive Problem Solving.* The concept of recursion is central in computer science. It plays an important role in compiler construction, data structures, artificial intelligence, general problem-solving, language theory, database construction, graphics, operating systems, and many other areas. Recursive techniques often provide more elegant solutions to complex problems than their iterative counterparts, especially when combined with recursively defined data structures.

Structure and Interpretation of Computer Programs [**1**] by Abelson and Sussman pioneered the educational use of functional programming techniques for teaching recursion as a natural problem-solving tool. Building on this seminal work we have developed a sequence of three laboratories using a more intuitive functional language, Standard ML [**14, 3, 27**]. Most students understand the basics of mathematical functions primarily as graphs drawn on real valued X,Y axes and the corresponding notation such as $f(x) = x^2$.

However, functions are much more powerful and general. They may be recursive, use other functions, have many arguments of any reasonable value (e.g., lists, trees, or even functions themselves) or return values of any kind, even functions.

Students work with more complex recursively defined structures such as lists and binary trees, and learn to use recursion as a problem-solving tool [**17**]. In one laboratory students learn the basics of Standard ML (SML) and apply recursive problem-solving techniques to develop function definitions for these complex structures. For example, functions which count the number of items in a list or sum the values of all the nodes in a labeled binary tree structure.

Students are not taught Standard ML, nor do we provide them with extensive books or reference manuals. The beauty of SML is that students can learn the requisite features of the language by experimenting with a small collection of representative examples. This is discovery learning [**5, 24**]. The language supports a natural expressive power which permits students to create very concise function definitions. As an example, consider the recursive definition of a list-processing function *found* which determines whether or not a specified element is in a list. Observe that this definition is very concise, using the natural recursive definition of a list structure. In addition, it will work correctly for any type of list (e.g., integer, real, boolean, strings, characters, or lists of lists). Students are also required to give inductive proofs for the recursive functions they define. This further reinforces the important principle of mathematical induction.

```
fun found(element, list) =
   if list = []                     { is 'list' an empty list? }
      then  false {'element' can't be found in an empty list}
      else if element = first(list)
                      {is 'element' found first in the list?}
            then  true
            else  found(element, tail(list));
                              {'element' found in rest of list?}
```

As a second example, consider the recursive definition of the binary tree processing function *preorder*. This function takes a labeled binary tree T as an argument and returns a list of node labels. It uses several other simple function definitions (e.g., functions for prepending an item to a list and for appending two lists) which students created in a previous list processing phase of the laboratory. The binary tree data structure definition and definitions of the functions *label*, *left_subtree* and *right_subtree* were provided by us.

```
fun preorder(tree) =
   if tree = empty          {is 'tree' an empty binary tree?}
     then  []  {if 'tree' is empty, return the empty list []}
     else prepend(label(tree),
                  appendlist(preorder(left_subtree(tree)),
                  preorder(right_subtree(tree)) ));
```

3. Conclusion and Discussion

For high school or college students considering a career in computer science or simply trying to understand more about computers, there is a need for more courses emphasizing mathematics, mathematical reasoning and problem-solving. Although computer programming courses are an option available to many students, these courses typically focus on the technical features of a specific programming language and a narrow range of algorithmic problem-solving strategies. Such courses build only slightly on prior mathematics experience, and do little to reinforce mathematical skills. Also, it has been shown that learning computer programming does not result in the improvement in general mathematical problem-solving skills [**21**].

Discrete mathematics courses provide a more valuable educational experience than traditional inward-looking technical courses. With its emphasis on logical reasoning and problem analysis and solution, discrete mathematics provides a catalyst for the general thinking and problem-solving skills that students need to be successful throughout their computer science studies and ultimately as computer science professionals. Moreover, an introductory course such as I've described appeals to a more diverse student population. The *Foundations of Computer Science* course at Stony Brook draws students from a wide range of majors, including psychology, philosophy, English, economics, music, art, engineering, mathematics, and the sciences. This course, and similar ones at other colleges and high schools, can prepare students not only for computer science, but for mathematics, engineering and science courses, and for reasoning in our complex modern world.

References

[1] Harold Abelson, Gerald Jay Sussman, and Julie Sussman, *Structure and Interpretation of Computer Programs.* The MIT Press McGraw-Hill, 1985.

[2] James L. Adams, *Conceptual Block Busting.* W.W. Norton, 1979.

[3] Åke Wikström, *Functional Programming using ML.* Prentice-Hall International, January 1987.

[4] A. Averbach and O. Chein, *Mathematics: Problem Solving Through Recreational Mathematics.* W.H. Freeman, 1980.

[5] John Seely Brown, *Learning-by-Doing Revisited for Electronic Learning Environments.* Lawrence Erlbaum Associates, 1983.

[6] W.F. Clockson and C.S. Mellish, *Programming in Prolog.* Springer-Verlag, 1984.

[7] E. Dubinsky N. Baxter and G. Levine, *Learning Discrete Mathematics with ISETL*. Springer-Verlag, 1988.

[8] E. Dubinsky J.T. Schwartz, R.B.K. Dewar and E. Schonberg, *Programming with Sets: An Introduction to SETL*. Springer-Verlag, 1986.

[9] Jennifer L. Dyck Richard E. Mayer and William Vilberg, "Learning to program and learning to think: What's the connection?" *CACM*, 29(7):605–610, July 1986.

[10] Susanna S. Epp, *Discrete Mathematics with Applications*. Wadsworth Publishing Company, 1990.

[11] ———, "Logic and Discrete Mathematics in the Schools," this volume.

[12] David Gries, "Teaching calculation and discrimination early in the curriculum." In *Proceedings of the IFIP WG 3.2 Workshop on Informatics Curricula for the 1990s*, April 1990.

[13] David Gries and Fred Schneider, *A Logical Approach to Discrete Mathematics*. Springer-Verlag, 1993.

[14] Robert Harper, "Introduction to Standard ML." Technical Report ECS-LFCS-86-14, University of Edinburgh, LFCS, Department of Computer Science, University Of Edinburgh, The King's Buildings, Edinburgh EH9 3JZ, November 1986.

[15] Peter B. Henderson, "Courseware for introductory foundations of computer science." In David L. Ferguson, editor, *Proceedings of the NATO Workshop on Advanced Technologies for the Teaching of Mathematics and Science*. Springer-Verlag, 1990.

[16] ———, "Discrete mathematics as a precursor to computer programming." *ACM SIGCSE Bulletin*, 22(1):17–21, February 1990.

[17] Peter B. Henderson and Francisco Romero, "Teaching recursion as a problem-solving tool using Standard ML." *ACM SIGCSE Bulletin*, 20(1):27–30, February 1989.

[18] Marvin Levine, *Principles of Effective Problem Solving*. Prentice-Hall, 1988.

[19] J. Mason and L. Burton, *Thinking Mathematically*. Addison-Wesley, 1982.

[20] Richard E. Mayer, *Thinking, Problem Solving, Cognition*. W.H. Freeman, 1983.

[21] ———, *Teaching and Learning Computer Programming*. Lawrence Erlbaum Associates, 1988.

[22] J.P. Meyers, "The central role of mathematical logic in computer science." *ACM SIGCSE Bulletin*, 22(1):22–26, 1990.

[23] Alan Newell and Herb Simon, *Human Problem Solving*. Prentice-Hall, 1972.

[24] Peter Pirolli and John R. Anderson, "The role of learning from examples in the acquisition of recursive programming skills." *Canadian Journal of Psychology*, 39:240–272, 1985.

[25] G. Polya, *How to Solve It*. Princeton University Press, 1973.

[26] Moshe F. Rubinstein, *Tools for Thinking and Problem Solving*. Prentice-Hall, 1986.

[27] Ryan D. Stansifer, *ML Primer*. Prentice-Hall, 1992.

[28] Charles W. Trigg, *Mathematical Quickies*. Dover Publications, 1967.

[29] Wayne Wickelgren, *How to Solve Problems*. W.H. Freeman and Co., 1974.

Department of Computer Science, SUNY Stony Brook, Stony Brook, NY 11794

E-mail address: `pbh@cs.sunysb.edu`

DIMACS Series in Discrete Mathematics
and Theoretical Computer Science
Volume **36**, 1997

The Role of Computer Science and Discrete Mathematics in the High School Curriculum

Viera K. Proulx

Computer science as a subject of study is rarely included in the high school curriculum. In some states there is no certification for teachers in computer science. This paper identifies those fundamental ideas of computer science that should be learned by every high school student. Based on the report in [**1**], it first identifies six main themes that present concepts that transcend computer science and promote critical thinking in the context of the modern complex world. Then it provides suggestions for simple activities students can do to explore problems related to each of the main themes. The paper concludes with some general notes about pedagogy. My goal is to implement an experimental, exploratory approach, with a great emphasis on creativity and on practice in expressing one's ideas in writing.

Studying computer science, and the related discipline of discrete mathematics, gives students an opportunity to learn about the representation and meaning of information, to learn the language needed to express logical ideas, and to practice the formal description of dynamic processes (algorithms). Studying computer science also introduces ideas related to managing complex systems — such as encapsulation, abstraction, and information hiding — that apply to systems outside the world of computing. Problem-solving techniques studied in discrete mathematics and computer science promote critical thinking and carry over to other disciplines. Finally, a measure of complexity and a sense of scale makes students aware of the fact that some problems are indeed hard, while many other apparently complex problems can be decomposed into manageable components.

The *ACM Model High School Computer Science Curriculum* [**1**] published by the Association for Computing Machinery (ACM) in 1993, developed by the Task Force on the High School Computer Science Curriculum of the Pre-College Committee of the Education Board of the ACM (formed in 1989) in which I participated, presents a curriculum that is focused on the fundamental concepts of computer science, presented as much as possible

1991 *Mathematics Subject Classification.* Primary 00A05, 00A35.

through experiments and concrete applications. Its curriculum appendices include several model implementations that have been tested in different high schools. In each case the emphasis is on practical experience as the basis for understanding each concept. The report also contains a list of "outcomes" — concepts, ideas, and skills — students should master in a computer science course. By outlining the curriculum as a list of concepts that should be covered, as opposed to writing a weekly course schedule, the Task Force wanted to make sure that the main emphasis is on the underlying concepts, not on a detailed list of topics and isolated skills. The aim is to let the students see the power of computers, introduce them to the algorithmic approach to problem solving, explain the basic stored-program computer model and its hardware implementation, and look at different aspects of information organization.

Although the report concentrates on computer science, many concepts and ideas introduced in such a course could just as naturally fit into a discrete mathematics curriculum. Indeed, computer applications are a rich source of examples of uses of discrete mathematics, as well as tools for exploring problems in discrete mathematics.

I will not not paraphrase the curriculum report; instead, I will identify the main themes that should recur throughout a computer science course and explain why they should become an integral part of every high school curriculum.

1. Main themes of a high school computer science curriculum

All high school students need to understand certain basic computer science concepts to become informed citizens of the information world we live in today. They need to know about the different ways in which information can be represented, organized, and processed. They need to learn that algorithms are descriptions of dynamic processes which can be represented as programs. They need to know that a computer is a machine capable of carrying out instructions encoded as a program. Furthermore, students need to understand how new algorithms are created. Only by understanding the ideas listed above can students begin to comprehend both the power and the limitations of computers. They need to learn what kind of problems computers can or cannot solve and why. It is also immensely important that students learn about the effects of computer technology — beneficial as well as detrimental — on today's society.

In the following subsections, I develop these ideas further, describing six main themes and the concepts students learn while studying them. In the next section I revisit each theme and describe classroom activities that can be used to explore these ideas.

1.1. Representation and Organization of Information. The world today is a world of information. Information affects our lives every day through news about the economy, politics, demographics, weather, sports, and travel. Computers allow access to a wealth of information, which also has its dangers.

By learning how data is represented in a computer, students can see the limitations imposed by computer systems — with respect to accuracy, privacy protection, speed of access, ability to find the desired information, etc. By learning about the binary system and the encoding of numbers, students gain a basis for understanding other aspects of data representation in computers.

There are also problems arising from storing large amounts of data. Students need to learn about the hierarchical organization of data banks, the structure of database systems, and the need for physical security and backups. They need to understand the differences between raw data, processed information, and knowledge gained from analyzing data.

Students will begin to understand the enormous advantages we have gained through easy access to large amounts of data and information — through modeling, analysis, visualization of data, computer simulations, and image processing. They will also be able to comprehend the problems our society faces in the new computer age — maintaining security, privacy, and equity of access to computer-based information.

1.2. Algorithms and Their Representation: Describing Dynamic Processes. The study of algorithms is the study of dynamic processes that arise in our daily lives. Looking at the underlying concepts promotes critical thinking and improves logical reasoning abilities.

To understand how computers perform a prescribed sequence of operations, students need to learn about algorithms as descriptions of dynamic processes and about programs as representations of algorithms in languages that both humans and computers understand. Students should concentrate not only on studying those well-defined algorithms that they can program with their newly learned programming skills. The selection is paltry, and the effort required for implementing them is typically mostly wasted on working out the finicky (albeit often self-inflicted) quirks of the system. The most important task here is to learn that, in order for a computer to perform any task, the task has to be defined precisely, with all possibilities accounted for. The task has to be represented as a series of instructions written in a formal language that is a communications tool between the computer and the programmer.

The concept of different levels of abstraction arises naturally from looking at the representation of problems in different languages. A comparison with other algorithm systems from daily life can help illustrate this point (maps, recipes, operating instructions for appliances, building plans, wiring instructions, etc.). The different levels of languages and their representations should be discussed, from the machine level (even Turing machines or finite automata) to the high level applications programs such as database or spreadsheet programs.

1.3. Computer Organization: Managing Complex Systems. The study of computer organization is the study of complex systems. The operating system itself is a complex resource-management system. The user's view of the computer is an abstraction designed to hide the underlying implementation. At the lowest level, the computer is an implementation of an algorithmic engine similar to a finite state automaton. One can again look at different levels of abstraction and observe again that complex systems are composed from components that act like building blocks. However, the most important point that needs to be made is that a computer will do only what someone tells it to do, and will carry out its task without any understanding of what the programmer intended.

Students need to learn about the basic physical components of a typical computer system (CPU, memory, I/O devices, network interfaces). They also need to learn that users interact with computers at several different levels of abstractions — from hardware designers who worry about the placement of a single wire, to naive users who type in the data, select from a menu of commands, and receive the results. Students need to learn that an operating system is a manager of computer resources which gives the user an ability to interact with the system at a higher level of abstraction. Students need to work with at least one concrete operating system and understand its functionality. Here the study of languages reappears from a different perspective — as a command language or a menu system used to control the operating system. Students should also learn about networks that connect computer systems all over the world and how is it possible to navigate through this wealth of information.

1.4. Problem Solving Techniques. Students should engage in the process of designing new algorithms. By this I do not mean programming algorithms that have already been thoroughly discussed and explained in a textbook. A translation of an algorithm from a description in a pseudocode into a programming language is also not what is meant here. Rather, in conceptualizing and developing an algorithm, the student should experiment with examples, observe patterns, make conjectures about what may be the possible solution, and verify this by using some type of formal reasoning. Similar techniques are often employed in innovative teaching of discrete mathematics. Students need to learn about standard problem-solving techniques: iteration, induction, recursion, divide-and-conquer (division of a problem into subtasks), simulations, and random sampling. They should begin to see the importance of creating the right abstract model of a given problem.

1.5. A Sense of Scale. It is very important that, while learning about different algorithms, looking at the representation of data, or examining the organization of a computer, students also develop some feeling for the scale of problems that are being handled. From the physical point of view, students need to have a sense of the speed at which computers operate, the speed of information transfer between different components or over computer networks, the size of the computer chip, and the depth of the hierarchical organization of computer systems. At the other end of the spectrum, they also need to see the large-scale view, with the whole world connected via electronic networks of all kinds.

To see both the power and the limitations of computers, students need to learn about the complexity of different algorithms. They need to understand that some problems cannot be solved no matter what computer power one may have. Students need to understand that the Tower of Hanoi problem, while it can be solved, cannot have the solution implemented within the lifetime of the universe. They also need to see that an apparently complex problem can often be reduced and solved quite quickly by using the divide-and-conquer method. Binary search is, of course, the prime example.

The immense complexity of computer systems illustrates another scale of difficulty. Here students need to see how creating numerous well-defined levels of abstractions makes possible the task of managing the complexity.

1.6. The Place of Computers in Today's Society. Computers today affect almost every aspect of daily life in our country. Students need to read, discuss, observe, and learn about both beneficial and harmful effects of the computer revolution on today's society. The power gained from access to data and information, as well as new devices, machines, discoveries, and applications, are changing our lives as we speak. Computers allow us to see the unseen, to perform experiments humans cannot do, to find patterns in vast arrays of data, and to respond immediately to observed changes. They automate factories, fly airplanes, connect telephone callers to each other, predict weather, and monitor patients in hospitals.

Students also need to examine some of the problems brought upon us by the widespread use of computers. The loss of privacy, an increased need for uninterrupted service of some computers, the loss of jobs to automation, the dehumanization of communications, and the loss of empowerment for those without access to computers are all new ills in our lives. Only through understanding computers and the way they are controlled can students comprehend the nature of the problems computers can cause or the empowerment they can bring.

2. Sample Activities.

In this section I discuss several ways in which students can explore the concepts presented above. I focus on activities that promote exploration and problem solving — not necessarily with the use of a computer. The software available today, especially spreadsheet applications and some of the simple new languages, makes it possible to implement exercises in each of these areas. These should be used once students understand the underlying concepts presented above.

2.1. Representation and Organization of Information. The exploration of data representation should start with encoding schemes students are likely to be familiar with — Roman numerals, Morse code, foreign alphabets, or number representations used in ancient times or in other cultures. Clocks, odometers, rulers, protractors, and numerous electrical or mechanical meters and measuring devices add to the variety. Make the exploration exciting. Give students a two-paragraph biography of Abraham Lincoln that uses Roman numerals for dates (including months and days), and ask students to identify the person. Ask them to encode messages and have their friend decode them.

You can introduce the binary number system through a game that appeared on cereal boxes a couple of years ago. A student is given a set of six cards as shown:

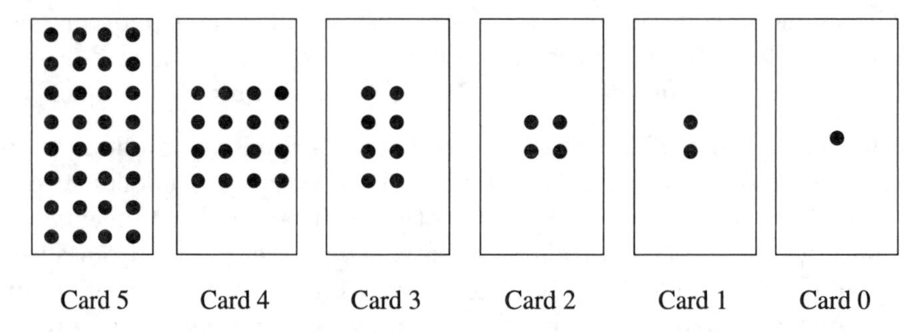

The goal is to figure out how to select the right set of cards, so that the total number of dots equals a given number. Choosing cards 3 and 4 gives us 24 dots, representing the number 24. What cards do we need to pick to represent number 37? How can we encode which cards have been used? What is the largest number we can represent? The game leads naturally to the binary representation of numbers. Number 24 in binary is 011000 representing our selection of cards 3 and 4.

The game can also be used to illustrate binary addition, including the carry. To see what happens, select all cards representing the numbers to be added, then look for duplicates, starting with the low end. For example, if card 3 is needed twice, replace it with one copy of card 4. Overflow occurs if card 5 is needed twice.

At this point it is very easy to introduce binary encoding of numbers, so that, for example, $37_{10} = 011011_2$. Repeat the game using ternary cards — or ask the students to make them. Ask them to rewrite Lincoln's biography using octal representation of numbers.

To learn about hierarchical and indexed ways of representing data, stucents can explore the organization of an encyclopedia — with indexes, tables of contents, brief synopses at the beginning, etc. They can explore an organizational chart of their school — teachers, students, classes, rooms. They can look at the way we search for information in the Yellow Pages. Some entries are listed under multiple headings, while some headings just reference other headings (e.g. Radio Repair — see Television and Radio Repair). Students can build a simple database using one of the standard application programs. They can also use data downloaded from the Internet — for example from the U.S. Census Bureau tables (`http://www.census.gov/`).

For examples of data visualization, you can start with maps of all kinds — geographical or political maps, maps for tourists which show key buildings, city planner's maps, or drawings describing how to get to a friend's house. It is also interesting to look at weather data and the representation of radar images from satellites. Students can look at different charts and graphs appearing daily in the press — illustrating projected growth, distribution of resources, and other information.

2.2. Algorithms and Their Representation: Describing Dynamic Processes. When talking about algorithms and problem solving, students need to explore many different ways in which algorithms can be described. They need to see the need for a good language that allows for an accurate description of the process. Turing machines, automata, or simple programming languages all provide interesting examples.

One way to start exploring different aspects of algorithms and their design is to ask students to describe as carefully as they can the rules for their favorite board game, card game, or team sport. Often, each game or sport has its own language, which students will have to explain. The rules may have many cases (e.g., you move double the distance if both dice have the same value, except when ...). Interestingly, algorithms describing the rules of many common games and sports are more complex than most programming exercises in typical introductory texts. One can let students play the game according to the given rules to discover omissions and ambiguities. This is a very good exercises that shows how difficult it is to cover all alternatives and to avoid ambiguities.

At this point, it will be clear that a language is needed to represent an algorithm. Students can then explore the different languages used in daily life to represent algorithms. Typically, these are used in instructions to perform certain actions. Ask students to look for such examples and explain their favorite. As mentioned earlier, cooking recipes, wiring diagrams, knitting patterns, musical scores, choreography scripts, or pictographs illustrating airplane emergency procedures all use languages that describe algorithms. Discuss the effectiveness of these languages, the semantics (meaning) of the symbols, and even the grammar rules. Some of these languages also illustrate levels of abstraction. A cookbook for the absolute beginner may define some basic rules (simmer, sear, fold in, check if done, ...) while others assume the user understands such terms.

With this preparation, students are ready to learn about finite automata. Start by asking them to create an automaton that describes a part of their daily routine (getting up, coming home from school, etc.). Next, ask students to create an automaton that represents VCR controls. A more abstract task is to describe a finite state automaton that performs long division. Many other ideas about exploring different types of automata, and related exciting problems can be found in [**8**]. Hypercard automata simulation is described in [**10**].

2.3. Computer Organization: Managing Complex Systems. There are many simple computer-based simulations that illustrate basic computer components and their behavior, especially the execution of a sequence of instructions. If possible, these should be used, especially if they give the user some degree of control and an opportunity for experimentation. The description of two such systems can be found in [**4, 12**]. Typically, they run on a particular hardware/software platform. To choose the right one, one should browse through the recent issues of the *ACM SIGCSE Bulletin.*

To illustrate the management tasks performed by an operating system we can use analogies from real life. For example, one can compare a computer operating system to the operation of a restaurant. The menu provides the command language; the waiter provides the user interface, and acts as a command interpreter. The cook represents the CPU, and the ingredients are the inputs. We may look at a particular order as a program for the cook, which then has to be translated into a recipe (machine language) before the cooking program is executed. The prepared meals are the outputs.

As an alternative, one can compare the computer to an office, with a clerk keeping track of all files and folders, and a boss requesting certain files from a secretary, who marks them "in use" until they are returned. The secretary also schedules appointments, decides when to interrupt a meeting with an important message, reserves meeting rooms, etc. Another example of a complex management system that can be compared to an operating system is an emergency room in a hospital. The operation of the Internet can be explained by comparing it to a post office or to a telephone company.

Students can simulate other organizational structures — either by role-playing or by using good computer-based simulation models. Of course, they can also explore alternatives using a pencil and a paper. In addition, to learn about some of the internal parts of an operating system, students can explore different scheduling algorithms (e.g., first-in-first-out which is like a grocery checkout line, first-in-last-out that represents getting in and out of a crowded bus, or priority first scheduling which is like the scheduling of treatments in an emergency room). They can also talk about queue management, routing protocols, and error detecting and correcting codes.

Of course, students should also gain experience working with a real computer and interacting with a real operating system.

2.4. Problem Solving Techniques. Many areas of discrete mathematics involve algorithms that are simple to explain, yet which represent a range of problem-solving techniques, such as backtracking, recursion, divide-and-conquer, and iteration.[1] These are also the algorithms that have the widest practical application — optimization, scheduling, forecasting, modeling, and communications, for example. Select examples that illustrate as many of these techniques as possible. Let students discover the techniques, then reflect on the underlying principles.

Let students explore maze-searching strategies, or try to solve the N-queens problem, to learn about backtracking. Use different minimum-spanning-tree algorithms to illustrate greedy strategies. Then try to fill a knapsack using a greedy strategy, showing that it does not always provide the best packing. One can illustrate that there is no simple algorithm for finding an optimum vertex cover — students can discover this by trying to determine the optimal location of fire stations.

Other examples are paper folding to generate the dragon curve, the traveling salesman problem, finding the shortest path in a network, exploring routing algorithms, graph coloring, and sorting.

2.5. A Sense of Scale. Ask students to solve the Towers of Hanoi problem by working with four or five different disks cut out of paper. Have them count the moves, deduce the strategy, and explain the recursive solution. Count the number of moves needed to move the 64 disks, and estimate the length of time for completion of the task (e.g., if each move takes one second).

Ask students to guess a number you have selected between one and one million. Ask how many questions they think they will need to guess the number. Let them discover why at most twenty questions will lead to a solution. The difference between exponential time, linear time, and logarithmic time should be explored. Let students think about how to manage Internet searches for information on a given topic.

2.6. The Place of Computers in Today's Society. This topic is best studied by reading the daily press, popular magazines, or some of the books that deal with issues related to social responsibility in computing. Students can present case studies, debate an issue, or write a position paper addressed to their congressional representative. Some discussion of these issues can be found in [**13**]. A comprehensive collection of risks related to computing [**14**] makes for fascinating reading and provides a variety of topics for discussions.

Weekly science supplements in major daily newspapers present many examples where the use of computers advances knowledge. There are examples from medicine, molecular biology and genetics, archaeology, space exploration, and ecology. Discuss the computational techniques that support these discoveries. Let students reflect on the need for speed, reliability,

[1] Editors' note: See Henderson's article in this volume for examples of activities that illuminate these concepts.

ability to manage large amount of data, and the new algorithms that lie behind these discoveries.

Ask students to write a paper on a day in the world without computers. What are the changes that affect us daily? Are they all positive? What are the drawbacks? What are the predictions for the future?

3. Issues of Pedagogy.

In a good English writing class students create their own stories. They read works of famous authors to learn technique and style, but nobody asks them to rewrite a scene from Hamlet. Their writing is evaluated and critiqued, and they learn by experimentation. In contrast, when studying mathematics, students often repeat proofs of theorems and solve problems using techniques that are given to them.

The advantage of starting with a new curriculum is that we can define from the beginning what are the most important experiences students should have. In computer science these are critical thinking, discovering algorithms, experimenting with alternatives, and representing one's ideas in different ways. We need to learn the language and the grammar — just as in English classes, yet we should always have ample opportunities to write our own stories, make our own discoveries, and learn from them.[2] We need to learn for ourselves what is a good poem, an effective essay, or a powerful argument. And, as in English classes, where we study the masters as examples of good writing, we should look at classical algorithms and solutions as examples to learn from. The world of a computer scientist or a mathematician is a creative one — we should strive to open up this world to our students.

To make this kind of teaching spread beyond the experimental stage, a number of issues need to be addressed by different segments of the educational establishment. Administrators have to be convinced of a need for computer science education (of the kind described here). Colleges and universities need to recognize the importance of courses of this type by including them in admission requirements. Computer science professionals, whether in industry or in academia, have to lend support by developing curriculum materials, assisting in teacher training (and retraining), and helping schools acquire better computing facilities — including hardware, software, and maintenance.

A major effort should go into the development of computer-based teaching tools for studying computer science and discrete mathematics concepts. Although the use of computer modeling is widespread in physics, calculus, the study of foreign languages, social studies, and other fields, computer science is still largely taught in classrooms where a chalkboard offers the only graphic display in the room. Simple computer animations of algorithms —

[2]Editors' note: The article by Casey and Fellows in this volume makes a similar argument for the K–4 mathematics curriculum.

of models of finite state automata, of data paths in query processing, of dynamic changes in data structures as an algorithm progresses — all of these can help students build new mental images and representations of these dynamic concepts. These animations should require a minimal hardware and software base, and, whenever possible, should be built in such a way that the students can use them interactively, stopping and restarting them at will, so they can serve as basis for meaningful laboratory experiments.

4. Conclusion.

While the computer revolution is changing the basic structure of modern society, the nature of this revolution is largely ignored in high schools. I have presented an argument for including a course in computer science in a regular high school curriculum and described the fundamental concepts this curriculum should cover. I have also proposed a new way of looking at the computer science curriculum, that is modeled after a typical English curriculum, namely, as a study of a language and concepts of computing, with a lot of room for creative thinking and expression.

5. Acknowledgments.

Numerous discussions with colleagues helped in shaping the ideas presented in this paper. I am indebted to all members of the ACM Task Force on Education — Charlie Bruen, J. Philip East, Darlene Grantham, Susan Merritt, Chuck Rice, Gerry Segal, and Carol Wolf. The collaboration with my colleagues at Northeastern University, Richard Rasala, Harriet Fell, and Cynthia Brown on developing curriculum materials for computer science has made me look constantly at different ways in which students learn and get engaged in learning. Discussions with Erich Neuwirth, Allen B. Tucker, Alan Biermann, A. Joe Turner, Peter Gloor, Wally Feuerzeig, Mike Fellows, and many others have helped in clarifying some of the issues. Their contributions are greatly appreciated.

Special thanks to the referee of this paper whose suggestions helped greatly in organizing this paper and focusing on the essential ideas.

References

[1] ACM Task Force, "ACM Model High School Computer Science Curriculum", *Report of the Task Force on High School Curriculum of the ACM Pre-College Committee*, ACM Press, 1993.

[2] Bezenet, L. P., "The Teaching of Arithmetic I, II, III: The Story of An Experiment," *The Journal of the National Educational Association*, November (1935).

[3] Biermann, A. W., *Great Ideas in Computer Science*, MIT Press. 1990.

[4] Biermann, A. W., A. F. Fahmy, C. Guinn, D. Penncock, D. Ramm, P. Wu, "Teaching a Hierarchical Model of Computation with Animation Software in the First Course," *SIGCSE Bulletin*, 26(1), pp. 295-299, 1994.

[5] Brown, C., H. J. Fell, V. K. Proulx, and R. Rasala, "Programming by experimentation and Example," in I. Tomek, ed., *Computer Assisted Learning, Proceedings of the 4th*

International Conference ICCAL '92, Wolfville, Nova Scotia, June 1992 (Springer Verlag 1992), 136-147.

[6] Carter, R., W. Feurzeig, J. Richards, and N. Roberts, "Intelligent Tools for Mathematical Inquiry," in *Proceedings of the 9th Annual National Education Computer Conference*, Dallas TX, (1988).

[7] Casey, N., *The Whole Language Connection*, Washington State Mathematics Council (1991).

[8] Dewdney, A. K., *The Armchair Universe*, W. H. Freeman and Co., New York, 1988.

[9] Fellows, M. R., "Computer Science and Mathematics in the Elementary Schools", *CBMS Issues in Mathematics Education*, Volume 3, pp. 143-163, 1993.

[10] Hannay, D., "Hypercard Automata Simulation: Finite-State, Pushdown, and Turing Machines," *ACM SIGCSE Bulletin*, 24(2), p. 55, 1992.

[11] Latour, B., "Visualization and Cognition: Thinking with Eyes and Hands," in *Knowledge and Society: Studies in the Sociology of Culture Past and Present,* 6 (1986) 1-40.

[12] Magagnosc, D., "Simulation in Computer Organization: A Goals Based Study," *SIGCSE Bulletin*, 26(1), pp. 178-182, 1994.

[13] Meyer, M., and R. Bauer, *Computers in Your Future*, Que Corporation, 1995.

[14] Neumann, P. G., *Computer Related Risks*, Addison Wesley, 1995.

[15] Proulx, V. K., "Computer Science in Elementary and Secondary Schools," in *Informatics and Changes in Learning,* Proceedings of the IFIP TC3/WG3.1/WG3.5 Open Conference on Informatics and Changes in Learning, Gmunden, Austria, 7-11 June 1993, D. C. Johnson, B. Samways, eds., North Holland, 1993, pp. 95-101.

[16] ———, "Suggested Exercises for College Preparatory Curriculum in Computer Science," College of Computer Science Technical Report NU-CCS-91-9, Northeastern University, 1991.

[17] Tinsley, D., ed. Proceedings of the IFIP WC3.1 Working Conference *Impacts of Informatics on the Organization of Education*, Santa Barbara, USA, August 1991, (Elsevier, North Holland, 1992).

College of Computer Science, Northeastern University, Boston, MA 02115
E-mail address: vkp@ccs.neu.edu

Section 8

Resources for Teachers

Discrete Mathematics Software for K–12 Education
NATHANIEL DEAN AND YANXI LIU
Page 357

Recommended Resources for Teaching Discrete Mathematics
DEBORAH S. FRANZBLAU AND JANICE C. KOWALCZYK
Page 373

The Leadership Program in Discrete Mathematics
JOSEPH G. ROSENSTEIN AND VALERIE A. DEBELLIS
Page 415

Computer Software for the Teaching of Discrete Mathematics in the Schools
MARIO VASSALLO AND ANTHONY RALSTON
Page 433

DIMACS Series in Discrete Mathematics
and Theoretical Computer Science
Volume **36**, 1997

Discrete Mathematics Software for K-12 Education

Nathaniel Dean and Yanxi Liu

1. Introduction

A significant number of researchers are developing general-purpose software and integrated software systems for domains in discrete mathematics, including graph theory, combinatorics, combinatorial optimization, and sets [**16**]. The main goal of such software is to provide effective computational tools for research, applications prototyping, and teaching in these domains. Some of the systems are being used for teaching courses on algorithms at the college level (for example, in Computer Science, Discrete Mathematics, or Operations Research) and to explain concepts that might otherwise be difficult to comprehend. They are used as part of labs for experimentation or in conjunction with projects for students to gain hands-on experience with algorithms. The visual and interactive nature of many of these systems tends to stir some enthusiasm in students who would otherwise have little or no interest in the course. Some systems are also being used at the high school level to motivate students to pursue careers in mathematics and computer science.

In this article we start with our report on two workshops with teachers and researchers using NETPAD (Section 2.1) and Combinatorica (Section 2.2) for K-12 mathematics education. To motivate the discussion, we raise and try to answer (Section 3, Section 3.3) the following questions, which should probably be considered in any evaluation of the use of computers in teaching:

1. *What are the desirable features of educational mathematics software from an educator's point of view?*
2. *To what extent can research-oriented mathematics software be used directly for education?*

1991 *Mathematics Subject Classification.* Primary 00A05, 00A35, 68N99.

3. *What do we gain by using computers in teaching discrete mathematics?*
4. *What discrete mathematics software is available?*

We hope that this article will help to facilitate collaboration among a diverse group of researchers and educators who are concerned with the development of software for various areas of discrete mathematics, and especially for K-12 mathematics education.

The results of the two workshops are presented in Section 2. We describe our observations and conclusions in Section 3. The Appendix summarizes and comments on the features of selected, currently available mathematics software which may be useful for educational purposes. The section also includes a list of other relevant software.

2. Two workshops involving researchers and educators

Many K-12 mathematics teachers have included discrete mathematics into their curriculum. The teachers and their students are having fun with it (see other articles in this volume). Some teachers suggest that certain students do very well on discrete mathematics even though they do not do well on traditional math (the students' main complaint about traditional math: too much rule-remembering). However, one can rarely find any K-12 teachers who use computers in their discrete mathematics classes. Why not? According to a high school teacher from New Jersey, all the software systems he has encounted are "not friendly enough". Within the special setting of the Center for Discrete Mathematics and Theoretical Computer Science (DIMACS), we had the opportunity to bring together, in two different one-day sessions, a group of middle and high school teachers[1] with researchers who developed NETPAD [**14, 15**] and Combinatorica [**37**]. In the following we report on

- our effort to obtain answers from the educators to some basic questions on educational software development, as described in the introduction;
- our observations of the interactions between the researchers and the educators; and
- our conclusions based on these observations.

2.1. NETPAD workshop.

2.1.1. *Introduction to NETPAD.* NETPAD is a software system with a menu-driven user interface, which allows the user to draw graphs, such as those on the left of Figure 1, including cycles, wheels, chains, and complete graphs with a specified number of vertices. It also permits the user to construct graphs with a mouse by placing vertices at arbitrary locations, then drawing lines connecting any specified pair of vertices; thus, for example, the

[1] Judy A. Brown, Pat Cline, Pat Johnson, Jan Kowalczyk, Jim Morris, Susan Picker, Susan Simon, and Ken Gittleson, all participants in the Leadership Program in Discrete Mathematics.

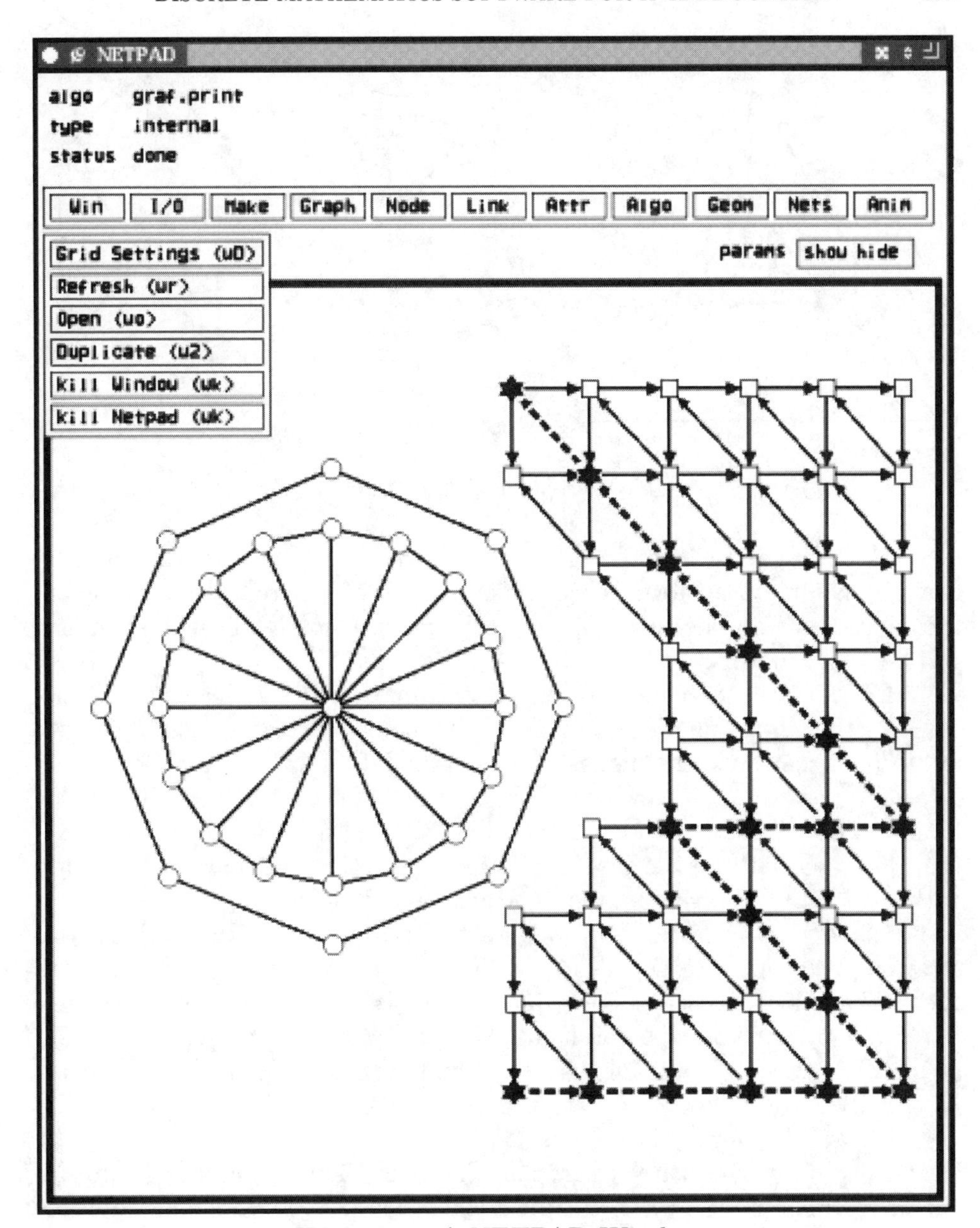

FIGURE 1. A NETPAD Window

user can easily draw any of the graphs in Figures 2, 3, and 4. The user can also convert these graphs into directed graphs or weighted graphs, and can associate labels with vertices or edges. Using the menus, the user can also apply any one of a number of algorithms to a graph, then display or even animate the result; for example, the directed graph on the right of Figure 1 includes a directed path consisting of dotted edges, found by an algorithm, from the lower left corner to the upper left corner. The user can transform graphs by dragging a vertex (and the edges incident with it) from its original

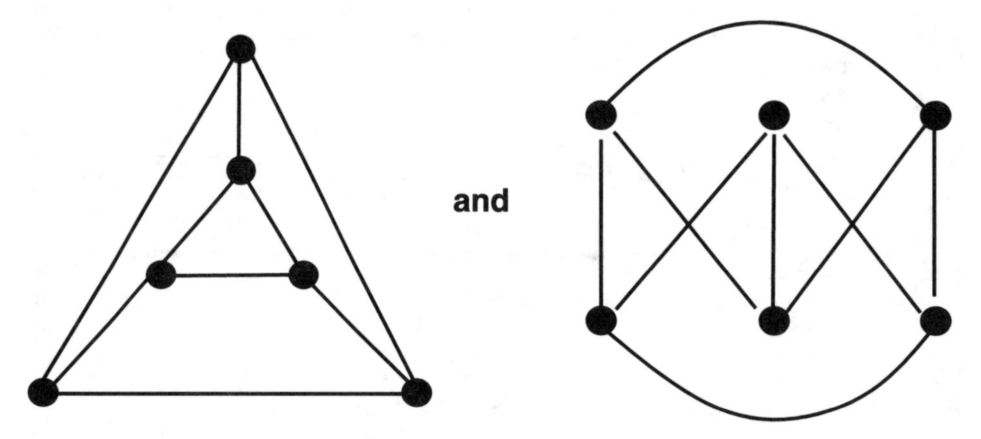

FIGURE 2. Example one

location to any other location. By using a sequence of such moves the user can, for example, transform the graph on the left of Figure 2 to the graph on the right. A comprehensive manual with exercises for NETPAD beginners can be found in [**30**].

NETPAD can also be customized and expanded by the user. Users can add their own functions to NETPAD, change various system defaults, and edit or rearrange the menus. There is no optional user-friendly textual interface; however, one can bypass the graphical user interface and call the functions or algorithms from a library directly (see the documentation [**14**], [**15**], [**32**], and [**33**] for details). NETPAD currently runs only on machines with the UNIX operating system, and not on personal computers.

2.1.2. *Session with teachers.* We started the day with a NETPAD demonstration given by Nate Dean, one of the developers of NETPAD, then let the teachers do two discrete mathematics exercises using NETPAD. We prepared a list of questions beforehand. No questionnaire was given, instead, at intervals we set a topic for discussion and recorded teachers' responses.

The two exercises we did with the teachers are:

1. Graph Isomorphism Exercise[2]:

Given a pair of pictures of graphs (Figure 2, Figure 3, Figure 4), create the first graph on NETPAD, then use NETPAD (i.e., modify the graph by moving its vertices) to transform it into the second graph.

2. Rectilinear Crossing Number Exercise:

For complete graphs K_n with $n = 3, 4, 5, \ldots$ try to determine the *rectilinear crossing number* (a crossing is an intersection of two edges not at a vertex); i.e., find a drawing of K_n with least number of crossings where each edge is drawn as a straight line segment. Note: The problem is unsolved for $n > 9$. (See [**21**] for more information on this problem.)

---

[2]This exercise, and the examples in Figures 2, 3, and 4 were suggested and developed by Joseph G. Rosenstein.

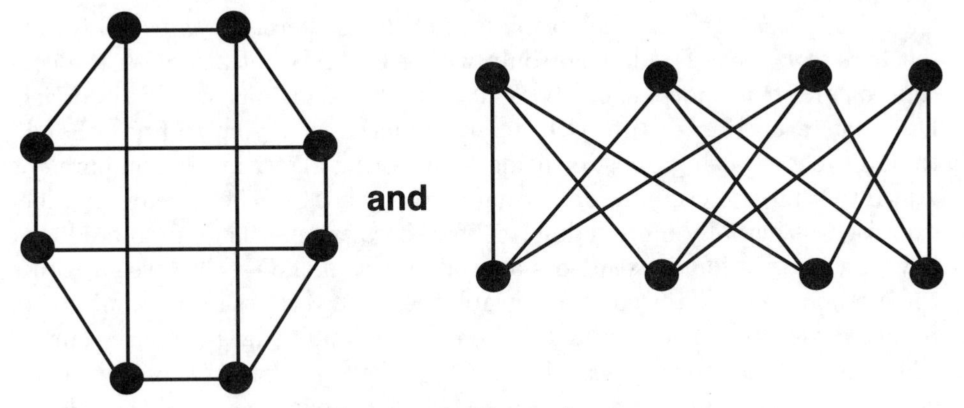

FIGURE 3. Example two

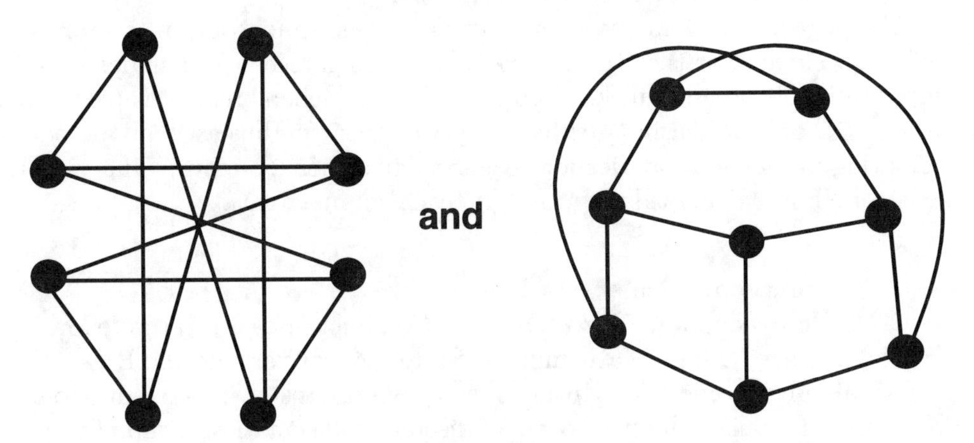

FIGURE 4. Example three

Teacher's comments on the graph isomorphism exercise:

> It is always important to start with simple, basic problems. This exercise is such a good problem for getting started with NETPAD, and it is harder and richer than it appears at first glance. For this problem, using computers is obviously easier than using pencil-and-paper. This exercise is very helpful for teaching students the idea that distance between vertices is irrelevant in a graph (a common misconception).
>
> One needs a strategy to tackle the problem, just playing around does not work. Knowing more graph theory changes what you learn from the exercise and it remains interesting at many different levels.

Teacher's comments on the crossing number exercise:

> The exercise needs to be motivated by some applications, although the exercise itself is very engrossing, "I could hardly wait to get back from lunch to continue!". This would be a very good problem for students.

2.1.3. *Observations.* The graph isomorphism exercise was useful for the teachers to become familiar not only with NETPAD but also with the concept of graph isomorphism. It is not at all boring, as we first worried. Different teachers came up with different methods: trying to find a hamiltonian cycle in both graphs, or using a 2-coloring algorithm to compare one graph to a bipartite graph, or "visualize-then-check". The crossing number exercise was actually quite 'addictive' for the teachers. They predicted that it would be true for the students as well. Most of the teachers seem to use the method: "create the complete graph then pull vertices into the middle to reduce crossings". They wanted to be able to "pull" edges (requires curved edges) as well as move nodes. The most positive impact of these two exercises is that they forced the teachers to look for patterns or methodologies in solving the problems, using NETPAD as a tool.

We began with some teachers working alone, and some in pairs. It quickly became clear that those working in pairs gained better understanding of the problem and quicker solutions. From teachers' reactions (see also Section 3), one can easily see that they were very enthusiastic about participating in all the activities and eager to give their opinions. The visual effect of NETPAD helped to bring out teachers' imaginations.

2.2. Combinatorica.

2.2.1. *Introduction to Combinatorica.* Combinatorica is a Discrete Mathematics package built on Mathematica. Section 4.1 of Vassallo and Ralston's article [**39**] gives a good description of some operations of the system. Combinatorica specializes in two aspects of discrete mathematics: combinatorics and graph theory. In [**37**] Skiena provides a problem-based approach to help the reader learn to use Combinatorica and discrete math. One advantage of Combinatorica over some software systems is the fact that it is associated with a large software system (Mathematica) and a publisher (Addison Wesley), so that the system gets distributed and publicized widely. On the other hand, the distribution of NETPAD depends on its developers.

2.2.2. *Session with teachers.* We started the day with a demonstration of Mathematica and Combinatorica by Steven Skiena, who developed Combinatorica. He showed for example how to do numerical calculations and solve polynomial equations (Mathematica), as well as how to display graphs and results of operations on graphs (Combinatorica). Participants were then presented with the following problem from [**25**], (pp. 34–48):

> Any permutation of numbers can be broken into 'runs', each of which is a maximal sequence of ascending, increasing, left-to-right consecutive elements. For example, the permutation (5,2,3,1,4) defines the runs (5), (2,3), (1,4).
>
> Questions:
>
> 1. What is the expected length of the first run in a random permutation of n elements, for large n?

2. Is the expected length of the second run longer, shorter or the same as that of the first run?

In contrast with NETPAD, Combinatorica's interface is strongly textual (although the display is graphical); one cannot use the mouse to move nodes around in a graph, and commands have to be typed in. The exercise was carried out by letting teachers request Combinatorica's functions through Steven Skiena, who typed in the relevant commands. Someone commented that we need Mathematica "plus an intelligent slave".

The teachers tried to solve this problem in several ways: all utilized exhaustive searches of all permutations; this made the calculation time unbearably long (the one teacher who tried to do this by hand could only get to $n = 4$, while those using Combinatorica couldn't get past $n = 7$). However, based on the results obtained for small n, the teachers started to make hypotheses and tried to justify them. Steven Skiena suggested using Combinatorica on large random permutations to try to justify answers, which led to an interesting comment (below).

Some of the teachers' comments on the permutation exercise were:

- "I'm really learning. But this problem is too hard for my students."
- "I'm not sure about the randomness, how many of those random permutations are enough for representing the real thing?" (This question came up after Steven Skiena suggested that instead of checking the average length of the 1st and 2nd runs of all the permutations, we can just do it for a random sample.) After seeing the variability of the results, the teachers were not willing to trust the answer of a random sample. This question had everyone thinking: "*When can random samples be trusted?*"
- While the above discussion was going on, a teacher commented: "It is hard to teach students that the procedure (method) of finding a solution is at least as important as a correct answer".
- One cannot do this exercise using NETPAD.

Teachers' general comments on Combinatorica/Mathematica (C/M):

- The functionalities of C/M presented here would be quite useful for teaching middle school students permutations (e.g., displaying permutations in an organized way), and high school students probability. Lots of other interesting things can be done such as testing graph isomorphisms (using IsomorphicQ).
- The graphics for 3D surfaces is very impressive.
- It is very useful to have a book and on-line help in Mathematica.
- However, you do not want students to be bothered by the syntax, since that will distract them from learning mathematics. We need predefined operators, and as little typing as possible in order to get into the mathematics quickly. *We need a Combinatorica in a NETPAD format.* (Although even the interface of NETPAD was not considered entirely friendly by the teachers.)

2.2.3. *Observations.*

- We witnessed a session where mathematics was treated as science in the sense that we first observed specific cases, then made hypotheses, and finally verified the hypotheses using Combinatorica. Here the computer software was used as a tool for exploration in an active learning process.
- The exploration of mathematics can be done using computer software or manually (as one of the teachers did). Obviously, computer software is much less error-prone and more convenient for generating large sets of examples. Only one example was completed manually during the session and many mistakes were made along the way.
- The question *when to trust random samples* raised by a teacher is a very good question that one has to address during the teaching or learning of probability. This episode showed us that although computer software can be very helpful in mathematics education, it cannot replace teacher's careful guidance on how to interpret the results generated by computers.
- As teachers pointed out, the user interface for computer software is crucial to its usefulness and functionality (it is the difference between being a really helpful tool and an imagination-limiting factor).
- Whether the software is widely distributable is an important yet practical issue for the users as well as for the software developers.

3. Discussion and Conclusions

3.1. Teachers' software wish-list. Based on our studies we have made a number of observations that deserve some discussion. Let us start with a summary of a software wish-list from the teachers in our workshops:

1. They want on-screen guided tutorials and an on-line help facility as well as written documentation. The tutorial should give just enough information to get started, but should also have a demo to "wow" students and hint at capabilities. One must be able to get in and out of the tutorial at any point.
2. They would like to be able to open a "journal" or "notebook" window to write down observations while working on exercises (e.g., as in Logo).
3. They want software that can be used at many levels, i.e., a tool that young students can use to explore, but which becomes even more useful as you learn more mathematics.
4. They want software to help students become familiar with concrete examples of geometric figures, graphs, and sets, so that they will be interested in learning about the more abstract theorems and algorithms later.

5. They consider topological concepts captured by graphs as important as geometric concepts. They would like something like "Geometer's Sketchpad" [**19**] for graph theory and combinatorics.
6. They want the software interface to be user friendly. The user should be able to give instructions by clicking the mouse, i.e., using a graphical interface. This is considered much more important than having lots of features.
7. They want the software to run on machines that they already have (PCs or Macs with 4mb internal and 80mb hard disk seemed realistic; a 20mb program sounded much too large), to use PC-windows conventions or Mac conventions (i.e., to use an interface that students and teachers already know), to be as small and flexible as possible, and to be fast, so that no one spends lots of time waiting for results.
8. They like the idea of modular software, starting with a basic program, then being able to add things like algorithms later.
9. They feel strongly that both teachers and students should be involved in all phases of the development of educational mathematics software.
10. They feel that software development and materials development need to proceed in parallel.

In general, the teachers desire a "user-friendly" and sound software system (points 1, 3, 4 and 6 above), which runs at a reasonable speed. Visualization and easy handling of the objects to be investigated are very important to the teachers. They prefer to let the students explore by themselves rather than explain every detail for them. They seem to want to provide students something to 'play' with that also stimulates the students to learn. This attitude is reflected by the teachers liking the graph isomorphism and crossing number exercises and their somewhat unwilling acceptance of using random samples.

The hardware requirement proposed by the teachers (point 7 above), however, is worth further discussion in light of recent technological advances and software developments. The computer industry is advancing at a fast rate, and today's PCs have features that have already surpassed yesterday's workstations. If one aims at the computers that teachers have today, the software may well be outdated when the final product is produced. Our recommendation is to build software to run *at least* on a 486 PC or comparable Macintosh with 8mb internal memory and 200mb hard disk. More demanding hardware requirements are surely foreseeable.

3.2. Responses to initial questions. Now let's return to the questions raised at the beginning of this article and see whether we can answer each of them:

1. *What are the desirable features of educational mathematics software from an educator's point of view?*

   See the teacher's software wish list and discussions above.

2. *To what extent can research-oriented mathematics software be used directly for education?*

   NETPAD and Combinatorica were both developed as tools for researchers, but evidently the teachers consider both of them to be useful for teaching discrete mathematics. Two of the key elements for using math software directly for educational purposes are the user-friendliness of the software and the availability of appropriate course materials.

   For user-friendliness, a graphical interface seemed much more attractive to the teachers than a textual interface (NETPAD over Combinatorica); on-line help is essential as well (Combinatorica over NETPAD).

   The course material to go with each software package needs to be carefully chosen in order for students to learn successfully. For example, one cannot do the estimations of runs in a string using NETPAD, nor can one enjoy doing graph isomorphism or crossing number exercises as much using Combinatorica, since one cannot drag vertices around with the mouse.

   This reflects both the potential and limitation of research-oriented mathematics software when it is used for educational purposes.

3. *What do we gain by using computers in teaching mathematics?*

   This is an important and broad topic. We do not intend to give a complete answer here. However one thing that is obvious in our study is the gained computing power. The computer allows us to access greater amounts of information, manipulate larger examples, and to do so quickly and correctly. This can be seen especially in the second workshop with Combinatorica, where only one example was worked out by hand by the teachers with a lot of backtracking to make corrections, while many examples were correctly displayed by Combinatorica. This computational power gave the teachers the opportunity to see the 'patterns' in the problem. Another observation is that an interesting question combined with the appropriate software can stimulate good discussions on mathematics and scientific methods to approach a mathematical problem (Section 2.2).

   More studies should be done on this topic for qualitative as well as quantitative measurements, and to evaluate the effects of using versus not using computers in teaching mathematics.

4. *What discrete mathematics software is available?*

   The answers to this question will be provided in the Appendix.

3.3. Conclusion. Besides what we learned from the teachers about their desires for educational software, we found that this kind of researcher-teacher session effectively facilitates two-way communication, and is beneficial for both the software developers and the teachers who are the potential users of the software. Teachers want to be involved in the software development process (points 9 and 10 of teacher's software wish list in Section 3). Teachers can provide concrete specifications on what subset of the existing software they need and what they would like to add specifically for educational purposes. For the teachers, as soon as they start to use the software, they immediately start to think about what course material can go with it; the developers/researchers learn unexpected weak points of the system, and receive feedback on the spot, which can have a positive impact on their future system design. Based on these facts we highly recommend this interactive approach for any educational software development project.

Due to the limitations of current software systems for discrete mathematics education, there seems to be an obvious demand for a math software system that covers a wide spectrum of discrete math topics *and* suits the educational needs of K-12 teachers and students. The only comprehensive discrete mathematics software project that the authors know of is LINK [**6**], where the goal is to build a software system for combinatorial computing and experimental discrete mathematics that is efficient, portable, and extensible. However, the educational requirement is not addressed. We are certainly interested in building an educational version of LINK.[3] We are encouraged by the enthusiasm and knowledge of the teachers, and we encourage our readers to join us in our efforts to develop better tools for discrete mathematics education.

Appendix: Other Existing Systems

The first few sections of this appendix give a sampling of other existing discrete mathematics systems that may be of interest to users or potential users, and that satisfy some of the desirable characteristics of educational software described by the teachers. A larger list is given in A.4 (without comments), but even that list is not close to being exhaustive. See [**16**] for more details on some of these and other packages. None of the packages described below contain the full range of features available in NETPAD (see Section 2.1.1). In particular, none of them contain built-in tools for manipulating attributes or for animating network algorithms. Several software libraries are omitted from this discussion including LEDA and the Stanford GraphBase [**26**]. The programs for the Macintosh and the IBM PC have an obvious computational disadvantage imposed by the hardware (i.e., speed and memory) but the necessary hardware for the UNIX-based systems is more expensive and, therefore, not as readily available for K-12 education (recall our discussions with teachers in Section 3).

A.1. IBM PC-based programs. INDS (Interactive Network Design System) is a software package developed at Bellcore by C. L. Monma and D.

[3]Since the writing of this paper LINK has been made publicly available. The software and documentation can be downloaded from the Web at `http://dimacs.rutgers.edu/Projects/LINK.html`.

F. Shallcross [**34**] for the IBM PC. It is used to solve certain network design and optimization problems. In particular, it uses an interactive graphical interface for applying heuristics to solve the traveling salesman problem and to find minimum cost 2-connected networks, problems that arise in the construction of survivable communications networks. This tool could be used to introduce students to telecommunications applications.

CARDD (Computer-Aided Representative graph Determiner and Drawer) is an expert system that constructs a graph with properties defined by the user. It was developed by T. Haynes, L. M. Lawson and M. W. Powell [**24**] and uses a forward chaining inference algorithm; i.e., once an invariant is resolved, it is never eliminated. The properties are specified by setting values for any subset of the available set of eight invariants: number of nodes, number of edges, maximum degree, minimum degree, independence number, maximum clique size, chromatic number and domination number. For students, this could be viewed as an interesting application of artificial intelligence techniques.

Dotty is a graph editor developed at Bell Labs [**28**]. It contains a programmable viewer and facilities for automatic graph layout, actually the most sophisticated graph layout algorithms of any package we have seen. It can be controlled through a graphical or a textual interface. The graphics and operating system details (i.e., fonts, menus, color maps) are hidden from the user, and so the same scripts that run in the Microsoft Windows version also run in the UNIX/X11 version. Dotty has already been incorporated into some software engineering applications at Bell Labs, and it could easily be used as a graphical front end for educational applications.

A.2. Macintosh-based programs. In [**13**], three versions of a program called CABRI are mentioned, one running on a Macintosh, another on a PC-compatible, and a third version for workstations that uses the BWE window management toolset from Brown University. (Only the Macintosh version was available to us.) It contains several network editing and analysis functions similar to NETPAD.

Groups & Graphs [**27**] is a program for manipulating graphs and groups. It contains various group theoretic algorithms of interest to graph theorists such as computing the automorphism group of a graph and determining whether two graphs are isomorphic. This program may be helpful to teachers in introducing and demonstrating group theoretic concepts and symmetries associated with graphs or geometric representations of graphs.

A.3. UNIX-based programs. The programs GMP/X by Esfahanian (personal communication with some of its users), GraphPack [**22**] and Combinatorica [**37**] all run under UNIX. The first author used an older version of GMP/X called GMP which used SunView. The new version GMP/X and GraphPack are based on X Windows and so are more portable. Combinatorica is actually a collection of programs written in Mathematica which must be purchased but runs on a variety of computers. GraphPack includes a language called *LiLa* which is based on the C programming language with additional primitives (i.e., *set, graph, tree*, etc.) to simplify the coding of new algorithms. The computer hardware required for these programs generally exceeds the limits of K-12 computing facilities, but for those who have access these programs certainly have a lot of potential for educational use. All of these and others mentioned below in Section A.4 have been used in undergraduate classes or labs.

Package	Type	H/W, S/W Constraints	Graphical or Textual?
AGE [**1**]	general	UNIX/X11	graphical
BINKY [**12**]	optimization	IBM PC	graphical
CABRI [**13**]	general	BWE toolset Mac or IBM PC	graphical
Cabri-Géomètre [**19**]	geometry	Mac or PC	graphical
CallCo [**17**]	enumeration	UNIX	both
CARDD [**24**]	graph parameters	IBM PC	textual
CATboxII [**3**]		IBM PC	graphical
Colbourn's program	reliability	Mathematica	textual
Combinatorica [**37**]	general	Mathematica	textual
Dotty [**28**]	graph layout	UNIX/X11 or IBM PC	both
EDGE [**35**]	graph layout		graphical
GAP [**18**]	general	IBM PC	graphical
GDR [**10**]	general	UNIX/X11	graphical
Geometer's Sketchpad [**19**]	geometry	Mac or IBM PC	graphical
Geometric Supposer [**38**]	geometry	Mac or IBM PC	graphical
Geomview [**29**]	geometry	SGI	graphical
GMP/X [**20**]	general	UNIX/X11	graphical
GraphLab [**7**]	general	NeXT	both
GraphPack [**22**]	general	UNIX/X11	both
GraphTool [**8**]	general	UNIX/X11 or SunView	graphical
Groups & Graphs [**27**]	general	Mac	
INDS [**34**]	optimization	IBM PC	textual
Maple's Package [**31**]	general	UNIX/X11	textual
METANET [**23**]	general	UNIX/X11	textual
NETPAD [**14**]	general	UNIX/X11	graphical
Network Assistant [**16**]	general	UNIX	textual
NPDA [**9**]	automata	UNIX/X11	graphical
SetPlayer	sets/hypergraphs	UNIX/X11	both
TRAVEL [**2**]	TSP	IBM PC	textual
xPAC [**16**]	backtracking	UNIX/X11	textual
XTANGO	animation	UNIX/X11	neither
XYZ GeoBench [**36**]	geometric algs	Mac	both

TABLE 1. Some Software Systems for Discrete Math

A.4. A list of other software systems (see Table 1). So many software packages for Discrete Mathematics are materializing in so many different places that it is hard to keep track of them all. At the DIMACS Workshop on Computational Support for Discrete Mathematics many of the experts demonstrated and presented papers on their programs [**16**]. From this experience it seems that there are other software tools comparable to NETPAD and Combinatorica that could be of interest to educators depending on their mathematical specialty, level of computer proficiency, the course subject area, hardware configuration, teaching style, or other factors. For example, AGE is similar to NETPAD and it emphasizes animation, but Network Assistant [**16**] and Maple's Networks Package [**31**] use a high level textual interface, as does Combinatorica, to manipulate graphs and compute graph invariants.

We were not able to evaluate these packages in great detail. (Both GraphPack [**22**] and SetPlayer are reviewed in this volume [**39**], however.) Since the teachers, both K-12 and beyond, have repeatedly asked for more information, we decided to assemble the list in Table 1 focusing on features that seem to be important to teachers and giving pointers to sources of more information. Most of the systems have some sort of graphical display or output, but to qualify as "graphical" here we insist that the user be able to interact with and manipulate the relevant objects (i.e., graphs, sets, polyominoes, etc.) with a pointer such as a mouse. To be "textual" we expect essentially the same functionality, but the objects are manipulated using the keyboard, possibly with no graphical display.

Acknowledgments

The authors would like to thank the teacher participants and Deborah Franzblau and Steven Skiena for their contributions to the workshops.

References

[1] J. Abello, S. Sudarsky, T. Veatch and J. Waller, "AGE: An animated graph environment." In [**16**].

[2] S. C. Boyd, W. R. Pulleyblank, G. Cornuéjols, "TRAVEL," Dept. of Mathematics and Statistics, Carleton Univ., Ottawa, Ontario K1S 5B6.

[3] A. Bachem, "CATbox II: Combinatorial Algorithm Toolbox for IBM Computer and Compatible," Presentation at the DIMACS Workshop on Computational Support for Discrete Math., Piscataway, NJ (March 1992).

[4] A. Badre, M. Beranek, J. Morgan Morris, and J. Stasko, "Assessing program visualization systems as instructional aids", Technical Report GIT-GVU-91-23, College of Computing, Georgia Institute of Technology, Atlanta, Georgia (October 1991).

[5] D. Berque, R. Cecchini, M. Goldberg, and R. Rivenburgh, "The SetPlayer system: An overview and a user manual," RPI Technical Report 90-20 (1991).

[6] J. Berry, N. Dean, P. Fasel, M. Goldberg, E. Johnson, J. MacCuish, G. Shannon, S. Skiena, "Link: a combinatorics and graph theory workbench for applications and research," DIMACS Technical Report 95-15 (June 1995).

[7] B. Birgisson and G. Shannon, "Graphview: An extensible interactive platform for manipulating and displaying graphs," Technical Report 295, Computer Science Dept., Indiana Univ., Bloomington, Indiana (December 1989).
[8] A. L. Bliss, M. B. Dillencourt and V. J. Leung, "GraphTool: A tool for interactive design and manipulation of graphs and graph algorithms." In [**16**].
[9] D. Caugherty and S. H. Rodger, "NPDA: A tool for visualizing and simulating non-deterministic pushdown automata." In [**16**].
[10] R. Cleaveland, P. Hebbar and M. Stallmann, "GDR: A visualization tool for graph algorithms (overview)." In [**16**].
[11] C. J. Colbourn, J. S. Devitt, and D. D. Harms, "Networks and Reliability in Maple." In [**16**].
[12] W. Cook, "Teaching combinatorial optimization with BINKY," Presentation at the DIMACS Workshop on Computational Support for Discrete Math., Piscataway, NJ (March 1992).
[13] M. Dao, M. Habib, J. P. Richard and D. Tallot, "CABRI, an interactive system for graph manipulation," Presentation at the DIMACS Workshop on Computational Support for Discrete Math., Piscataway, NJ (March 1992).
[14] N. Dean, M. Mevenkamp and C. L. Monma, "NETPAD: An interactive graphics system for network modeling and optimization," Computer Science and Operations Research, Pergamon Press (1992) 231-243.
[15] ———, "NETPAD Version 1 - User's Guide," Bellcore Technical Memorandum TM-ARH-022264, Nov. 17, 1992.
[16] N. Dean and G. Shannon, eds., *Computational Support for Discrete Mathematics*, Am. Math. Soc., DIMACS Series, V. 15. Proc. of DIMACS Workshop, March 1992.
[17] M. Delest and N. Rouillon, "CalICo: Software for Combinatorics," In [**16**].
[18] D. S. Dillon and F. Smietana, "An interactive, graphical, educationally oriented graph analysis package." In [**16**].
[19] J. W. Emert and W. V. Habegger, "Cabri-Géomètre vs. the Geometer's Sketchpad: A comparison of two dynamic geometry systems," *Notices of the AMS* **40** (Oct 1993) 988-992.
[20] A. H. Esfahanian, G. Zimmerman and D. Vasquez, "GMP/X, an X-Windows Based Graph Manipulation Package" In [**16**].
[21] M. R. Garey and D. S. Johnson, "Crossing number is NP-complete," *Siam J. Alg. Disc. Meth.* **4** (1983) 312-316.
[22] M. Goldberg, E. Kaltofen, S. Kim, M. Krishnamoorthy and T. Spencer, "GraphPack: a software system for computations on graphs and sets," manuscript.
[23] C. Gomez and M. Goursat, "METANET: A System for Network Analysis." In [**16**].
[24] T. Haynes, L. M. Lawson and M. W. Powell, "CARDD (Computer-Aided Representative-graph Determiner and Drawer)," *Congressus Numerantium* **77** (1990) 163-168.
[25] D. E. Knuth, *The Art of Computer Programming, vol 3, Sorting and Searching*, Addison-Wesley, Reading MA, 1973.
[26] ———, *The Stanford Graphbase: a platform for combinatorial computing*, Addison-Wesley, Reading MA, 1993.
[27] W. Kocay, "Groups & graphs, a Macintosh application for Graph Theory," *JCMCC* **3** (1988) 195-206.
[28] E. Koutsofios and S. C. North, "Applications of graph visualization," manuscript.
[29] S. Levy, T. Munzner and M. Phillips, "Geomview: An interactive geometry viewer," *Notices of the AMS* **40** (Oct 1993) 985-988.
[30] Y. Liu, "A Comprehensive Manual for NETPAD Users with Exercises on Discrete Mathematics," October 1993, DIMACS technical report 94–36.
[31] Maple's Networks Package, Maple V Release 2 help pages (1993).

[32] M. Mevenkamp, "NETPAD Version 1 - Programmer's Guide," Bellcore Technical Memorandum TM-ARH-022265, Nov. 17, 1992.

[33] ———, "NETPAD Version 1 - Reference Guide," Bellcore Technical Memorandum TM-ARH-022266, Nov. 17, 1992.

[34] C. L. Monma and D. F. Shallcross, "A Graphical Interactive Network Design System for the IBM PC: User's Manual", Bellcore, TM ARH-008041 (1986).

[35] F. N. Paulisch, "EDGE: An extendible directed graph editor," Technical Report 8/88, University of Karlsruhe, Institute for Informatics, June 1988.

[36] P. Schorn, "The XYZ GeoBench for the experimental evaluation of geometric algorithms." In [**16**].

[37] S. S. Skiena, *Implementing Discrete Mathematics: Combinatorics and Graph Theory with Mathematica*, Addison-Wesley (1990).

[38] Sunburst, Catalog from Sunburst Communications, Pleasantville NY (1994) 36.

[39] M. Vassallo and A. Ralston, "Computer Software for the Teaching of Discrete Mathematics in the Schools," this volume.

BELL LABS
nate@research.bell-labs.com

THE ROBOTICS INSTITUTE, CARNEGIE MELLON UNIVERSITY
(PREVIOUS ADDRESS: DIMACS, RUTGERS UNIVERSITY)
yanxi@cs.cmu.edu

DIMACS Series in Discrete Mathematics
and Theoretical Computer Science
Volume **36**, 1997

Recommended Resources for Teaching Discrete Mathematics

Deborah S. Franzblau and Janice C. Kowalczyk

1. Introduction

If you are teaching or planning to teach discrete mathematics, or if you are a mathematics supervisor assisting teachers in using discrete mathematics, you may be wondering what sorts of classroom materials are available. What books should you own or ask your school to buy for the library? Where can you get ideas for interesting student activities? Are there good videos or software? What have other teachers used successfully? This article addresses these questions, and provides recommendations for a selection of the best resources known to us. This is not a comprehensive listing, but rather a description of a core library recommended for anyone teaching discrete mathematics in grades K-12. Several of the resources mentioned here (especially videos or stand-alone activities) are also useful for convincing parents, school boards, or administrators of the value of teaching discrete mathematics

We have collected these recommendations from a number of sources. The most important of these are teachers and instructors from the Leadership Program in Discrete Mathematics (LP).[1] Both of us have been involved in the LP for over two years, Janice Kowalczyk as an alumna, lead teacher, workshop leader, and program evaluator; Deborah Franzblau as instructor and summer institute organizer. Kowalczyk maintains a resource list for the program, and edits a resource column, "The Discrete Reviewer," in the LP newsletter, *In Discrete Mathematics: Using Discrete Mathematics in the*

1991 *Mathematics Subject Classification.* Primary 00A35.

[1]This NSF-funded program has been in operation since 1989, beginning with high school teachers. Middle school and K-8 teachers have joined more recently. Further details on the program can be found in [**32**]. Suggestions for and comments on resources are gathered at alumni sharing sessions, through email to a group bulletin board, and individual contacts.

Classroom (see section 3), which Franzblau has edited. Several of the resources described here have been reviewed in the newsletter. Kowalczyk has taught mathematics in middle school. Franzblau has taught mathematics at Vassar College and is currently teaching at the College of Staten Island; she also edits the "Education Forum" of the *SIAM Activity Group on Discrete Mathematics Newsletter*.

Both of us believe strongly in activity-centered teaching methods which focus on student exploration and discovery first, and abstraction and precision later. Although we agree that hard work is essential for understanding, and drill and practice are useful at the right time, we believe that student motivation and problem-solving must always come first. Thus, our focus is on materials and activities which either provide motivation for learning or lead to interesting mathematical discoveries. For this type of teaching, it is essential that the teacher have sufficient background and confidence to recognize and help students articulate their discoveries, so we have included resources providing background and breadth for teachers.

We believe that teachers are the best judges of what is appropriate for the ages or grade-level of their students. In our experience, the best activities and ideas work at many levels, and skillful teachers can adapt them to the level of their own classroom. We do however give suggested grade-level ranges for each resource, where appropriate, based on comments from teachers who have used the materials.

An important resource to teachers at all grade levels is [**31**], which presents the discrete mathematics chapter from the *New Jersey Mathematics Curriculum Framework.* This chapter is the first comprehensive attempt to describe what activities and topics from discrete mathematics are appropriate at each grade level cluster from K-2 to 9-12. The materials in the *Framework* are also based on the experiences of LP teachers.

Many of the resources listed here were developed for grades 7-12 (or the college level). Until recently, there has not been much available material that is labeled "discrete mathematics" at the elementary level. Nevertheless discrete mathematics often appears in publications such as *Wonderful Ideas* or *The Elementary Mathematician* (see Section 3). Moreover, there are many activities and topics that can be adapted from materials written for a higher level, as well as some excellent children's literature (see Section 4) that can be used to to introduce these activities.

In addition, we recommend the following catalogs for browsing; a number of the resources mentioned here can be found in them. The first three are good sources for physical models ("manipulatives") and software. Addresses of the publishers are given in the Appendix.

- Creative Publications
- Cuisenaire
- Dale Seymour
- Key Curriculum Press
- Mimosa Publications (Primarily K-8)

- COMAP (*Consortium for Mathematics and its Applications*) (Primarily 9-college)

The resource descriptions are organized by category, as follows.

Section 2: **general textbooks** and curriculum materials, suitable for high school courses or as teacher resources;
Section 3: other sources for **activities** which can be integrated into new or existing courses;
Section 4: literature and periodicals that are recommended for **student reading**;
Section 5: **supplementary reference** works primarily for teachers;
Sections 6, 7, and 8: **videos**, **software**, and **World-Wide Web** (Internet) sites for supplementary activities and materials.

Each title is followed by a suggested grade-level range and the resource topic is given (in ***bold italics***) if it is not clear from the title or the context. Within each subsection, unless stated otherwise, resources are arranged by approximate grade-level.

For each resource, we list the publisher and/or a distributor (as of 1995/96). Print resources with no distributor listed can usually be found at large bookstores. The prices given are approximate retail cost, based on 1995 or 1996 catalogs or bookstore quotes (prices may vary a few dollars between different sources).

The article is accompanied by several appendices to assist the reader in locating resources:

Appendix A: addresses and contact information for most of the publishers and distributors;
Appendix B: an index of all resource titles (except those mentioned in passing), arranged alphabetically, by type;
Appendix C: an index of titles appropriate for each of the grade-level ranges K–2, 3–5, 6–8, and 9–12;
Appendix D: a list of major topics accompanied by recommended titles.

2. Discrete Mathematics Textbooks

When teachers name the texts they like best for teaching discrete mathematics, there are four titles that are mentioned over and over.[2] In roughly descending order of popularity, the texts are as follows: *Mathematics, a Human Endeavor*; *For All Practical Purposes*; *Excursions in Modern Mathematics*; and *Discrete Mathematics through Applications.* These have been

[2]Since separate courses on discrete mathematics don't make much sense at levels K-8, recommendations in this section are primarily from high school teachers; however, middle and elementary school teachers have enjoyed using these texts as resources. At the elementary level, the *Everyday Math Program* from the Everyday Learning Corp. (Janson), *Math Their Way* from Creative Publications, or materials from Mimosa Publications can be used to create a mathematics curriculum which incorporates many discrete concepts.

used not only as texts for discrete mathematics courses, but as supplementary reading for teachers who are introducing discrete topics in traditional courses. In the remainder of this section we give further details on content, describe how teachers have used the books, and give teachers' comments.

There are also several mathematics curriculum development projects at the high school level which contain many discrete topics. One is the NSF-funded Core-Plus Mathematics Project, which is discussed in [**15**]; these materials are being published under the title *Contemporary Mathematics in Context* by Everyday Learning Corporation. Another is the NSF-funded ARISE project, being developed by the Consortium for Mathematics and its Applications (COMAP); these materials are being published under the title *Mathematics: Modeling Our World* by South-Western Educational Publishing Company.

Mathematics, a Human Endeavor *(7–12)*
Harold Jacobs; W.H. Freeman, 3rd Ed., 1994; $49.

Mathematics, A Human Endeavor was clearly a textbook ahead of its time. After going through its third revision in 1994, it is as popular today as a classroom text and a teacher resource as it was back in the late 60s. It is a recommended favorite of teachers in the Leadership Program in Discrete Mathematics. Much of the mathematics is connected to real world applications, and there are many science and mathematics connections. The book is jammed with problem-solving activities and "What if ...?" questions, and emphasizes mathematical thinking. It is also full of photos, drawings (e.g., from Escher), and mathematical cartoons (e.g., from *Peanuts*, *BC*, and *The New Yorker*). The following background is adapted from information provided to us by the publisher:

> In the late 1960s, Harold Jacobs, teaching at Grant High School in Southern California, began exploring various mathematical applications for use in the classroom. He decided to write a textbook, something original, for those students who had not done well in mathematics. Neither the author or the publisher anticipated the response of instructors to the book when it was first published. Much to his surprise, Harold Jacobs found that he had taken the mathematical community by storm. And the author is still teaching students and finding ways to introduce them to the beauty of mathematics—to motivate them to see beyond its apparent difficulty. (W. H. Freeman, private communication.)

Content. Technically, this is not a "discrete mathematics" text, but many of the topics included are in fact discrete. The book has an eclectic list of topics, including the following.

- inductive and deductive reasoning;
- patterns and sequences (e.g., Fibonacci numbers);

- symmetry and regular polygons;
- combinations, permutations, and Pascal's triangle;
- probability and statistics;
- logic and puzzles;
- network (graph) problems—including Euler paths and trees.

Classroom Use. This book is most often used as a supplementary text or resource for bright 7-10th grade students, in a variety of courses, including mainstream 7th- or 8th-grade mathematics. (One 8th-grade teacher's students requested that she use the book all the time instead of their regular text.) One teacher used it in an Applied Mathematics course, which also included computing. Another used it for students off the "calculus track"; the course proved to be popular enough that "on-track" students took it as well. Other teachers have used it as a resource for themselves.

Comments. This book invariably draws enthusiastic comments from teachers. One comment was especially strong:

> I've been using this book for over 20 years. I've used it for a senior elective for college-bound students who have had a rough time with the traditional SAT curriculum. I've used several topics in a general math class (inductive reasoning, sequences, combinatorics, elementary probability). I've used activities from it in traditional courses (functions, conics, logs). No matter what school I teach in, I make sure that I have a classroom set of this book. (Marilyn Goldfarb LP '93), private communication.)

The following are comments from other teachers who have used this text.

- A wonderful all-around reference. Perfect for self-teaching. Good combinatorics section.
- This is my favorite math text book.
- A "must" for every library.
- It was written with "people who don't like math" in mind. It has a nice tone, develops the topics nicely, and goes into some depth.
- The supplemental transparency package is a definite "keeper." The material in it is excellent.

For All Practical Purposes (FAPP) *(9–College)*
Solomon Garfunkel (Project Director), et al.; W.H. Freeman (for COMAP), 3rd Ed., 1994 (4th Ed., 1997); $45.

Although written for a college-level audience, the material in this book has been used successfully at many grade levels. It is a great reference that is fun to read; everyone should have this book on their shelf. It has a wealth of contemporary topics, and each edition has added new topics. Of special interest are *Spotlights*—profiles of mathematicians or computer scientists who have contributed substantially to the solutions of problems mentioned in the book. There is an accompanying series of 26 half-hour video tapes which give overviews and applications of the topics; text supplements and an instructor's guide are available.

Content. This book is packed with topics and problems; there is enough material for a year-long course on the uses of mathematics. Many different semester courses could also be designed around this book. The third edition has five parts, as follows:

1. *Management Science*: graph models and problems (Eulerian paths, Traveling Salesman Problem, spanning trees, and scheduling), linear programming;
2. *Statistics/Data*: data collection and representation, probability and statistics;
3. *Coding Information*: identification and error-correcting codes;
4. *Social Choice*: voting, fair division, apportionment, game theory;
5. *Size and Shape*: growth models, measurement, astronomy, fractals, symmetry, tilings.

The focus is on real-world problems and models. Note that a number of standard discrete mathematics topics are not covered, such as shortest path problems, graph coloring, and permutations, and some of the mathematics used is continuous rather than discrete.

Classroom Use. This book was intended for introductory college courses, but teachers have reported using it successfully with a wide range of younger students, from gifted 7th- and 8th-graders in an enrichment program, to 11th- and 12th-graders (including non-college-bound students). Most high school teachers find the reading level too high for their students; instead, they use it as a resource for themselves, adapting the problems for their classes. Specific examples of the use of the text are given in this volume [**17**, **34**].

Comments. The following is an excerpt from a review of the first edition (1988), by Anthony Piccolino, on the mathematics faculty at Montclair State University, who used the book for three years in teaching an elective mathematics course to high school seniors [**28**].

> [This is] a textbook which addresses real-life situations, emphasizes mathematical modeling, encourages students to make mathematical connections, and devotes extraordinary efforts to changing students' narrow view of mathematics. ...
>
> I found this book to be most effective when used in conjunction with the 26-program video series of the same name. The animation and the delivery style are wonderfully motivating and give students an excellent sense of the "big picture" for each chapter before having students delve into various detailed aspects of the chapter. ...
>
> I recommend this book with great enthusiasm!

Joseph Malkevitch, on the mathematics faculty of York College of CUNY, and one of the authors of the section on Management Science, believes the text has played a key role in the evolution of college "mathematics-for-liberal-arts" courses [**24**]. He wrote,

> [Its] approach underlines the importance for students of analyzing and understanding real-world situations and building mathematical models rather than gaining facility with solving exercises based on the models.
>
> [Texts like this can] show how the ideas being developed in discrete mathematics are essential for emerging new technologies such as robotics, computer vision, data compression, and medical imaging.

Excursions in Modern Mathematics *(9–College)*
Peter Tannenbaum and Robert Arnold; Prentice-Hall, 2nd Ed., 1995; $57.

This book, first published in 1992, was evidently inspired by *For All Practical Purposes* (FAPP), and, like FAPP, is intended for an introductory college-level course. The title conveys its goal of providing interesting "trips" through the realm of contemporary mathematics. The style is less cluttered than FAPP, and many find that it makes a better textbook than FAPP (which is a better reference, however). Many high school teachers use it as a resource for themselves, finding the reading level too high for their students (as with FAPP). The publisher also offers a supplement of *New York Times* articles connected to the text topics. The following description and comments are based on the first (1992) edition.

Content. Excursions covers a subset of the topics of FAPP (there is no section on codes, and fewer topics in each section), and introduces topics in a different order. The topics covered are divided into four parts, as follows:

1. *Social Choice*: voting theory, fair division, apportionment;
2. *Management Science*: Eulerian tours, traveling salesman problem, minimum-cost spanning trees, scheduling;
3. *Growth and Symmetry*: Fibonacci numbers and sequences, population growth, transformation geometry, fractals;
4. *Statistics*: collecting and describing data, probability, and the normal distribution.

Classroom Use. Teachers have reported using this as a text for a discrete mathematics elective at the 12th-grade level, and as a resource or supplement for discrete mathematics courses for grades 9–10 and 10–11. It has also been used to introduce discrete topics in calculus at grade 12.

Comments. One teacher, who used the book as supplementary reading and as a resource for herself, felt it was the most useful of the four books in terms of the topic coverage and quantity of material. She found the material on fair division and voting very strong, but was disappointed that graph coloring and codes were not included.

The following comments are taken from a review [**2**] by high school teacher and mathematics supervisor, Ethel Breuche LP '91, who used the book as a resource.

> I feel that this text is wonderfully *rich*. It is rich with examples and simple and thorough explanations. It is rich in discussion and explorations. Most of all it is rich in exercises at the end of each chapter ..., which increase in level of difficulty and creative problem solving.
>
> Some chapters are followed by an additional appendix. ... For example, in the voting theory section, the voting scheme for the nominations for the Academy Awards is described in detail. Every chapter offers references for further research and readings. The text is written with deliberate thoughtfulness with regard to racial and gender equity. ...
>
> Whether using the book as a resource or as a classroom text, I cannot praise it enough. Although written for a college-level liberal-arts math course, [college-bound high-school] juniors and seniors whose reading ability is average or above can use this text as well.

Discrete Mathematics Through Applications *(7–12)*
Nancy Crisler, Patience Fisher, and Gary Froelich;
W. H. Freeman (for COMAP), 1994; $36.

This book was developed to fill the need for a high-school level discrete mathematics text [**8**]. It is addressed to the student, and is less sophisticated mathematically than either FAPP or *Excursions*. Teachers of students in grades 7–9 (or average students in grades 10–12) have been very pleased with it; but teachers of more advanced students have found it less useful. Overall, the comments on this book have been more mixed than those on the previous three. However, the book can provide a good introduction for a teacher who has not seen discrete mathematics previously.

Content. The book includes the following chapters:

1. Election Theory;
2. Fair Division;
3. Matrix operations and applications;
4. Graphs and applications;
5. Recursion.

Each chapter begins with a group exploration and a set of exercises to introduce the topic. This is followed by short lessons and exercises. In another article in this volume [**8**], the authors give more detail on the content and development of the text.

Classroom Use. One teacher used it as a text for a course called "Math in the Real World", for students at the 11th- and 12th-grade level with poor mathematics backgrounds. Another used it as a text for a discrete mathematics course for grades 9–10, while another used a draft version for an 11th-grade discrete mathematics course.

Comments. A teacher who used it as a text, who felt that FAPP and *Excursions* were too abstract and notation-heavy for her students, said:

> It's an excellent book. It's very concrete, interesting, and to the point. Starting with the first chapter, where there's an election activity on soda preferences, there are many good activities that involve the students. The exercises are conducive to group work, and range from very basic to challenging. (Diane DePriest LP '93, private communication.)

She supplemented the text with logic puzzles, and material on permutations and combinations which she created.

A teacher who chose *not* to use the book for a 12th-grade course felt that it "looked too easy," and feared it could have a "bad effect on the reputation of the course." A college faculty member was disappointed that the book "does not point out where the mathematical structure is," and does not do more "summing up" of the mathematics topics learned.

On the other hand, an elementary school teacher advises teachers of elementary or middle grades to use the text:

> It lends itself best of all the texts to adapting ideas to the elementary and middle grades. Also, the instructor's manual is user-friendly for teachers who are not familiar with discrete math, and would otherwise be afraid to try it. I also encourage them to order the COMAP modules on graphs.[3] (Penni Ross LP '94, private communication.)

3. Sources for Student Activities

This section contains recommended newsletters and books which provide activities on a diverse collection of topics, as well as modules or books providing resources on special topics. The resources within each subsection are listed in order of suggested grade levels.

3.1. Newsletters and General Collections.

Math Their Way (*Book, K–3*)
Mary Baretta-Lorton; Addison-Wesley, 1976; $39 (available from Creative Publications); blackline masters are included.

An old favorite. This book contains an activity-centered mathematics program which focuses on patterns. It can be used as a self-contained curriculum or for enrichment.

Wonderful Ideas (*Newsletter, K–6*)
Wonderful Ideas, 8/yr; $26/yr (individuals), $38/yr (schools).

[3]These modules are *Drawing Pictures with One Line*, *The Mathematician's Coloring Book*, and *Problem Solving with Graphs*, described in Section 3.

This newsletter provides interesting elementary-level activities, often developed around a single theme. Sample topics include the "Monty Hall" problem (probability) and making a pinwheel (geometry and origami).

The Elementary Mathematician (*Newsletter, K–6*)
COMAP, 4/yr; only available with $16/yr membership (which includes other publications and discounts).

This is another good source of classroom activities, many of which integrate mathematics with science, health, history, and other areas. The pull-out section in each issue contains a complete classroom-ready lesson. Also available is **The Elementary Mathematician Pull-out Book** (Laurie Aragon, Ed.; $9), a collection of pull-out lessons from past newsletters.

INsides, OUTsides, LOOPS and LINES (*Book, K–8*)
Herbert Kohl, W.H. Freeman, 1995; $13.

One of us (Kowalczyk) recently discovered this excellent book, which introduces concepts from discrete mathematics and topology.

> If I had to recommend one resource book to elementary and middle grade teachers in discrete mathematics this would be it. ... [T]his book provides playful introductory activities [based on concepts in graph theory and topology]. The five chapter titles give you the best flavor for its contents: *Lost in the Garden*—simple closed curves; *Map Coloring*—figuring out the rules; *Tracings*—simple beginnings lead to complicated patterns; *Stretching, Bending and Twisting*—a new way to look at shapes; and *Mobius Strips*—some thoughts on doing mathematics with a twist. [**21**]

Math on the Wall (*Looseleaf collection, 1–6*)
Linda Holden; Creative Publications, 1987; three sets are available, for grades 1–2, 3–4, and 5–6, $8.50/set.

This is a collection of problems and puzzles designed to decorate a classroom bulletin board. Blackline masters for cut-out designs are included. It has lots of combinatorial puzzles, posing such problems as "How many different bouquets can you make with these flowers?" (This includes pictures of flowers and vases to cut out and color).

Discrete Mathematics Across the Curriculum, K–12
(*Collected Articles*) NCTM, 1991; see [**19**].

This edited collection is one of the NCTM *Yearbook* series, written to supplement the Discrete Mathematics Standard in the 1989 NCTM *Standards* [**27**]. Most of the articles focus on a single topic, such as graphs or recursion, illustrated with activities suggested for specific grade levels.

Teaching Children Mathematics (*Journal, K-4*)
Mathematics Teaching in the Middle School (*Journal, 5-8*)
Mathematics Teacher (*Journal, 9-12*)
National Council of Teachers of Mathematics, available with membership (which includes other publications and discounts).

These journals often publish articles or include pullout sections with discrete mathematics activities. The newsletters of local mathematics teachers associations may also publish useful activities.

In Discrete Mathematics: Using Discrete Mathematics in the Classroom (*Newsletter, K–12*)
DIMACS DM-Newsletter, 2/yr; for a subscription (no cost — supported by DIMACS), write to DIMACS-DM Newsletter (address in Appendix A).

This newsletter features articles by teachers (primarily participants in the Leadership Program in Discrete Mathematics) describing their experiences in creating and/or using discrete mathematics topics in the classroom. There are articles on special topics, such as elections or game theory, along with bibliographies for background reading, as well as resource recommendations. Puzzles and drawings by readers (or their students) also appear. The newsletter was previously edited by Joseph Rosenstein and Franzblau (and currently by Robert Hochberg); Kowalczyk edits a resource review column. Judy Brown (see Section 8) has introduced a new Internet resource column.

Discrete Mathematics in the Schools *(Book, K-12)*; see [**33**].

This edited collection of articles (which includes this article on "Recommended Resources") is based on a conference which took place in October 1992 at Rutgers University, under the sponsorship of DIMACS, the NSF-funded Center for Discrete Mathematics and Theoretical Computer Science. The articles present different perspectives of discrete mathematics and how it can be reflected in the school curricula. Many activities and topics are presented as illustrations of the authors' perspectives.

A Comprehensive View of Discrete Mathematics: Chapter 14 of the New Jersey Mathematics Curriculum Framework *(Article, K-12)*; in [**33**]

This chapter from the *New Jersey Mathematics Curriculum Framework* presents a comprehensive discussion of discrete mathematics topics and the grade levels at which they can be appropriately presented. The grade-level overviews are illustrated by several hundred activities appropriate for various grade levels.

HiMap and HistoMap Series (*Topic modules, 4–12*)
Various authors; COMAP; $9 - $12 (catalog available).

These are self-contained supplements on specific topics that can be used in a number of courses. They contain background reading, as well as blackline masters for student activity sheets and transparencies. A few of the modules are accompanied by special-purpose software, such as an implementation of the *Moving Knife* strategy to accompany a module on fair division. Suzanne Foley LP '95 (private communication.) mentioned that, "One useful aspect of the HiMap books that I enjoy is the history behind the mathematics," and said that she appreciates the practical activities. There are many discrete topics in the large collection, including voting, fair division, graph problems, recurrence relations, and codes. (See also the specific modules recommended in Section 3.2.)

Consortium (*Newsletter, 9–College*)
COMAP, 4/yr; only available with $32/yr membership (which includes other publications and discounts).

This newsletter is intended primarily for teachers of undergraduates, but continues to add more items of interest to high-school teachers. It focuses on applications of mathematics, and features pull-out modeling activities for the classroom. Lists of current and future HiMap module titles are given in each issue.

Problem Solving Strategies: Crossing the River with Dogs
(*Book, 9–College*)
Ted Herr and Ken Johnson; Key Curriculum Press, 1993; $25 (teacher's resource book, $20).

This book, written to be used as a course text, uses problem-solving to encourage reasoning skills across the curriculum. Used mainly as a teacher resource, the problems have been used with students ranging from gifted fifth-graders to undergraduates.

3.2. Activities on Special Topics: This section lists books and modules which each give activities related to one special topic. The activities can usually be integrated into a wide variety of courses, on many levels.

The Mathematician's Coloring Book (*Module, 4–10*)
Richard L. Francis; COMAP, HiMAP Module **13**, 1989; $12.

This entertaining unit, developed around the Four-color Theorem, is especially appropriate for general math or geometry students. It includes maps, pictures, and worksheets to explore map coloring (which have been used in the LP K–8 programs). Many of the coloring activities are appropriate for young students. At the end there are challenging problems on coloring maps on doughnut-shaped and other surfaces, suitable for advanced students.

Drawing Pictures with One Line: Exploring Graph Theory (*Module, 4-10*)
Darrah Chavey; COMAP, HistoMAP Module **21**, 1992; $12.

This popular module introduces Eulerian tours and other topics. Here is an excerpt from a teacher's review [**29**]:

> [The book takes the reader] through the historical beginnings of graph theory as recreational puzzles, to the array of applications for which graph theory is used today. Included are multicultural aspects[4] of graph theory as it exists in cultures in Africa and Oceania as part of a heritage of sophisticated story-telling.

Problem Solving Using Graphs (*Module, 6–10*)
Margaret B. Cozzens and Richard Porter; COMAP,
HiMAP Module **6**, 1987; $10.

This is an excellent source of problems on a variety of graph topics, including Eulerian tours, shortest paths, minimum spanning trees, and the Traveling Salesperson Problem. The problems can be used for teaching modeling, problem-solving, or algorithm design. Many of these problems have been used successfully in the Leadership Program, and subsequently by the participating teachers when they return to their classrooms.

Fractals for the Classroom: Strategic Activities (*Workbook, 7–12*)
Heinz-Otto Peitgen, Hartmut Jürgens, Dietmar Saupe, Evan Maletsky, Terry Perciante, and Lee Yunker;
Springer-Verlag, 1991; two volumes, $20/vol. (available from NCTM).
Vol. 1 includes slides of 2- and 3-D fractal images.

This is a set of ready-to-use classroom activities, designed to introduce the concepts of *self-similarity*, fractal generation via the *Chaos Game*, and fractal *complexity* (or *dimension*). The book is the result of a collaboration between researchers on the mathematics of fractals (the first three authors) and specialists in mathematics education.

Authors Maletsky (see [**23**]) and Perciante have each used some of these activities successfully with teachers at all levels in the Leadership Program. William Bowdish LP '92 reported using the material in a classroom activity in which his 9th-grade Honors Algebra students estimated the fractal dimension of the coastline of Martha's Vineyard [**1**].

Codes Galore (*Module, 9–12*)
Joseph Malkevitch, Gary Froelich, and Daniel Froelich;
COMAP, HistoMAP Module **18**, 1991; $12.

This module begins with a historical account of codebreaking and its impact on WWII. Its main focus is on common error-detecting and error-correcting codes (such as the zipcode and ISBN numbers), and their many

[4] See also **Ethnomathematics** in Section 5.3.

applications (such as compact discs, or transmitting images from spacecraft to Earth). These codes use the idea of *check digits*, and are a good introduction to modular arithmetic. There are many activities, many of which require a calculator. A number of the activities lend themselves to cooperative group work—like decoding a simulated picture from space. A good complement to this module is the segment on codes in the COMAP *Geometry* video, described in Section 6.

4. The Students' Bookshelf

4.1. Literature. A number of teachers have found that one way to interest and engage students in mathematical thinking is through literature. At the primary level, weaving together literature and mathematics may be a necessity, since so much classroom time at this level is needed to develop literacy. This section contains a short list of books that teachers have found connect well to topics in discrete mathematics. There are a number of publications that provide more complete lists of literature/mathematics connections, such as [**4**]; however we are not aware of any that highlight discrete math topics.

Dr. Seuss books (*pre-K–2*) ***Iteration and Recursion***
"Dr. Seuss"; Random House; $8–$15.

A number of Dr. Seuss books contain the seeds for thinking about iteration and recursion. **The Cat in the Hat** (1957) and **Green Eggs and Ham** (1960) are just two that come to mind. In these stories, events or activities are repeated over and over but in each repetition a new event is added. Another Dr. Seuss book, **The Lorax** (1971), is a variation on this pattern, with an environmental theme. As events occur in *The Lorax* they trigger other events, eventually throwing the environment out of balance.

One Hundred Hungry Ants (*K–3*) ***Listing and Counting***
Elinor Pinczes, Houghton-Mifflin, 1993; $15 (avail. from Creative Pub.).

This is a simple story told in verse about one hundred hungry ants heading towards a picnic. Different formations of 100 ants are tried in order to speed their way to the food; illustrations provide visual patterns for counting by twos, fives, and tens. In the process the author introduces the factors of 100 and the problem of counting factors.

Grandfather Tang's Story (*K–3*) ***Visual Problem Solving***
Ann Tompert; Crown, 1990; $15 (avail. from Cuisenaire or Creative Pub.).

This is a Chinese folktale told with tangrams, a set of seven simple shapes cut from a square. Using tangrams, Grandfather Tang tells a story to his granddaughter: two shape-changing fox fairies try to best each other until a hunter brings danger to both of them. With tangrams students can be encouraged to investigate geometrical concepts and to use their visual imaginations to retell or invent their own stories.

The Tangram Magician (*K–3*) ***Visual Problem Solving***
Lisa Campbell Ernst and Lee Ernst, Harry N. Abrams, NY, 1990; $20.

This is another good book for developing visual imagination for geometric problem solving. The story involves a magician who can change shape, illustrated with tangrams. The reader is asked to supply the end of the story by creating the next shape that the magician will take on. High school teacher Eric Simonian LP '94 reports on an interesting classroom experience using this book involving students in grades 2, 4, and 10, in [**35**].

A Three Hat Day (*K–3*) ***Listing and Counting***
Laura Geringer; Harper-Collins, 1987; $5 (paper).

This amusing tale about a hat collector and his search for a perfect wife provides an opportunity for teachers to ask "How many different ways are there to ...?" The story can be used to introduce the concept of combinatorics, i.e., systematic counting.

Sam Johnson and The Blue Ribbon Quilt (*K–4*)
Tessellations and Geometric Transformations
Lisa Campbell Ernst, Wm. Morrow and Co., 1983 (paper, 1992), $5 (paper).

Sam Johnson and the Blue Ribbon Quilt raises both mathematical and social questions. While mending the awning over the pig pen, Sam discovers that he enjoys sewing patches of cloth together. However, when he asks his wife if he can join her quilting club, he meets at first with scorn and ridicule. The borders of the pages feature traditional American quilt patterns which can be used to introduce mathematical concepts such as symmetry, tessellations, and transformational geometry.

A Cloak for the Dreamer (*K–5*) ***Tessellations***
A. Friedman, Scholastic, 1995; $15 (available from Creative Pub.).

A tailor asks each of his three sons to sew a colorful cloak that will keep out wind and rain. The first two sons sew watertight cloaks made of rectangles and triangles, but the third son, the dreamer, makes a cloak out of circles, which is full of holes. This book gives a good introduction to tessellations, and contains (for parents and teachers) a section on the underlying mathematical concepts.

Two of Everything (*1-4*) ***Iteration; Exponential Growth***
Lilly Toy Hong, Albert Whitman & Co., 1993; $15 (available from Creative Pub.).

A Chinese folktale about a couple who find a magic pot that doubles everything that is put into it. The story is retold and illustrated by the author. This is a good starting point for thinking about iteration and exponential growth.

Anno's Mysterious Multiplying Jar (*1-6*) ***Listing and Counting***
Masaichiro and Mistumasa Anno; Putnam Publishing, 1983;
$17 (available from Creative Pub.).

Simple text and beautiful illustrations tell a tale about a porcelain jar with a sea inside. In the sea is one island, and on the island are two countries, and in each country are three mountains .... This book provides a rich introduction to the concept of counting by multiplication. One of our favorites!

Math Curse (*4-8*) ***General Problem Solving***
J. Scieszká and L. Smith; Viking/Penguin Group, 1995; $11.

A popular new book. A young girl wakes up one day afflicted with the "math curse"—she sees math in everything she does. The book has a wonderful crazy-quilt graphical style, and is full of engaging problems (mainly on algebra, geometry, and numbers—but with a few discrete math problems).

Jurassic Park (*6–up*) ***Fractals and Iteration***
Michael Crichton; Random House, 1990; $6 (paper).

Jurassic Park, which connects mathematics, biotechnology, and prehistoric legend, has proved to be a "student magnet" in a number of mathematics classrooms and can serve as a stepping stone to the introduction of fractals. In the beginning of each chapter, the "dragon-curve" fractal is constructed to one more level, as a mathematical foreshadowing of the events to come. An 8th-grade teacher's students asked her to teach them more about fractals after reading the book, leading to some of the best learning she'd seen in her career—sparked by a real "need to know."

4.2. Other Good Reading. You may want to keep the following periodicals and books in the classroom for students to browse through.

Aha! Insight (*Book, 6–10*)
Martin Gardner; W. H. Freeman (for *Scientific American*), 1978; $15 (paper).

This book contains short, fun problems that Gardner has collected, that can all be solved easily—if one chooses a good approach. Many of the problems involve combinatorics (counting), algorithms, or logic. Illustrated with cartoons.

Math EQUALS (*Book, 6–10*)
Teri Perl, Addison-Wesley, 1978; $20.

Each chapter contains a biography of a woman mathematician, followed by mathematics activities related to her work. There are a few discrete mathematics activities, including those in a chapter on Lady Ada Byron Lovelace (1815-1852), who wrote about Charles Babbage's design for a machine that anticipated the modern computer.

QUANTUM: The student magazine of math and science
(*Magazine, 10–College*) Springer-Verlag, 6/yr.; $20/yr.

This is a new magazine on mathematics and physics for sophisticated high school students. It has well-written, challenging articles about current math and science topics and also has a problem section. It is a collaborative effort of the NCTM and the American Association of Physics Teachers, along with their counterparts in Russia. (*Quantum* was inspired by the Russian journal *Kvant*, which in turn may have been inspired by an older Hungarian mathematics newsletter for students.)

Math Horizons (*Magazine, 11–College*)
MAA, 4/yr; $35/yr ($20/yr, MAA members).

Written for undergraduates, this also makes interesting reading for high-school students, and teachers at all levels. Articles include profiles of famous contemporary mathematicians, including John Conway (tiling, knots, and game theory), Fan Chung (graph theory), Persi Diaconis (probability and magic), and Jean Taylor (soap-film surfaces). There are articles giving information on careers in mathematics, as well as a problem section.

5. The Teachers's Bookshelf: Reference and Reading

This section focuses on print resources for teachers that are appropriate for reference, background reading, or as sources for ideas. Many of these resources are written for undergraduates and/or high-school and college faculty, although mature students may also be able to use them.

5.1. College Textbooks on Discrete Mathematics. Here are several sources for getting more background or detail on discrete math topics, as well as many applications and examples. Many of the problems, although intended for the college level, have been used successfully at lower levels.

Discrete Mathematics and its Applications (*Book, College*)
Kenneth Rosen, McGraw Hill, 2nd Ed., 1991; $70.

Moshe Vardi, a computer science professor at Rice University, asked college/university faculty which book they used to teach their introductory course on discrete mathematics. Over a dozen faculty responded: Rosen's book was the most popular text in the survey.[5] The book has a companion volume, with further applications, written by various authors: **Applications of Discrete Mathematics**, J. Michaels and K. Rosen, Editors (McGraw Hill).

[5]The survey was conducted informally using the electronic mailing list **TheoryNet**, in May 1993. Rosen's book was mentioned by faculty from U. Mass (Amherst), Columbia, U.C. Santa Cruz, U.C. Berkeley, RPI, SUNY Brockport, and U. Kansas. On the other hand, when I (Franzblau) did a similar informal survey of colleagues (mainly in mathematics departments), I found no clear favorite text, though Rosen's book was mentioned favorably [**13**].

Other texts mentioned several times in Vardi's survey were **Discrete Mathematics**, by Ken Ross and Charles Wright, Prentice-Hall, 3rd Ed., 1992; and **Discrete Mathematics with Applications**, by Susanna Epp, Wadsworth, 1990. (In [**12**], the author discusses her experience teaching basic logic in the course at DePaul University that led to this text.)

Applied Combinatorics (*Book, College–Graduate*)
Fred S. Roberts, Prentice-Hall, 1984; $70.

This is a good source of applications of discrete mathematics.[6] The book also describes many classic algorithms, such as those for finding shortest paths, minimum spanning trees, Eulerian tours, and maximum flow. I (Franzblau) have used the text several times for an intermediate-level undergraduate course; students find it somewhat difficult, so I prefer using it as a resource for myself.

Graph Theory Applications (*Book, College*)
L.R. Foulds, Springer-Verlag, 1992; paper, $49.

This text was recommended by Susan Picker LP '90 (private communication):

> The distinctive thing about this text is the variety of applications, including social sciences, economics, physics, biology, chemistry, civil engineering, operations research, circuit design, matrices, algorthms, architecture, and industrial engineering.

Graphs: An Introductory Approach (*Book, College*)
Robin Wilson and John Watkins, Wiley, NY, 1990.

One of the referees of this article strongly recommended this text, as a "great introduction to graphs."

Graphs, Models, and Finite Mathematics (*Book, College*)
Joseph Malkevitch and Walter Meyer, Prentice-Hall, 1974.

This book is the source of some of the material in the HiMAP module **Problem Solving Using Graphs** (described above in Section 3).

5.2. Periodicals. The following are high-quality newspapers and magazines that you may want to read regularly. They are good sources for current discoveries and applications of both discrete and continuous mathematics.

New York Times (*Newspaper*)
NY Times Co., daily; cost varies.

---

[6]In [**30**], the author provides examples of applications of discrete topics and discusses how he uses them in the classroom.

See especially the *Science Times* (Tuesdays): Gina Kolata, who recently won an award from the Joint Policy Board for Mathematics, often writes articles in this section on mathematical topics, such as recent breakthroughs on factoring large primes, the Traveling Salesperson Problem, and "DNA computing." The economics and patents columns in the business section, as well as financial and stock market data, are good sources for both applications and problems.

What's Happening in the Mathematical Sciences (*Booklet*)
Barry Cipra; AMS, annual (since 1993); $7 (ea.).

A well-written review of "hot" and accessible mathematics topics on which progress was made during the year. The author writes frequently on mathematics topics for the newsletter of the Society for Industrial and Applied Mathematics (SIAM).

Mathematical Intelligencer (*Magazine*)
Springer-Verlag, 4/yr; $33/yr.

This is an unusual journal with lively expository articles for a general mathematical audience. It has book reviews and articles about mathematicians and the history of mathematics. The text is sprinkled with intriguing and humorous quotations and pictures. A special feature is the *Mathematical Tourist* column, in which readers report on mathematically interesting sites (such as buildings or sculpture that incorporate interesting geometry or topology, or home towns of famous mathematicians), illustrated with photographs.

Scientific American (*Magazine*)
Scientific American, 12/yr; $36/yr.

This classic journal has a variety of articles on current science topics aimed at a broad, educated audience. There are often good articles involving mathematical modeling, some suitable for student reading. The column *Mathematical Recreations*, edited by Ian Stewart, which has undergone several incarnations since Martin Gardner's *Mathematical Games* column, is still a great source of fun problems. Many of Gardner's columns can be found in books, such as [**14**]. His successor, Douglas Hofstadter also published his columns [**16**], which can be used for classroom activities [**34**].

American Scientist (*Magazine*)
American Scientist, 6/yr; $28/yr.

This is the magazine of Sigma Xi, the scientific honor society. It has well-written articles on a range of scientific topics, as in *Scientific American.* It contains an extensive section of book reviews, and has good science cartoons.

5.3. Books on Special Topics . The following sources give interesting background material not usually found in textbooks. We include grade-level suggestions for use of the material (or for student reading), where appropriate.

Ethnomathematics: A Multicultural View of Mathematical Ideas ***Graph Theory and Applications***
Marcia Ascher; Chapman and Hall, 1991; $42.

This is a fascinating book, which will give you a new perspective on mathematical ideas that one often takes for granted, such as the words used for counting. Of special interest is a chapter exploring Eulerian paths, a standard topic in discrete mathematics (e.g., see [**17**]), as an artistic aid to story-telling in the South Pacific islands. This is the source of some of the material in the module *Drawing Pictures with One Line*, described on page 13. The following is an excerpt from a review by Susan Picker LP '90.

> But *Ethnomathematics* explores more than the topic of graph theory as it presents the mathematical ideas of number, kin relations, games of chance and strategy, and symmetric strip decorations. ... It provides a comprehensive look at the meaning and use of similar mathematical ideas in different cultures, illuminating both the mathematics and the culture in which it appears, and through this showing the value of the study of mathematics in a multicultural setting.

Alan Turing: The Enigma ***Codebreaking***
Andrew Hodges, Simon and Schuster, NY, 1983.

This is a popular book that tells the story of Alan Turing and his role in breaking the German ENIGMA code during WWII. It makes good background reading for teaching cryptography or computer science.

Visions of Symmetry *(7–up)* ***Escher's Tessellations***
Doris Schattschneider; W.H. Freeman, 1990; $25.

This excellent book on the work of M.C. Escher provides both historical and mathematical perspective on the evolution of his tessellation prints. Beginning with his sketches of mosaics in the Alhambra, the author guides the reader through Escher's notebooks, explaining how he developed an understanding of the mathematics of tessellation (as well as his own system of classification) in order to create his work. The book is illustrated with many color reproductions from his notebooks.

An Introduction to Tessellations *(4–12)*
Dale Seymour and Jill Britton; Dale Seymour, 1989; $21.

This is a good source of ideas and activities for students. It illustrates basic tessellation concepts and gives a variety of ways to create interesting artwork via geometric transformations of simpler tessellations.

Teaching Tessellating Art *(K–12)*
Jill Britton and Walter Britton; Dale Seymour, 1992; $21.

This book contains transparency masters and activities that are very useful for teaching Escher-style tessellation techniques, as well as transformation geometry. I (Franzblau) have used these (as well as pictures from

Visions of Symmetry) for hands-on workshops for K-8 teachers and high school students.

Orderly Tangles (*8–up*) ***Knots***
Alan Holden; Columbia U. Press, 1983; $32.

This is an excellent, accessible introduction to knots. The author begins with a discussion of highway interchanges, and goes on to discuss knots in general. It is illustrated with photographs of wonderful knot models made out of wooden dowels.

The Knot Book (*College–up*)
Colin Adams; W. H. Freeman, 1994; $33.

Adams has led many undergraduate research projects on knots; this book grew out of his notes. It develops the mathematics of knot theory, and can be read at many levels. A good source of problems about knots.

Chaos, Fractals, and Dynamics: Computer Experiments in Mathematics
Robert Devaney, Addison-Wesley, Menlo Park, NJ, 1989.

Devaney is one of the pioneers in the applications of fractals, as well as the introduction of dynamical systems and chaos in the undergraduate and high school curriculum. See also Devaney's article in this volume [**10**], his video (section 6), and the Web site he developed (section 8).

Fractals, The Patterns of Chaos
John Briggs; Simon and Schuster, 1992; $20 (paper).

The following is a recommendation from a Leadership Program participant:

> When I try to explain what fractals are and how they relate to our world, I sometimes come up short. I was wandering through my favorite bookstore and discovered this book. It's terrific. In addition to the pretty fractal pictures, he shows fractals in the real world—decaying leaves, weather systems, lightning, cauliflower, the human body, etc. He even relates fractals to art and architecture. In the appendix, the author lists and evaluates fractal software both for the IBM and the Mac (both shareware and commercial software). He also lists several fractal publications. (Edward Polakowski LP '92, email posting.)

Game Theory and Strategy (*12–College*)
P. Straffin; MAA, 1993; $34.

This book was recommended by Joe Malkevitch [**25**]:

> This is a wonderfully rich book about the theory of games. It covers most of the major ideas in a motivated and succinct way, and has many examples.

Fair Division: From Cake Cutting to Conflict Resolution
Steven Brams and Alan Taylor, Cambridge Univ. Press, 1996. $18.

If you're interested in estate settlement, voting, auctions, and similar topics, this book comes highly recommended by high-school teacher L. Charles Biehl LP '90 (email posting). It was written by the authors of a well-known recent work on "envy-free" cake-cutting.

Basic Geometry of Voting
Donald G. Saari, Springer-Verlag, NY, 1995; $39.

This recent book provides additional mathematical background for teaching election theory and voting paradoxes, topics which are covered in the books *For All Practical Purposes*, *Excursions in Modern Mathematics*, and *Discrete Mathematics through Applications*, discussed in Section 2.

Unit Origami (*7–up*) ***Geometry of Polyhedra; Graph Problems***
Tomoko Fusé; Japan Publications, 1990; $19 (available from Cuisenaire or Dale Seymour; a classroom guide is also available).

This is an introduction to a style of origami that involves building complex shapes, including regular polyhedra, by fitting together simple (usually identical) units. It is often recommended by teachers. Many of the constructions are suitable for teaching to students, and can provide inspiration for exploring the geometry of polyhedra. Creating the shapes out of small units also leads one naturally to discrete problems: counting vertices, edges, and faces (as a prelude to Euler's formula), as well as to questions on coloring edges and faces.

Build Your Own Polyhedra (*5–up*)
Peter Hilton and Jean Pederson; Addison-Wesley, 1988; $28
(available from Dale Seymour).

This is a beautiful book with illustrated directions for constructing polyhedra. It also introduces some of the mathematics of polyhedra, such as Euler's formula. This is a good source for classroom projects at almost all grade levels. See [**18**] for a discussion of the value of introducing students to polygons and polyhedra long before beginning the formal study of geometry.

6. Video Tapes

In addition to the COMAP Videos which accompany the text FAPP (see section 2), we found only two video tapes that have been widely used and recommended, *Geometry: New Tools for Technologies* and *Powers of Ten*, which are described below. If you are not able to purchase them, you may be able to borrow them through a school or regional library.

For All Practical Purposes (*9-College*)
Twenty-six half-hour programs on 15 cassettes;
Available from Annenberg/CPB Multimedia Collection (address in Appendix A); $39.95 per cassette (or $85 per module of three cassettes)

These video tapes accompany the text FAPP discussed in Section 2; there is an introductory tape, plus five half-hour programs for each of the five sections of FAPP.

Geometry: New Tools for Technologies *(5–12)*
Graph Applications; Codes
Five units, 10-15 minutes each (1 hr. total); COMAP, 1992; $70 (user's guide, $10).

The following description is from a teacher's recommendation:

> This well-done ... video, complete with user's guide, illustrates the geometry of the 20th century: motion planning, error-correcting codes, Euler circuits, vertex coloring, and tomography. (Ethel Breuche LP '91 [**3**])

The five unit titles are as follows:

- *A Hero's Jouney: Motion Planning for Robots*
- *Connecting the Dots: Vertex Coloring*
- *Picture Perfect: Error-correcting Codes*
- *X-ray Vision: the CAT Scan*
- *Snowbound: Euler Circuits*

This videotape was conceived and directed by Joseph Malkevitch who says (private communication) that he "still finds parts of the tape very robust in attracting interest. For example, the piece on CAT scans shows people that math is behind tomography as well as computing and engineering and physics." (See also Malkevitch's article [**26**] on the value of addressing real-world problems in teaching mathematics.)

Several teachers have also mentioned that the tape appeals to a broad audience. For example, the segment entitled "Snowbound" begins with a charming cut of young children trying to draw a house without lifting a pencil, before getting into more practical applications. Another cut in this segment uses interesting graphics to let you "ride along" a graph. The segment "Connect the Dots", shows an application of graph coloring to creating zoo habitats. Both of these segments have been very successful as a follow-up to problems on Eulerian paths and graph coloring with all teachers in the Leadership Program in Discrete Mathematics. One of us (Franzblau) likes to use the segment on error-correcting codes as part of teaching a unit on codes, using the first half to motivate the concept of error correction.

Powers of Ten *(6–12)* ***Iteration; Exponential Growth and Decay***
10-15 minutes; W.H. Freeman; $40.

Teachers are very enthusiastic about this short video. It has been used at a number of grade levels, as well as in teacher-training programs, especially for number sense, positive and negative exponents, and scientific notation. Most find the graphics very powerful, and show it two or more times with discussion in between. It could be a good resource when discussing the exponentially growing time needed to run many exhaustive search algorithms.

Here are some comments made by teachers who have used it:

- I show the first half, go over exponents, scientific notation, etc., and then show the whole thing, introducing negative exponents.
- It's also a really great visual experience for the "trip" back down the powers of ten.
- One of the amazing things is how outer space looks so much like inner space with the vast distances between things. It's a great link to science because through the use of the negative exponents you travel to the molecular level.
- It's one of those *ahhhhhh* films!

There are other good videos available that are less well known. We list below a few that have been especially recommended.

Mathematical Eye Series (*9–12*) ***Graphs; Probability; Logic***
20 min. each (approx.); Journal Films; $270 each.

This series was recommended enthusiastically by high-school teacher Diane DePriest LP '93 (who borrowed the films from a district library). She provided the following description.

This outstanding series includes 18 titles, including the following on discrete math topics:

> *Lines and Networks* (Euler paths, subway representation, isobars);
> *Fibonacci and Prime Numbers*;
> *Logic and problem solving* (flow charts, probability, truth tables);
> *Probability*; and
> *Shapes and Angles* (includes tessellations).

Lines and Networks works well in a geometry class, and shows, for example, how the tangled mess of a real subway system can be represented much more simply with an abstract network (graph) model.

Futures Series (*9–12*)
Narrated by Jaime Escalante; PBS; $450 for the entire 12-part series.

This series was also recommended by Diane DePriest LP '93 (private communication):

> Each part deals with a different aspect of math and how it is used by real people in the real world. There are many famous guests (such as Cindy Crawford, Arnold Schwarzenegger, Jackie Joyner Kersee, and Sally Ride). The titles with some discrete math content are *Statistics and Sports performance,* and *Water Engineering and Optics.* The series is *excellent*!

Fractals: The Colors of Infinity (*8–up*)
52 min.; Films for the Humanities, 1994; $149

This film is recommended by high-school teacher Edward Polakowski LP '92 (email posting):

In addition to beautiful images of the Mandelbrot set (the best I've seen), it includes interviews with Benoit Mandelbrot, who talks about his discovery, the rise of fractal geometry as a means of looking at the world, and the "practical" uses of fractals. It is narrated by Arthur C. Clarke in an understandable fashion and even includes an interview with Stephen Hawking: "Is the world infinitely small?"

Professor Devaney Explains the Fractal Geometry of the Mandelbrot Set *(10–College)*
Key Curriculum Press, $25.

This is another good video for teaching fractals. Although recommended for higher grades, one teacher (Erica Voolich LP '94, email posting) found that her 7th-grade students responded enthusiastically. The only background concepts needed are that of a function and multiplication of complex numbers. (An article by Devaney on chaos is included in this volume [**10**] and in [**19**].)

NOVA Series *(6–up)*
NOVA/WGBH; some of the videos in the series can be purchased or rented; otherwise, check local televion listings or your library (teacher's guides and transcripts are also available).

This television series, shown regularly on PBS, has occasional programs in areas relevant to discrete mathematics. For example, "The Man Who Loved Numbers" (1988) is about the self-taught Indian mathematician Srinivasa Ramanujan, who developed many remarkable facts in number theory. More recently, a program on codes used in WWII, "The Codebreakers" (1994), was shown.

7. Software

In this section we list the programs and software that teachers have found most useful in teaching discrete mathematics. Overall, general-purpose programs, such as spreadsheets and drawing programs, or programs that allow open-ended explorations are much more useful than the many limited-use or "drill-and-kill" programs that dominate the market. As one can see in the articles [**9, 36**], there is essentially no general-purpose software designed for discrete mathematics that is appropriate in grades K-12; nevertheless software intended for other purposes has proven useful.

Teachers interested in using software may be interested in contacting **CLIME**, the *Council for Logo and Technology In Mathematics Education* (see the Appendix). The organization, which is affiliated with the NCTM, publishes a newsletter; issues have included a "top-ten mathematics-education software list" (based on a survey of teachers in a mentorship program at Stevens Institute of Technology), as well as a complementary "top-ten lessons-with-software list". In our recommendations below, we refer to the software survey as the "CLIME survey."

Spreadsheets (Excel, ClarisWorks, Lotus 123, etc.) (*4–up*)

Spreadsheets contain rows and columns of *cells* in which one can enter either numbers or algebraic formulas. Although one doesn't normally think of spreadsheets as educational software, they are in fact one of the most useful tools to have in the middle- or high-school classroom, especially for Algebra or Pre-Algebra students. Spreadsheets were rated "No. 2" in the CLIME Survey ("No. 1" was the *Geometer's Sketchpad*, described below).

Spreadsheets are ideal for explorations involving iteration and recursion, such as experimentation with population growth and other models of change, such as described in [**11, 15**]. Of course, calculators can perform many of the same functions, but spreadsheets can do many calculations in parallel, and so can be much faster and easier to use. Also, data from spreadsheets can often be exported to graphics programs to be displayed in different formats.

Classroom Use. The following are some examples of projects in which students used spreadsheets.

- (*6th grade*) Students acted as landscape consultants to students from other countries (communicating over the Internet), doing their planning, estimates, and budgeting with the spreadsheet. (Sr. Diane Mollica LP '95, private communication.)
- (*High school*) As preparation for finding probabilities mathematically, students rolled dice and entered their data on a spreadsheet; they then estimated the probabilities of various outcomes, such as rolling "7" with two dice. (Br. Patrick Carney LP '91, private communication.)
- (*High school*) In a unit on understanding the electoral college, students used a spreadsheet to compute the number of electoral votes each state would get using various methods of apportionment. Using their data, they showed how Hayes won over Tilden in 1876 in the electoral college—even though Tilden won the popular vote. (William Bowdish LP '92 and David Fogle LP '93, private communications.)

Geometer's Sketchpad (*7–up*)

Key Curriculum Press; Windows or Macintosh, $170 (plus shipping); site licenses are available.

Sketchpad is a drawing program that allows one to do precise geometric constructions. Teachers can use it to create demonstrations and examples; students can experiment with examples to make conjectures or verify theorems. It is one of the all-time favorites of mathematics teachers everywhere. It was rated "No. 1" in the CLIME survey. In addition to being a great experimental tool for traditional geometry, it is also excellent for demonstrating tessellation concepts, or creating tessellations. One can create animations easily by saving "scripts". Sketchpad can also be used to generate fractal patterns, such as the Sierpinski Triangle or Koch Snowflake (see *Fractals for the Classroom*, discussed in section 3.2). It is also possible to create activities that illustrate graph (network) concepts, such as those described for *Netpad* in [**9**].

A similar program, which uses slightly different conventions, but is not as well known, is **Cabri** (Texas Instuments), a limited version of which is installed on the new TI-92 graphing calculator. Another recommended though less-popular program is the **Geometric Supposer** (*8–up*) (Sunburst Communications; Mac, $129; Apple/Windows, $99), developed by Dr. Judah Schwartz of the Educational Development Corporation. For lower grades (5–up), there is the **Geometric PreSupposer** (Sunburst; Apple/Windows, $99). Standard **drawing programs** such as in ClarisWorks may also be used effectively (e.g., for tessellations and fractals).

Logo (*3–up*)
Logo Computer Systems Inc. (LCSI); $70-$200 (Windows, Mac, Apple IIe); (see also the references in [**22**] for other versions).

The best description of Logo is perhaps the one given by Hal Abelson of MIT: "Logo is the name of a philosophy of education and a family of computer languages that aid in its realization." The idea behind Logo is that we learn by constucting our own knowledge. Logo, as conceived by Seymour Papert, was intended to be the educational clay that would facilitate the building of mathematical knowledge.

A Logo program is a set of commands to a "turtle" that moves and draws on the computer screen in response to those commands. Students can construct new commands, and in turn use them as building blocks for even more elaborate creations. For example, students can write Logo programs to generate fractal curves, such as the Koch Snowflake. Logo's structure makes it especially easy to implement iteration and recursion.

Classroom Use. I (Kowalczyk) have extensive experience using and teaching with Logo. For example, I used it to simulate "gnomon growth", which leads to Fibonacci spirals, as described in [**20**]. (The text *Excursions in Modern Mathematics* described in Section 2 has good material on this topic.) Charlie Hennessy LP '95, who teaches 6-8th grades, reports that it is a good tool for getting students comfortable with the concept of a variable, since students can change parameters and see the effects immediately. Philip Lewis, a high school teacher, describes success using Logo to teach vector algebra and elementary algebra with an algorithmic approach [**22**].

Further Information. One of the most popular versions of Logo over the last 10 years has been **Logowriter** (LCSI; Mac/Apple IIe, $199). This version is slowly being replaced by its more sophisticated relative, **Microworlds** (LCSI; Mac, $99), which features an unlimited number of turtles, parallel processing, a melody editor, drawing tools, and the ability to create hypermedia links. **MathLinks** (LCSI; $79) (rated "No. 4" in the CLIME "top-ten lessons with software" list) is an interactive set of activities designed to help students develop mathematical thinking and become mathematical problem solvers. Activities are organized around the topics of polygons, repeating patterns, permutations and combinations and transformations. **Turtle Math** (LCSI; Mac, $69),[7] another version of Logo, was developed to address the needs of the elementary classroom. It is accompanied by 36 classroom-tested activities and materials geared to grades 3–6, which involve students in such discrete math topics as geometry, similarity, patterns, and probability.

Tesselmania! *(3–12)*
Dale Seymour; $48 (Mac), $69 (Windows), $79 (CD-ROM).

This program is designed for demonstrating and creating tessellations. *Tesselmania* is easy to use and good for sparking interest. Judy Nesbit LP '94 reports using the program for a student tessellation contest: "Students have been very enthusiastic about this project. They learn a lot and many of them are quite creative!" The program includes animation that is good for illustrating the different types of polygons that tile the plane and the transformations (reflections, translations, and rotations) involved in creating the tiling. The program comes with a resource guide, which has a series of lesson plans for teaching tessellations and using the program. Judy Brown LP '92 comments: "*Tesselmania* is the perfect tool for teaching teachers about tessellations. It gives teachers the ability to experiment with a number of different types of transformations and tessellations in a quick and accurate manner."

Many, including myself (Franzblau), have found that in teaching tessellations it's much better to start by having students create tessellations by hand; but once they understand the concepts, the program may be useful as a demonstration, or to inspire student creativity. Introducing any software too early may give students the misconception that a computer is necessary to create the tessellations.

I find that paper and pencil, drawing programs, or *Geometer's Sketchpad* are better tools for creating interesting tessellation designs. Although *Tesselmania* allows one to add decorations easily, and does the transformations automatically, its drawing capabilities are limited.

8. World Wide Web Sites

Teachers with access to the Internet can now find a variety of resources through the World Wide Web. There is no way to keep up with the constant growth and change on the Web, but we have identified several well-established Web sites that are useful for teachers or students of discrete mathematics, which we expect to continue and grow. The Web addresses (URLs) that we give are current as of July, 1996, but are subject to change.

Special thanks are due to Judy Ann Brown LP '92, a mathematics teacher at Pleasant Valley Middle school in Pennsylvania, who has used the

[7] *Turtle Math* was created by Doug Clements and Julie Sarama [**7**], and is available through both Dale Seymour and LCSI.

Web extensively in her teaching over the last few years, and who provided us with much of the information in this section. She has been a lead teacher in the Leadership Program in Discrete Mathematics, as well as an adjunct professor at Allentown College, teaching a discrete mathematics course for educators. Judy's home page, which has links to many of the resources mentioned here, is at `http://dimacs.rutgers.edu/~judyann/`; it also has links to other resources that she uses.

General Resources. We first mention several addresses with general information for mathematics and/or science teachers, which also have links to further resources.

Eisenhower National Clearinghouse for Math and Science Teachers (Ohio State University) `http://www.enc.org/`

This umbrella site is a gold mine for mathematics and math education. You can read newsletters, specialized catalogs, and topical publications, or connect to other education sites. An *Online Documents* service allows you to find articles on curriculum issues in mathematics that are available in an electronic format.

A *Lessons and Activities* page includes classroom activities with projects and supporting materials. When we last checked, the page contained sixteen different activities; the following three titles give a flavor for the kind of things that can be found here.

- The CHANCE Database contains materials to help teach a CHANCE probability course or a standard introductory probability or statistics course.
- The *Good News Bears Stock Market Project* is an interdisciplinary project designed for middle school students. It is an interactive stock market competition between classmates using the New York Stock Exchange and NASDAQ.
- The *Spanky Fractal Database* is a collection of fractal images, documents, and software available for free distribution. It links to several other fractal sites on the Internet.

Mathematics Forum (Swarthmore College)
`http://forum.swarthmore.edu/`

This is a new site, extending the well-known **Geometry Forum**, which was developed several years ago at Swarthmore. The site is intended to create an on-line mathematics community; it hosts discussion groups on teaching mathematics, provides useful information and interesting math problems, and has a search engine for mathematics resources.

One of the useful features at this site is **Ask Dr. Math**. K-12 students can get answers to their mathematical questions by linking to the *Dr. Math Archive*, which is a database of previously and frequently asked questions in mathematics. Students can also send a message directly to *Dr. Math* (actually a team of mathematics students at Swarthmore) using the email address dr.math@forum.swarthmore.edu. Replies are usually sent within 24 hours.

NCTM Standards (National Council of Teachers of Mathematics)
http://www.enc.org/online/NCTM/280dtoc1.html

A complete copy of the NCTM *Curriculum and Evaluation Standards for School Mathematics* can be found here, which includes a standard for discrete mathematics in grades 9-12. This excellent resource includes a discussion of each standard at its designated grade level groups: K-4, 5-8, or 9-12. These discussions contain a wealth of ideas and materials that could be adapted for use in the classroom. This is a great starting place for anyone who is interested in mathematics education.

Special Topics. Our next set of recommendations is for Web sites which have proven valuable sources for enrichment material for students or teachers on specific topics, or for classroom activities.

MegaMath (Los Alamos National Laboratory)
http://www.c3.lanl.gov/mega-math/

The MegaMath site brings important mathematical ideas to elementary school classrooms with unique activities. The site was developed by Nancy Casey and Michael Fellows [**5, 6**]. It includes many discrete math activities including the following:
The Most Colorful Math of All (map and graph coloring);
Games on Graphs;
Untangling the Mathematics of Knots;
Algorithms and Ice Cream for All (an interesting problem on graphs); and
A Usual Day at Unusual School (logic and paradoxes).
This site offers teachers the opportunity to bring discrete mathematics into their classroom with engaging stories.

MacTutor for Math History Information
http://www-groups.dcs.st-and.ac.uk:80/~history/Mathematical_MacTutor.html

MacTutor offers the history of mathematics on the Web. The Welcome Page for the site includes a Famous Curves Index, a Biographical Index, Chronologies, a History Topics Index, a Birthplace Map, the Mathematicians of the Day, Anniversaries for the Year, a Search Form, and Search Suggestions. I (Kowalczyk) used MacTutor to find a wealth of information about Fibonacci and the Fibonacci numbers (see also the article [**20**], which I wrote before I found this site).

Dynamical Systems (Boston University)
http://math.bu.edu/DYSYS/dysys.html

This site is designed to help teachers bring contemporary mathematics topics—chaos, fractals, and dynamics—into the classroom and to illustrate how to use technology effectively in the process. The interactive activities at this site can also help teachers understand the mathematics behind these topics. This site is well worth exploring—especially for high school teachers. This site was developed by Robert Devaney, who has used these activities in teaching calculus [**10**] and differential equations.

After exploring the site, Judy Brown LP '92 sent a note which is excerpted below.

> I spent a lot of time with the "chaos for the classroom" section. This is in such an easy-to-read digestible format that I really can't wait to go back and investigate the Mandelbrot and Julia set information. I don't know exactly how to express my feelings, except that some of the "neat stuff" that I've done before now has taken on a more mathematical tinge. There are also a few fractal "movies" that you have got to see [under the heading *Rotations and Animation*]. My favorite is the dancing Sierpinski triangle. I'll never be able to look at a Sierpinski triangle again without imagining it dancing.

Fractal Frequently Asked Questions and Answers (Ohio State)
`http://www.cis.ohio-state.edu/hypertext/faq/usenet/fractal-faq/faq.html`.

This is another good source for those interested in fractals, with hyperlinks to many other interesting and useful fractal sites on the Web. A sample of questions addressed are as follows. "I want to learn about fractals. What should I read first?" "What is a fractal?" "What are some examples of fractals?" "What is chaos?" This site is recommended for both teachers and students who want to begin learning about fractals.

AIMS Puzzle Page
`http://204.161.33.100/puzzle/puzzlelist.html`

Here is a delightful site with excellent classroom-ready activities, which are based primarily on discrete mathematics. Each month a new and challenging puzzle is posted, complete with student worksheets, which can be downloaded. As I (Kowalczyk) viewed the puzzles for the first four months of 1996, I could hardly wait to print them so I could get started.

The World of MC Escher
`http://www.texas.net/users/escher`

At this site you will come to know this fascinating artist (and mathematician) through stories, his tessellations and other art works, as well as his insights into these works. The site also offers high-quality (commercial) products featuring Escher's designs. If you are already familiar with Escher you'll have a great time just looking around, otherwise, it's time to explore and be captivated by his work. (See also the books *Teaching Tessellating Art* and *Visions of Symmetry*, described in Section 5.)

Acknowledgments

We wish to thank the many people who contributed to this article. In addition to those we have already mentioned or quoted in the text, many others provided titles, information, or comments: Doris Abraskin (*LP '95, PS 233, NY*), Jan Amenhauser (*LP '92, Perry L. Drew Sch, NJ*), Charlie Anderson (*LP '94, Arvada West HS, CO*), Jeremy Avigad (*DIMACS Postdoc*), Chuck Biehl (*LP '90, Math/Science Acad, DE*), Judy Brown (*LP '92, Pleasant Valley MS, PA*), Ethel Breuche (*LP '91, Freehold HS, NJ*), Br. Pat Carney (*LP '91, Bishop Walsh MS/HS, MD*), Connie Cunningham (*LP '93, Rocky Grove HS, PA*), Val DeBellis (*Rutgers U., NJ*), Diane DePriest (*LP '93, Purnell Sch, NJ*), Suzanne Foley (*LP '95, Olney Cluster, PA*), Carol Giesing (*LP '92, Hoover HS, CA*), Marilyn Goldfarb (*LP '93, State College Alternative, PA*), Charlie Hennessy (*LP '95, Holy Trinity, Washington, DC*), Laura Holland (*LP '95, Bowne-Munro Sch, NJ*), Art Kalish (*Syosset HS, NY*), Virginia Kostisin (*LP '92, Haledon PS, NJ*), Joe Malkevitch (*York College, CUNY*), Sr. Dianne Mollica (*LP '95, Immaculate Conception Sch, NJ*), Judy Nesbit (*LP '94, Montclair Kimberly Acad, NJ*), Tony Piccolino (*Montclair State U, NJ*), Susan Picker (*LP '90, Manhattan Sch Dist, NY*), Ed Polakowski (*LP '92, Manalapan HS, NJ*), Janice Ricks (*LP '91, Marple Newtown HS, PA*), Penni Ross (*LP '94, The Langley Sch, VA*), Joe Rosenstein (*Rutgers U., NJ*), Reuben Settergren (*Grad Student, Rutgers U., NJ*), Barbara Stapleton (*LP '95, Bowne-Munro Sch, NJ*), Marylu Tyndell (*LP '94, Wall Township HS, NJ*) Dave VanSchaick (*LP '93, Gowana Jr. HS, NY*), and Erica Voolich (*LP '94, Solomon Schecter Day Sch, MA*).

References

[1] William L. Bowdish, "Finding the Fractal Complexity of a Coastline", *In Discrete Mathematics: Using Discrete Mathematics in the Classroom* **5** (Nov 1994), p. 3.

[2] Ethel Breuche, Book Review, *In Discrete Mathematics: Using Discrete Mathematics in the Classroom* **2** (October 1992), p. 10.

[3] ———, in "The Discrete Reviewer", *In Discrete Mathematics: Using Discrete Mathematics in the Classroom* **6** (Spring/Summer 1995), p. 7.

[4] Marilyn Burns and Stephanie Sheffield, *Math and Literature, K–3,* Math Solutions, Sausalito, CA, 1992. (Avail. from Cuisinaire.)

[5] Nancy Casey and Michael Fellows, "This is MEGA-Mathematics!", Los Alamos, 1993.

[6] ———, "Implementing the Standards: Let's Focus on the First Four", this volume.

[7] Doug Clements and Julie Sarama, "Turtle Math: Redesigning Logo for the Elementary Classroom", *Learning and Leading with Technology* (April, 1996), p. 10.

[8] Nancy Crisler, Patience Fisher, and Gary Froelich, "A Discrete Mathematics Textbook for High Schools", this volume.

[9] Nathaniel Dean and Yanxi Liu, "Discrete Mathematics Software for K-12 Education", this volume.

[10] Robert L. Devaney, "Putting Chaos into Calculus Courses", this volume.

[11] John A. Dossey, "Making a Difference with Difference Equations", this volume.

[12] Susanna S. Epp, "Logic and Discrete Mathematics in the Schools", this volume.

[13] Deborah Franzblau, "New Models for Courses in Discrete Mathematics", *Newsletter of the SIAM Activity group in Discrete Mathematics* **4** (Winter, 1993-94), p. 1-3.

[14] Martin Gardner, *The Scientific American book of Mathematical Puzzles and Diversions,* Simon and Schuster, NY, 1959. See also other books by author.

[15] Eric W. Hart, "Discrete Mathematical Modeling in the Secondary Curriculum: Rationale and Examples from the Core-Plus Mathematics Project", this volume.
[16] Douglas R. Hofstadter , *Metamagical Themas*, Basic Books, New York, 1985.
[17] Bret Hoyer, "A Discrete Mathematics Experience with General Mathematics Students", this volume.
[18] Robert E. Jamison, "Rhythm and Pattern: Discrete Mathematics with an Artistic Connection for Elementary School Teachers", this volume.
[19] Margaret J. Kenney and Christian R. Hirsch, Eds., *Discrete Mathematics Across the Curriculum*, 1991 Yearbook, NCTM, Reston, VA, 1991.
[20] Janice C. Kowalczyk, "Fibonacci Reflections: It's Elementary!", this volume.
[21] ———, resource review in "The Discrete Reviewer", *In Discrete Mathematics: Using Discrete Mathematics in the Classroom* **7** (Fall/Winter 1995), p. 7.
[22] Philip G. Lewis, "Algorithms, Algebra, and the Computer Lab", this volume.
[23] Evan Maletsky, "Discrete Mathematics Activities for Middle School", this volume.
[24] Joseph Malkevitch, "Applied Discrete Mathematics for Liberal Arts Students", *Newsletter of the SIAM Activity group in Discrete Mathematics* **4** (Summer, 1994), p. 1-2.
[25] ———, "Game Theory Bibliography", *In Discrete Mathematics: Using Discrete Mathematics in the Classroom* **6** (Spring/Summer 1995), p. 9.
[26] ———, "Discrete Mathematics and Public Perceptions of Mathematics", this volume.
[27] National Council of Teachers of Mathematics, *Curriculum and Evaluation Standards for School Mathematics*, NCTM, Reston, VA, 1989.
[28] Anthony Piccolino, Book Review, *In Discrete Mathematics: Using Discrete Mathematics in the Classroom* **2** (October 1992), p. 10.
[29] Susan Picker, Resource Review, *In Discrete Mathematics: Using Discrete Mathematics in the Classroom* **3** (August 1993), p. 4.
[30] Fred S. Roberts, "The Role of Applications in Teaching Discrete Mathematics", this volume.
[31] Joseph G. Rosenstein, "A Comprehensive View of Discrete Mathematics: Chapter 14 of the New Jersey Mathematics Curriculum Framework", this volume.
[32] Joseph G. Rosenstein and Valerie A. DeBellis, "The Leadership Program in Discrete Mathematics", this volume.
[33] Joseph G. Rosenstein, Deborah S. Franzblau, and Fred S. Roberts, Eds., *Discrete Mathematics in the Schools*, (this volume), American Mathematical Society, AMS/DIMACS series, 1997.
[34] Reuben J. Settergren,"What we've got here is a failure to cooperate", this volume.
[35] Eric Simonian, "The Tangram Magicians", *In Discrete Mathematics: Using Discrete Mathematics in the Classroom* **7** (Fall/Winter 1995), p. 3.
[36] Mario Vassallo and Anthony Ralston, "Computer Software for the Teaching of Discrete Mathematics in the Schools", this volume.

Dept. of Mathematics, CUNY/College of Staten Island, NY
(Previous address:) DIMACS, Rutgers University, Piscataway, NJ
E-mail address: `franzblau@postbox.csi.cuny.edu`

Rhode Island School of the Future, P.O. Box 4692 Middletown, RI 02842
E-mail address: `kowalcjn@ride.ri.net`

Appendix A. Publishers, producers, and distributors.

AMS (American Math. Society)
P.O. Box 6248
Providence, RI 02940-6248
800-321-4267, 401-455-4000

American Scientist
P.O. Box 18975
Research Triangle Park, NC 27709-9890
800-282-0444, 919-549-0097

Annenberg/CPB Multimedia Collection
Dept. CA95 P.O. Box 2345
S. Burlington, VT 05407-2345
800-LEARNER

CLIME (Council for Logo and technology In Mathematics Education)
Ihor Charischak
Stevens Institute of Technology – CIESE Center
Hoboken, NJ 07030
Email: `icharisc@stevens-tech.edu`

COMAP (Consortium for Mathematics and its Applications)
Suite 210, 57 Bedford Street
Lexington , MA 02173
800-772-6627, Fax: 617-863-1202,
Email: `orders@comap.com`

Creative Publications
5040 West 111th St.
Oak Lawn, IL 60453
800-624-0822, Fax: 708-425-9790

Cuisenaire
P.O. Box 5026
White Plains, NY 10602-5026
800-237-0338 (orders), 800-237-3142 (service), Fax: 800-551-RODS

Dale Seymour
P.O. Box 10888
Palo Alto, CA 94303-08799
800-872-1100

DIMACS-DM Newsletter
P.O. Box 10867
New Brunswick, NJ 08906
908-445-4065, Fax: 908-445-3477
Email: `estler@dimacs.rutgers.edu`

Everyday Learning Corporation
See Janson Publications.

Films for the Humanities
Box 2053
Princeton, NJ 08543-2053
800-257-5126

W. H. Freeman
New York, NY
800-877-5351, 212-576-9400

Janson Publications
800-382-1479 or 800-322-MATH (6284)
Fax: 312-540-5848

Journal Films
800-323-9084, 708-328-6700

Key Curriculum Press
P.O. Box 2304
Berkeley, CA 94702-0304
800-995-MATH (6284), Fax: 800-541-2442
Email: `info@keypress.com`,
WWW: `http://www.keypress.com`

LCSI (Logo Computer Systems, Inc.)
3300 Chemin Cote Vertu Road, Bureau/Suite 201
Montreal (Quebec) H4R 2B7, CANADA
800-321-LOGO, 514-331-7090, Fax: 514-331-1380

MAA (Mathematical Association of America)
1529 Eighteenth St., NW
Washington, DC 20036
800-331-1622, Fax: 202-265-2384

Email: maahq@maa.org,
WWW: http://www.maa.org

Mimosa Publications
P.O. Box 26609
San Francisco, CA 94126
800-MIMOSA-1, Fax: 415-995-7155

NCTM (National Council of Teachers of Mathematics)
1906 Association Drive
Reston, VA 22091-1593
800-235-7566 (orders),
703-620-9840 (info),
Fax: 703-476-2970
Email: nctm@nctm.org

New York Times Co.
229 W. 43rd St.
New York, NY 10036-3959
800-698-4637,
WWW: http://www.nytimes.com

NOVA/WGBH
125 Western Ave
Boston, MA 02134
800-255-9424, 617-492-2777
Note: Distributors of NOVA video tapes change often; contact WGBH for information on distributors of particular programs.

PBS/Futures Series
1320 Braddock Place
Alexandria, VA 22314-1698
800-344-3337

Prentice Hall
Englewood Cliffs, NJ 07632
800-848-9500, 201-592-2000

Scientific American, Inc.
415 Madison Ave.
New York, NY 10017-1111
800-333-1199, 515-247-7631

Simon and Schuster
(Same as Prentice-Hall)

Springer-Verlag NY
175 Fifth Ave.
New York, NY 10010
800-SPRINGER, 201-348-4033

Sunburst Communications
101 Castleton St.
P.O. Box 100
Pleasantville, NY 10570-0100
800-320-7511

Texas Instruments
800-TI-CARES
Email: ti-cares@ti.com
WWW: http://www.ti.com

WGBH (See NOVA/WGBH)

Wonderful Ideas
P. O. Box 64691
Burlington, VT 05406-4691
800-92-IDEAS

Appendix B. Index by title.

The following is a list of titles of all resources described in the text, each followed by its corresponding page number. Titles are arranged alphabetically within the main resource categories: Texts, Literature, Periodicals, Videos, Software, and Web Sites.

Texts.

Aha! Insight, 16
Applied Combinatorics, 18
Basic Geometry of Voting, 22
Build Your Own Polyhedra, 22
Chaos, Fractals, and Dynamics: Computer Experiments in Mathematics, 21
Codes Galore, 13
Discrete Mathematics Across the Curriculum, 10
Discrete Mathematics and its Applications, 17
Discrete Mathematics through Applications, 8
Drawing Pictures with One Line: Exploring Graph Theory, 13
Ethnomathematics: A Multicultural View of Mathematical Ideas, 20
Excursions in Modern Math, 7
Fair Division: From Cake Cutting to Conflict, 22
For All Practical Purposes, 5
Fractals for the Classroom: Strategic Activities, 13
Fractals: The Patterns of Chaos, 21
Game Theory and Strategy, 21
Graphs: An Introductory Approach, 18
Graphs, Models, and Finite Mathematics, 18
Graph Theory Applications, 18
HiMap/HistoMap Series, 12
Insides, Outsides, Loops, and Lines, 10
Intro to Tessellations, 20
Knot Book, 21
Mathematician's Coloring Book, 12
Mathematics, a Human Endeavor, 4
Math EQUALS, 16
Math on the Wall, 10
Math Their Way, 9
Orderly Tangles, 21
Problem Solving: Crossing the River with Dogs, 12
Problem Solving Using Graphs, 13
Teaching Tessellating Art, 20
Unit Origami, 22
Visions of Symmetry, 20

Literature.

Alan Turing: The Enigma, 20
Anno's Mysterious Multiplying Jar, 16
Cloak for the Dreamer, 15
Dr. Seuss Books, 14
Grandfather Tang's Story, 14
Jurassic Park, 16
Math Curse, 16
One Hundred Hungry Ants, 14
Sam Johnson and the Blue Ribbon Quilt, 15
The Tangram Magician, 15
Three Hat Day, 15
Two of Everything, 15

Periodicals.

American Scientist, 19
Consortium, 12
Elementary Mathematician, 10
In Discrete Mathematics: Using Discrete Mathematics in the Classroom, 11
Mathematical Intelligencer, 19
Mathematics Teacher, 11
Mathematics Teaching in the Middle School, 11
Math Horizons, 17
New York Times, 18
QUANTUM: The student magazine of math and science, 17
Scientific American, 19
Teaching Children Mathematics, 11
What's Happening in the Mathematical Sciences, 19
Wonderful Ideas, 9

Videos.

For All Practical Purposes, 22
Fractals: The Colors of Infinity, 24
Futures Series, 24
Geometry: New Tools for New Technologies, 23
Mathematical Eye Series, 24
NOVA Series, 25
Powers of Ten, 23
Professor Devaney Explains the Fractal Geometry of the Mandelbrot Set, 25

Software.

Geometer's Sketchpad, 26
Logo, 27
Spreadsheets, 26
Tesselmania, 28

Web Sites.

AIMS Puzzle Page, 31
Dynamical Systems, 30
Eisenhower Nat'l Clearinghouse for Math and Science Teachers, 29
Fractal Frequently Asked Questions and Answers, 31
MacTutor for Math History Information, 30
Mathematics Forum, 29
MegaMath, 30
NCTM Standards, 30
World of MC Escher, 31

Appendix C. Index by grade level.

In this section, for each of the grade-level ranges K–2, 3–5, 6–8, and 9–12, we list the titles of resources described in the text whose content seems especially appropriate. Within each subsection, titles are ordered (approximately) by reading level. Each title is followed by the number of the page that contains its description. See Appendix B for unabbreviated titles.

Grades K–2

- **Texts**:
  Math Their Way, 9
  Insides, Outsides, ... , 10
  Math on the Wall, 10
- **Literature**:
  Dr. Seuss books, 14
  One Hundred Hungry Ants, 14
  Grandfather Tang's Story, 14
  The Tangram Magician, 15
  Three Hat Day, 15
  Sam Johnson ... Quilt, 15
  Cloak for the Dreamer, 15
  Two of Everything, 15
  Anno's ... Jar, 16
- **Periodicals**:
  Teaching Children Math, 11
  Wonderful Ideas, 9
  Elementary Mathematician, 10
- **Web Sites**:
  MegaMath, 30

Grades 3–5

- **Texts**:
  Math on the Wall, 10
  Insides, Outsides, ... , 10
  Math Coloring Book, 12
  Intro to Tessellations, 20
  Build Your Own Polyhedra, 22
- **Literature**:
  Sam Johnson ... Quilt, 15
  Cloak for the Dreamer, 15
  Two of Everything, 15
  Anno's ... Jar, 16
  Math Curse, 16
- **Periodicals**:
  Wonderful Ideas, 9
  Elementary Mathematician, 10
- **Software**:
  Logo, 27
  Tesselmania, 28
- **Web Sites**:
  MegaMath, 30

Grades 6–8

- **Texts**:
  Insides, Outsides, ... , 10
  Math Coloring Book, 12
  Drawing Pictures/One Line, 13
  Intro to Tessellations, 20
  Build Your Own Polyhedra, 22
  Aha! Insight, 16
  Math EQUALS, 16
  Unit Origami, 22
  Fractals for the Classroom, 13
  Math/Human Endeavor, 4
  DM through Applications, 8
  HiMap/HistoMap Series, 12
  Prob Solving Using Graphs, 13
- **Literature**:
  Math Curse, 16
  Jurassic Park, 16
- **Periodicals**:
  Math Teaching in the Middle School, 11
  In Discrete Mathematics, 11
- **Videos**:
  Geometry: New Tools ... , 23
  Powers of Ten, 23
  NOVA Series, 25
  Fractals/Colors of Infinity, 24
- **Software**:
  Logo, 27
  Tesselmania, 28
  Spreadsheets, 26
  Geometer's Sketchpad, 26
- **Web Sites**:
  MacTutor/Math History, 30
  AIMS Puzzle Page, 31
  World of MC Escher, 31

Grades 9–12

- **Texts**:
  Math Coloring Book, 12
  Drawing Pictures/One Line, 13
  Intro to Tessellations, 20
  Build Your Own Polyhedra, 22
  Unit Origami, 22
  Fractals for the Classroom, 13
  Math/Human Endeavor, 4
  DM through Applications, 8
  Prob Solving Using Graphs, 13
  HiMap/HistoMap Series, 12
  Visions of Symmetry, 20
  Teaching Tessellating Art, 20
  Codes Galore, 13
  Orderly Tangles, 21
  Prob Solving/Crossing River with Dogs, 12
  Excursions in Modern Math, 7
  For All Practical Purposes, 5
  Ethnomathematics, 20
  Game Theory and Strategy, 21
  Fair Division, 22
- **Literature**:
  Jurassic Park, 16
  Alan Turing: The Enigma, 20
- **Periodicals**:
  Mathematics Teacher, 11
  In Discrete Mathematics, 11
  QUANTUM, 17
  Math Horizons, 17
  Consortium, 12
  New York Times, 18
  Mathematical Intelligencer, 19
  Scientific American, 19
  American Scientist, 19
- **Videos**:
  For All Practical Purposes, 22
  Geometry: New Tools ... , 23
  Powers of Ten, 23
  NOVA Series, 25
  Fractals/Colors of Infinity, 24
  Mathematical Eye Series, 24
  Futures Series, 24
  Prof. Devaney Explains Fractal Geometry, 25
- **Software**:
  Logo, 27
  Tesselmania, 28
  Spreadsheets, 26
  Geometer's Sketchpad, 26
- **Web Sites**:
  MacTutor/ Math History, 30
  AIMS Puzzle Page, 31
  World of MC Escher, 31
  Mathematics Forum, 29
  Fractal Q & A, 31
  Dynamical Systems, 30

Appendix D. Index by topic.

In this section, for each major topic, we list the titles of resources described in the text which are recommended either for background or as a source of activities. Resources are listed in order by increasing grade level within each subsection. The title of each resource is followed by the number of the page that contains its description. See Appendix B for unabbreviated titles.

Graph Problems and Applications

- **Print Resources**:
  Insides, Outsides, . . . , 10
  Math Coloring Book, 12
  Drawing Pictures/One Line, 13
  Math/Human Endeavor, 4
  Prob Solving Using Graphs, 13
  Excursions in Modern Math, 7
  For All Practical Purposes, 5
  Ethnomathematics, 20
  Applied Combinatorics, 18
  DM and its Applications, 17
  DM through Applications, 8
  Graphs, Models, . . . , 18
  Graph Theory Applications, 18
  Graphs: Intro Approach, 18
- **Videos**:
  For All Practical Purposes, 22
  Geometry: New Tools . . . , 23
  Mathematical Eye Series, 24
- **Web Sites**:
  MegaMath, 30

Iteration, Recursion, and Fractals

- **Print Resources**:
  Dr. Seuss Books, 14
  Two of Everything, 15
  Anno's . . . Jar, 16
  Jurassic Park, 16
  Fractals for the Classroom, 13
  Math/Human Endeavor, 4
  DM through Applications, 8
  Excursions in Modern Math, 7
  For All Practical Purposes, 5
  Fractals/Patterns of Chaos, 21
  Chaos, Fractals, Dynamics, 21
  DM and its Applications, 17
- **Videos**:
  For All Practical Purposes, 22
  Powers of Ten, 23
  Fractals/Colors of Infinity, 24
  Prof. Devaney Explains Fractal Geometry, 25
- **Software**:
  Logo, 27
  Spreadsheets, 26
  Geometer's Sketchpad, 26
- **Web Sites**:
  Fractal Q & A, 31
  Dynamical Systems, 30

Geometric Patterns and Transformations

- **Print Resources**:
  Grandfather Tang's Story, 14
  The Tangram Magician, 15
  Cloak for the Dreamer, 15
  Sam Johnson . . . Quilt, 15
  Build Your Own Polyhedra, 22
  Intro to Tessellations, 20
  Unit Origami, 22
  Math/Human Endeavor, 4
  HiMap/HistoMap Series, 12
  Visions of Symmetry, 20
  Teaching Tessellating Art, 20
  Excursions in Modern Math, 7
  For All Practical Purposes, 5
- **Software**:
  Logo, 27
  Tesselmania, 28
  Geometer's Sketchpad, 26
- **Web Sites**:
  World of MC Escher, 31
  Mathematics Forum, 29

Listing and Counting (Combinatorics)

- **Print Resources**:
  One Hundred Hungry Ants, 14
  Three Hat Day, 15
  Anno's ... Jar, 16
  Math on the Wall, 10
  Aha! Insight, 16
  Math/Human Endeavor, 4
  Applied Combinatorics, 18
  DM and its Applications, 17
- **Videos**:
  Mathematical Eye Series, 24

General Problem Solving (modeling, logic, etc.)

- **Print Resources**:
  Math Their Way, 9
  Grandfather Tang's Story, 14
  Math on the Wall, 10
  Math Curse, 16
  Aha! Insight, 16
  Math EQUALS, 16
  Math/Human Endeavor, 4
  Prob Solving/Crossing River with Dogs, 12
- **Videos**:
  Geometry: New Tools ... , 23
  Mathematical Eye Series, 24
  Futures Series, 24
- **Software**:
  Logo, 27
- **Web Sites**:
  MegaMath, 30
  AIMS Puzzle Page, 31
  Eisenhower Clearinghouse, 29
  Mathematics Forum, 29

Social Choice (fair division, voting, game theory)

- **Print Resources**:
  DM through Applications, 8
  HiMap/Histomap Series, 12
  Excursions in Modern Math, 7
  For All Practical Purposes, 5
  Game Theory and Strategy, 21
  Fair Division: From Cake Cutting to Conflict, 22
  Basic Geometry of Voting, 22
- **Videos**:
  For All Practical Purposes, 22

Codes and Information

- **Print Resources**:
  DM through Applications, 8
  Codes Galore, 13
  For All Practical Purposes, 5
  Alan Turing: The Enigma, 20
  Applied Combinatorics, 18
- **Videos**:
  For All Practical Purposes, 22
  Geometry: New Tools ... , 23
  NOVA Series, 25

Knots and Topology

- **Print Resources**:
  Insides, Outsides, ... , 10
  HiMap/HistoMap Series, 12
  Orderly Tangles, 21
  Knot Book, The, 21
- **Web Sites**:
  MegaMath, 30

History and People

- **Print Resources**:
  Math EQUALS, 16
  New York Times, 18
  For All Practical Purposes, 5
  Math Horizons, 17
  Ethnomathematics, 20
  Mathematical Intelligencer, 19
- **Videos**:
  Geometry: New Tools ... , 23
  NOVA Series, 25
  Futures Series, 24
- **Web Sites**:
  MacTutor/Math History, 30

DIMACS Series in Discrete Mathematics
and Theoretical Computer Science
Volume **36**, 1997

The Leadership Program in Discrete Mathematics

Joseph G. Rosenstein and Valerie A. DeBellis

1. Introduction to the Leadership Program

During the period from 1989 to 1998, the *Leadership Program in Discrete Mathematics* will have involved about 1000 K–12 teachers in an intensive and exciting introduction to discrete mathematics. In this article we will describe the *Leadership Program* (LP) and the lessons that we have learned from it. We will also describe the ways in which the LP serves as a continuing resource to teachers who have not participated in the program, as well as those who have.

A. History of the Leadership Program. The story of the *Leadership Program in Discrete Mathematics* begins with a proposal to the National Science Foundation in 1988 for funding the Center for Discrete Mathematics and Theoretical Computer Science (DIMACS)[1] as a Science and Technology Center (STC). The proposal included a provision that DIMACS would support programs for teachers and students in collaboration with the Rutgers University Center for Mathematics, Science, and Computer Education (CMSCE). Soon after NSF announced in February 1989 that DIMACS would receive the STC award, planning began for a summer program for teachers. This two-week program, entitled *Networks and Algorithms*, took place in the summer of 1989 with 27 high school teachers; it was funded entirely by DIMACS.

1991 *Mathematics Subject Classification.* Primary 00A05, 00A35.

Joseph G. Rosenstein has served as Director of the Leadership Program in Discrete Mathematics since its inception in 1989. Valerie A. DeBellis has also been associated with the LP since its inception, and has served as its Associate Director since 1992.

[1]DIMACS is an NSF-funded Science and Technology Center which was founded in 1989 as a consortium of Rutgers and Princeton Universities, AT&T Bell Laboratories, and Bellcore (Bell Communications Research). With the reorganization of AT&T Bell Laboratories in 1996, it was replaced in the DIMACS consortium by AT&T Labs and Bell Labs (part of Lucent Technologies). DIMACS is also funded by the New Jersey Commission on Science and Technology, its partner organizations, and numerous other agencies.

The 1989 *Networks and Algorithms* institute served as a pilot for a more ambitious program that was first funded by NSF the following year — the *Leadership Program in Discrete Mathematics.* Initially, the LP was funded by NSF for two years as an institute for high school teachers. Then, as the LP — Phase II, it was funded by NSF for three years as an institute for high school and middle school teachers. Finally, as the LP — Phase III, it was funded by NSF for four years as an institute for K–8 teachers. Each program also received financial support from DIMACS and from Rutgers University, and each program was co-sponsored by DIMACS and CMSCE.

B. Size of the Leadership Program. A clear progression over time has been the increase in the size of the program. During Phase I of the LP (of course, it was not referred to as Phase I before Phase II was funded), there were about thirty-five participants during each of 1990 and 1991. During Phase II, which involved parallel institutes for high school and middle school teachers, there were about eighty teachers during each of 1992, 1993, and 1994. During Phase III, there were three institutes for K–8 teachers with 120 participants in 1995, and, in each of 1996, 1997, and 1998, five institutes with 180 participants. Two of the Phase III institutes each year are residential institutes at Rutgers, and the other three are commuter institutes, one at Rutgers and two at other sites; the "off-site" institutes in 1996 were in Rhode Island and Virginia, and in 1997 will be in Rhode Island and Arizona. Including the pilot institute in 1989, the total number of participants is approximately 1000. The scope of Phase III of the LP can be seen by reviewing the schedule for the summer of 1997. There are five two-week institutes for teachers who are new to the program, five one-week institutes for teachers in the 1996 cohort who are returning for their second summer's activities, two one-week institutes for teachers from 1989–1995, and a two-day "crash course" for high school teachers. Altogether, there are over seventeen weeks of institutes during the summer, with an anticipated total attendance of over 400 teachers.

C. The evolving target audience of the Leadership Program. Another clear progression over time has been that the participants have been teachers of progressively younger and younger students. This was not the intention at the outset, but it reflected what we learned from the participants in the program about the suitability and the value of introducing discrete mathematics to students of all grade levels and all ability levels.

The pilot program was targeted to teachers of high achieving seniors, since we thought that it was for those students that discrete mathematics was most appropriate. Many of the participants in the pilot program were teaching Advanced Placement (AP) Computer Science courses in their schools, and one focus of the summer institute was writing Pascal computer programs to implement network algorithms.

We soon learned that other students were able to benefit from exposure to discrete mathematics as much, if not more than the high achieving students, and so our assumption became that teachers in the program would be introducing discrete mathematics to average high school students. Then we learned that students who had been unsuccessful with traditional mathematics could be successful with discrete mathematics — partly because of the visual, geometric component of the topics in discrete mathematics that were the focus of the program, and partly because discrete mathematics did not require a strong background in the mathematics with which they had been unsuccessful.

This realization, that discrete mathematics could provide a "new start" for students, led to the October 1992 conference *Discrete Mathematics in the Schools: How Can We Make an Impact?* which in turn led to the publication of this volume. It first appeared in a "concept document" developed by Joseph Rosenstein in January 1991, and was then incorporated into the charge to the conference, a revised version of which appears as the **Introduction**[2] to this volume.

We soon realized that many of the topics which we discussed in the program were equally appropriate for middle school students. As a result, Phase II of the LP included a middle school component. Each summer from 1992 to 1994, the LP involved two parallel institutes, one for forty high school teachers and one for forty middle school teachers. The program for middle school teachers was conducted by faculty from Montclair State College (now Montclair State University). As was the case earlier, we learned from the teachers in the program that discrete mathematics is also appropriate for students at earlier grade levels. Phase III accordingly is addressed to elementary school teachers as well as middle school teachers. We have found that most of the topics in discrete mathematics that are included in the program can indeed be introduced to students at all grade levels, although of course the way in which the topics are introduced may differ considerably between grade levels.

D. Participation in the Leadership Program. Participants in the LP have typically been expected to attend a two- or three-week institute in the first summer, up to four follow-up sessions during the following school year, and a one- or two-week institute the following year. Including the follow-up sessions, each fully-participating teacher has been involved in four to six weeks of discrete mathematics workshops during a period of a year (including two summers), five weeks for teachers in Phase I, six weeks for teachers in Phase II, and four weeks for teachers in Phase III. With the support of DIMACS, the LP has also provided opportunities for participants to

[2]See "Discrete Mathematics in the Schools: An Opportunity to Revitalize the Mathematics Curriculum", this volume [**5**], and the articles by Susan Picker [**4**] and L. Charles Biehl [**1**] for further information about how discrete mathematics has served as a new start for students.

continue to attend follow-up sessions after their year as official participants in the program, to attend "veterans institutes" in subsequent summers, and to communicate with each other on an active email network.

Participants in the LP are expected to introduce discrete mathematics in their classrooms, incorporate discrete mathematics into their schools' curricula, and introduce their colleagues, both locally and broadly, to topics in discrete mathematics. A substantial percentage of LP participants have fulfilled these expectations and have remained active in LP activities beyond their formal affiliation with the program.

E. Instructional Staff of the Leadership Program. An important feature of the program has been the participation in the instructional staff of college faculty, including both researchers and mathematics educators. This has enabled participants to be in contact with real-live mathematicians and computer scientists, to learn how people working in the field think about their subject, and to experience the mathematical sciences as living disciplines.

Among the faculty members who have conducted workshops extending a week or more have been Ravi Boppana (Rutgers), Margaret Cozzens (Northeastern), Valerie DeBellis (Rutgers), Deborah Franzblau (Rutgers), Robert Garfunkel (Montclair), Robert Hochberg (Rutgers), Glenn Hurlburt (Arizona State), Robert Jamison (Clemson), Kenneth Kaplan (Rutgers), Laura Kelleher (Massachusetts Maritime Academy), Rochelle Leibowitz (Wheaton College — MA), Evan Maletsky (Montclair), Joseph Malkevitch (York College — CUNY), Terence Perciante (Wheaton College — IL), Anthony Piccolino (Montclair), Fred Roberts (Rutgers), Joseph Rosenstein (Rutgers), Donald Smith (Rutgers), Diane Souvaine (Rutgers), Ann Trenk (Wellesley), Tom Trotter (Arizona State), and Kenneth Wolff (Montclair). (Note that affiliations given are those at the time of participation in the LP.)

Many other researchers and educators have visited and participated in the LP, including Steven Brams, Doug Clements, John Conway, Dannie Durand, Nate Dean, Ron Graham, Stuart Haber, Eric Hart, David Johnson, Stephen Maurer, Michael Merritt, Peter Winkler, and Ann Yasuhara. Of special note is Michael Fellows (University of British Columbia), who showed how research problems in computer science can often be brought down to a level which second-graders can understand.

Also of note is the interplay in the formulation and development of the LP between a mathematician (Joseph G. Rosenstein) and a mathematics educator (Valerie A. DeBellis). As Director and Associate Director of the LP, they together developed a program which presents the appropriate content in a form which both challenges the participants and enables them to meet the challenges, which both involves teachers in mathematical learning and activities and enables them to replicate that learning and those activities in their classrooms. We also acknowledge the efforts of Janice Kowalczyk, Bonnie Katz, and Stephanie Micale who have served the LP for many years as Evaluation Coordinator, Program Coordinator, and Secretary, respectively.

2. Broader Goals of the Leadership Program

The goals of the LP are not defined exclusively in terms of the accomplishments of the participants in the area of discrete mathematics, but also in terms of their attitudes and understandings toward mathematics and the teaching and learning of mathematics. By 1991, we had learned that discrete mathematics was not just another interesting area of mathematics which teachers could use in their classrooms, but that it was also an excellent vehicle for changing mathematics education. This was reflected in the "concept document" referred to above: "Discrete mathematics provides an opportunity to focus on *how* mathematics is taught, on giving teachers new ways of looking at mathematics and new ways of making it accessible to their students. *From this perspective, teaching discrete mathematics in the schools is not an end in itself, but a tool for reforming mathematics education.*" [**5**] The broader goals reflected in this passage are explored in the following paragraphs.

A. Changing participants' attitudes about mathematics. Teachers often view mathematics exclusively as a body of knowledge, as a set of facts, which it is their job to transmit to their students; this should not be surprising since, after all, this has likely been their experience in learning mathematics. It should also not be surprising that, as a result, many students attribute their lack of success in mathematics to their inability to remember all of the required facts, formulas, and techniques.[3]

We would like teachers to view mathematics in terms of reasoning and problem-solving; in order to do that we must expect teachers to reason and solve problems. We would like teachers to recognize the applications of mathematics to the world; in order to do that we must show them how to wear eyeglasses through which they can see the world mathematically.[4] Wrestling with a mathematical situation, what mathematicians would call "doing" mathematics, is not something with which many teachers are familiar; we need to introduce them to the idea of doing mathematics, and foster the idea that they themselves can function as mathematicians, as can their students.[5] And, as educators of teachers, we need to provide teachers with a

[3]When Rosenstein interviewed Rutgers students in his "Mathematics for Liberal Arts" class, many responded independently that they were unsuccessful in mathematics in high school because of their inability to memorize all the formulas and proofs.

[4]See also Joseph Malkevitch's article "Discrete Mathematics and Public Perceptions of Mathematics" [**3**] in this volume, and a children's book called *Math Curse* [**8**] which features a child who sees mathematics everywhere.

[5]One striking image of early LP institutes is Joe Malkevitch's asking the participants whether they thought of themselves as "mathematicians" and conveying to them that if they are doing mathematics, then it is entirely appropriate for them to refer to themselves, and their students, in that way.

supportive learning environment so that they will be comfortable with doing mathematics.

Discrete mathematics is a particularly appropriate environment for enabling teachers to function as mathematicians; as a result, all of these goals have been features of the LP. Participants in the LP are expected to solve problems. These problems go beyond the warm-up exercises that simply test their recall of the workshop topics, although such exercises facilitate a gradual transition from easier to more difficult problems. After each morning workshop, they spend an hour in study groups grappling with the problems, and discuss their solutions in the homework review session the next morning; the daily problems also are the focus of many evening discussions among participants in the residential programs. Discrete mathematics is an area where problems can be concisely stated and easily understood, no matter whether their solutions are simple or difficult, or even if their solutions are unknown. Since many topics in discrete mathematics are connected with real world situations, participants can learn to wear mathematical eyeglasses, seeing applications of graphs, counting, and algorithms all around them. Discrete mathematics lends itself readily to exploration, enabling participants to rediscover principles that mathematicians refer to as theorems. Moreover, since most teachers have had no exposure to topics in discrete mathematics such as graphs (the kind with vertices and edges) and since these topics have few mathematical prerequisities, all participants start on a "level playing field". This makes it possible for early elementary teachers with little mathematical background to work together in a supportive environment with middle school teachers who are more familiar with traditional topics in mathematics; despite their differences in background, it is not uncommon in this type of environment for primary teachers to grasp the essence of a situation before their middle school colleagues.

B. Learning mathematics. The high expectations that we have created for participants are reflected in the schedule of the institute itself. Half of each day is devoted to learning mathematics, and the other half to introducing that mathematics into K–8 classrooms. Each morning, participants are involved in a two-hour content-based workshop on new mathematical topics. This is followed by a one-hour study session in which participants work in small groups on a set of "homework" problems based on the topic of the workshop; before the morning workshop on the next day, they will present solutions to the entire group. The schedule of the follow-up sessions is similar.

Altogether there are twenty-one workshops for the K–8 LP participants; these include the ten workshops during the first summer, six workshops at the follow-up sessions, and five workshops during the second summer program.[6] The first week of the summer institute focuses on graphs and their

[6]There are six follow-up workshops because we conduct Saturday follow-up sessions in October and May, and Saturday-Sunday follow-up sessions in December and March;

applications, and the second week focuses on patterns in numbers and geometry. The second summer institute focuses on games and probability. The topics of the workshops at the follow-up sessions are a variety of significant topics in discrete mathematics which are independent of each other and of the topics of the summer workshops. Following are the titles of all the workshops:

First summer

1. Coloring Maps & Resolving Conflicts
2. Drawing Pictures with One Line: Euler Circuits
3. Hamilton Circuits & the Traveling Salesperson Problem
4. Making the Right Connections: Spanning Trees and Algorithms
5. Shortest Routes
6. Introduction to Systematic Counting
7. Combinatorics and Pascal's Triangle
8. Iteration and Recursion
9. Patterns in Geometry
10. Generating Fractals

Follow-up sessions

11. Voting: Consolidating Individual Preferences
12. Codes: Error Detection and Error Correction
13. Fair Division
14. Number Patterns in Nature (including Fibonacci numbers)
15. Directed Graphs and Tournaments
16. Alphabetizing and Sorting

Second summer

17. Paths and Matchings
18. Matchings and Games
19. Games and Strategies
20. Probability
21. Probability and Games

The workshops for high school and middle school teachers in Phase I and Phase II of the LP addressed similar topics, although because high school and middle school teachers typically have more experience in mathematics, these topics could be discussed at greater depth and additional topics could be introduced. During Phase II, when participants attended three weeks in the first summer and two weeks in the second summer, the five weeks focused on the following themes:

1. graphs and their applications
2. algorithms for graphs
3. combinatorics and probability
4. social choice
5. recursion and fractals

---

the two-day follow-up sessions make it easier for participants who must travel a distance to attend the expected four follow-up sessions.

The description of discrete mathematics presented in the article in this volume entitled "A Comprehensive View of Discrete Mathematics: Chapter 14 of the New Jersey Mathematics Curriculum Framework" [**6**] reflects in part the activities used by LP participants in their classrooms. The first draft of that article was drawn from a "content map" which participants in the 1994 veterans program were asked to help develop based on their classroom experiences; the "content map" was designed to indicate classroom activities appropriate for different grade levels and to trace the development of topics in discrete mathematics across the different grade levels.

C. Changing instructional practices. Although each morning at the LP is devoted to mathematical content, the way in which that content is delivered is designed to convey messages about mathematical instruction. We consciously model the behaviors that we would like teachers to carry into their own classrooms, the types of mathematics instruction recommended by the NCTM Standards. Some of these behaviors are described in the following paragraphs.

Using a variety of instructional formats. The morning workshops involve a mixture of whole-group instruction and small-group activity. The pattern that is repeated throughout each workshop involves introduction of new content material, participants' working on a problem, and discussion of the problem and the material. For homework review sessions, the whole group is divided into two smaller groups. Seating in workshop groups in the K–8 program is heterogeneous, with teachers from different grade levels and with different mathematical backgrounds working together; seating in classroom implementation groups in the afternoon is homogeneous, with teachers working with colleagues who deal with children at similar ages. Participants leave the institute with models of introducing new mathematical material which serve as alternatives to the lecture method.

Working in groups on problem-solving. Solving problems in groups provides powerful lessons for all participants, even for those who had been using groups in their own classrooms, because they typically had never themselves learned content material in a group setting. This is facilitated by having participants working at round tables which are conducive to small group interaction. Participants learn about the power of discussion in assisting mathematical learning. They learn about the advantages of working in a group where different participants bring different perspectives and strengths to the problem-solving process; as noted above, each group of teachers in the K–8 program typically includes teachers from all grade levels. They learn about how to achieve the goal of ensuring that everyone in the group has learned the material. And they learn about dealing with the difficulties that arise when some individuals tend to dominate the group and when others tend to withdraw from the group.

Peer Mentoring. An important aspect of the program is the role played by "lead teachers" during all activities. Lead teachers serve as coaches during the problem-solving activities, not providing answers but raising pertinent questions, suggesting possible directions, and reinforcing participants' confidence in their ability to solve problems themselves. They also conduct the homework review sessions, and the sessions on classroom implementation, including presentations of their own classroom activities with discrete mathematics. The presence of the lead teachers not only facilitates the learning, but also provides strong role models for the future achievement of program participants. Participants are very aware that the lead teachers were introduced to discrete mathematics only a few years ago, and that now they are serving in a leadership capacity. Over 40 teachers have served in this leadership role during the course of the program. Each lead teacher has used discrete mathematics extensively in his or her classroom, has made presentations on her or his classroom experiences to colleagues, including presentations during the summer and follow-up programs, and has served as a coach to participants in the program, both during the institute and subsequently.

Journal writing. A more recent addition to the program is the use of journal writing, with continued feedback from staff, to enhance the mathematical learning of participants. Participants are provided with a ten-page "journal" in which they make daily entries regarding their mathematical learning; this gives participants an opportunity to describe their understanding of the new material and to highlight areas where they are having difficulty with the material. Journals are collected near the end of each day and are reviewed by lead teachers, who respond daily (in writing) to the entries. Journals serve as a way of assessing the program as well as the learning of individual participants. If patterns are found among the journals, the lead teachers respond collectively to the group the following day, and convey the participants' areas of difficulty to the workshop leaders.

Providing opportunities for reflection. After modeling each instructional strategy, participants are asked to reflect on that strategy in organized discussions; this enables them to better understand the strategy and how it can be used. This mode of "modeling then reflecting"[7] on desired behavior is now utilized in a number of contexts, including group learning, problem-solving, journal writing, assessment, and developing an equitable learning environment. Regular opportunities are incorporated into the program to allow participants time to reflect on the institute experience. For example, after working in groups for several days, participants are asked to reflect on what it means to engage in group work and what are important aspects to remember about participating in group learning environments.

D. Challenging the participants. Many teachers who come to professional development activities are looking primarily for activities they can

[7]Use of this mode of "modeling then reflecting" in the LP was stimulated by Eric Hart, who often encouraged us to reflect on this mode as well as model it. As a result we expanded its use, which gave us more to reflect on.

do in their classrooms; they want to be given things that will interest, occupy, and challenge their students. Typically, they are not looking to be challenged themselves. The conflict between these two perspectives emerges by the third day of each institute, when some participants, feeling the challenge keenly, raise the question of "why do we need to learn this, we are only teaching at the x'th grade level", and others, beginning to yield to old negative attitudes about mathematics, decide that they will never overcome the challenges that the LP presents.

The LP is designed to challenge participants to learn mathematics by doing mathematics, that is, by solving mathematical problems whose answers and solution methods they do not know in advance. They need to understand that learning often involves dealing with situations that are challenging, and with concepts that appear inpenetrable; they need to experience frustration when a problem appears insoluble and excitement when it has finally been overcome. Not only do they better understand the mathematical themes and strategies involved in the problem and gain confidence in their mathematical abilities, they also understand the difficulties their students have with situations that are challenging and with concepts that appear inpenetrable. Many teachers have forgotten what it is like to be a learner — what it is like to be frustrated and what it is like to be successful.

Creating a program which provides frustration for its participants is a perilous undertaking. However, as a result of the environment in the LP, not one participant has yet left the program. An important reason for this is the presence of the lead teachers (see Section 2.C). One of their important roles is to identify participants who are experiencing difficulties, quickly provide additional assistance and counseling, and, where appropriate, refer them for further assistance and encouragement to the Program Directors (the authors).

Another strategy is that we enable participants to recognize that their frustration is an entirely valid component of problem solving. For many K–8 teachers, solving mathematical problems means memorize the rule, then apply it; often they are surprised to find that the problem-solving process is very different. While solving problems, they often experience negative emotions such as fear, frustration, uncertainty, and anger. On the third day of the program, participants are involved in a workshop on problem solving which deals explicitly with the role of affect in problem solving.[8] This workshop enables them to reflect on these emotions, recognize that they are normal, and apply problem-solving strategies to work through them to complete the problem. In order for teachers to model productive problem-solving behaviors in their classroom, they need to have a clear understanding of the process itself; this workshop not only provides insight into the participants' problem solving but also helps participants experience the difficulties their students have in solving problems.

[8] This workshop was designed by Valerie DeBellis based on research in mathematics education by her [**2**] and Gerald Goldin.

E. Empowering the participants. In addition to offering mathematical challenges to the participants and challenging them to bring LP materials to their classrooms, curricula, and colleagues, the LP strives to empower the participants to meet these educational challenges. As noted above, an important component of this empowerment is that participants are encouraged to see themselves as mathematicians when they are engaged in problem solving activities in mathematics. They are also encouraged to play leadership roles in introducing discrete mathematics into American classrooms. We describe them as experts in K–12 discrete mathematics, since in fact the teachers who have completed the LP are among the first teachers who have incorporated discrete mathematics in their classrooms, and are a substantial percentage of teachers who have done so. Their expertise becomes clear to them at the end of the first summer program when they are asked to reflect on what they have learned in their two-week encounter with the LP; they are amazed by what they have learned, by the amount they have learned, and by the fact that they have succeeded in learning mathematics which they never dreamed existed. Challenging teachers to learn mathematics has great risks, but there are also great rewards, because they learn that they can do mathematics, and pass on that sense of empowerment to their students.

The LP also empowers teachers to initiate mathematical explorations in their classrooms. Explorations imply that the class may travel to uncharted territory, where the teacher may not know what questions to ask, and what answers to give to students' questions. Because teachers are accustomed to be in a position of authority, they may be reluctant to ask any question whose answer they do not already know. We empower the teacher to overcome this barrier by modeling — showing the teacher that the workshop leader does not know the answers to all the questions — and by introducing mathematical questions to which no one knows the answer. In order for teachers to entertain questions whose answers they don't know, they have to become comfortable responding "I don't know the answer to that question; let me think about it overnight". This kind of response also lets the students know that there are problems which cannot be solved quickly, that some problems require thought and time. We often complain that students are unwilling to work on problems which require more than ten minutes (or even ten seconds) of their time. However, they don't see adults spending time on problem-solving. For the teacher to say "let me think about that one" sends a message that problems whose solutions are not obvious are worth considering.

F. Reflections. It should be noted that we have incorporated into the above discussion the lessons we have learned from directing the LP. Participants in the early institutes will recognize only some of these themes. As the program evolved, we learned what features we could incorporate into the program and how to make these features more meaningful to teachers. As we began to understand the power of discrete mathematics to facilitate change, we were able to introduce deliberately and intentionally various components of the program.

Many of these features can be introduced in programs that address other content areas of mathematics. However, the qualities of discrete mathematics make possible the inclusion of all of these components of education reform. Moreover, repeating an important point introduced in the 1991 "concept paper" (see Section 1.C), discrete mathematics offers a new start for teachers because they can incorporate these features into their discrete mathematics lessons, where they are not restricted by existing curriculum requirements, and once successful adapt them into other mathematics instruction.

3. The LP as a continuing resource

Discrete Mathematics Newsletter. The newsletter *In Discrete Mathematics: Introducing Discrete Mathematics in the Classroom* has been published for the past six years, and includes articles written by participants in the LP describing their classroom experiences with discrete mathematics. The newsletter is distributed at no charge to over 3000 teachers. Its founding editor was Joseph G. Rosenstein, and it has since been edited by Deborah S. Franzblau and Robert A. Hochberg. The publication of the newsletter has been funded by DIMACS and NSF.

Workshops in Your District. For the past four years, the LP has offered to send a team of experienced teachers to any district to conduct workshops in discrete mathematics for middle and high school teachers. Beginning in 1998, similar workshops will also be available for elementary school teachers. The contents of the high school and middle school workshops were developed at summer workshops (called "workshop workshops") during 1993 and 1994, and parallel workshops for elementary schools are currently being developed. In the workshop workshop, teachers who are experienced with the use of discrete mathematics in their own classrooms develop a series of one-and-a-half hour workshops, and receive training on how to deliver similar workshops. In the typical case of "Workshops in Your District", two teachers go to a district and together present four of these workshops during an inservice day. The district pays only the honoraria for the presenters and the cost of the materials; the publicity and administration of the "Workshops in Your District" project are paid for by DIMACS.

The Franchise Program. Beginning in the summer of 1998, it will be possible for any college teacher with a background in discrete mathematics to replicate locally the Phase III program in discrete mathematics for K–8 teachers, using the workshop materials developed for the LP and with the assistance of lead teachers from the LP. It is anticipated that the franchising arrangements will result in a number of commuter programs for K–8 teachers at different sites throughout the United States.

Curriculum Materials. A curriculum materials development project is currently underway, and it is anticipated that materials for K–8 teachers will be available, in print or electronic format, by 1999. These materials will be targeted to teachers who have no experience with discrete mathematics, and will be based on the experiences of the LP program. In addition, the LP Web site (accessed from http://dimacs.rutgers.edu), currently under development, will contain materials from the LP and resources developed by LP participants.

Conference Presentations. Participants of the LP regularly make presentations on discrete mathematics at NCTM regional and national conferences and at conferences of local and state organizations of mathematics teachers. (Some have even made presentations at international conferences on mathematics education.) These teachers serve as an ongoing resource for conference organizers throughout the country who wish to schedule sessions on discrete mathematics.[9]

4. Participants' Statements

We conclude this article with a number of statements prepared by participants in the *Leadership Program in Discrete Mathematics* in response to the simple question, "What has resulted from your participation in the LP?" Taken together, these statements (presented alphabetically) illustrate and highlight many of the features of the LP that were presented in this article.

As a result of my participation over the years in the LP, I have introduced my high school to discrete mathematics. Aided by the staff at Rutgers University, I have been successful in implementing a full year discrete math course. Also, discrete math is now an integral part of a topics in math course. I have also spread the word about discrete math by making presentations at conferences on the local, state, regional, and national level.

William Bowdish, LP '92, teaches in the Sharon (MA) High School.

I am a teacher and mathematics supervisor in one of five high schools in a regional district. Since the summer of 1991, at least two department meetings a year in my school have been solely dedicated to a presentation on a discrete math topic and encouraging teachers to infuse this material in their classes. Discrete math is now offered as a full year course in all five high schools and enrollment appears to be growing in the larger schools. Textbooks have also been selected with an eye toward how much discrete math they contain. Since the summer of 1992, when I participated in the Workshop Workshop program and helped author some of the workshops, I have presented discrete math workshops to over ten schools or districts. I have also presented workshops in discrete math topics every year since 1992 at the AMTNJ conference and every other year at the "Good Ideas in Teaching

[9]Another important resource for conference organizers are the teachers who participated in the "Implementation of the NCTM Standard in Discrete Mathematics Project" directed by Margaret Kenney of Boston College; during each of 1993-1996, summer institutes for high school teachers were conducted at six sites throughout the country.

Pre-Calculus and ... " conference at Rutgers. One of the most gratifying experiences I have had is coming back to follow-up sessions and hearing or seeing how some of the participants have used one of the workshops that I had presented as a lead teacher, and usually how they have improved upon it. I haven't even begun to describe how much I have learned about a topic I had no prior knowledge about before the summer of '91. My image of myself as a mathematician was first formed that summer. In my wildest dreams, I could not have imagined the personal, professional, and academic growth I have achieved. I have enjoyed learning and sharing this knowledge and enthusiasm for mathematics with my students, with the teachers I work with, and with the wonderful people I have met through the Leadership Program.

Ethel Breuche, LP '91, is a teacher and mathematics supervisor at the Freehold (NJ) High School.

The LPDM has made me a better teacher and has made my school a better place for mathematicians to grow and develop. As a result of my participation in 1993-1995 in summer programs and my ongoing conversations with colleagues through follow-ups and e-mail communication, I have been involved in the development of the AP Statistics Program and work as a consultant for the College Board. I have introduced the AP Statistics course with a unit on Discrete Mathematics and teach this course at my school. We have also introduced a year-long Discrete Mathematics course at the high school. On a more philosophical level, my students benefit from having a teacher who knows that there are many good approaches to problem solving, and often many good answers to the same problem.

Anne M. Carroll, LP '93, teaches at Kennett (PA) High School

The LP has enriched my ability to bring real world connections to my students. It has enhanced my professional portfolio by exposing me to cutting-edge mathematical thought and theory and providing me with a network of resources from both the educational and research communities. This exposure allows me to bring a new lease to the mathematical life in my classroom. My students look forward to exploring current situations with a mathematical eye and become empowered when they realize that they too can think and speak mathematically.

Carol Ann DiMauro, LP '92, is a consultant with the New York City Mathematics Project, Institute for Literacy Studies, Lehman College, CUNY.

Since participating in the LP in '95, I have seen how much discrete mathematics is already inherent in many of the text books and NCTM publications on the market today. Discrete math lends itself well to performance-based tasks and assessment and helps to build students' problem solving skills. Using discrete math allows the teacher and student to make more real life mathematical connections. I have shared discrete mathematics with other teachers through hands-on workshops and in class demonstrations. Students in my own classroom, my school and other classes in my cluster have participated in a variety of discrete math activities. Even at recess students solve problems and practice mathematical tasks by "playing" on specially developed discrete math activities that are painted in the school yard. If you're in a class of mine, you can't escape learning mathematics with a lot of discrete activities.

Suzanne Foley, LP '95, is K–8 Technology Coordinator for the Olney (PA) Cluster of districts.

My involvement in the LP has broadened my personal approach to math and problem-solving as well as given me new avenues to reach different math students in different ways. First, discrete math is a great forum for teaching problem-solving. Rich problems with open-ended solutions allow for many kids to dive in and explore. Students with different strengths and interests can use approaches that suit their learning styles or multiple intelligence strengths. Second, the LP gave me connections to interesting and dedicated people who are a rich resource for further growth and new applications that are age- and grade-level appropriate.

Charles G. Hennessey, LP '95, teaches middle school students at the Holy Trinity School in Washington, D.C.

Two of my 8th-grade students at opposite ends of the spectrum provide the best example of the impact of discrete mathematics on my classes. One student, in the "learning disabled" program, hates school. He has no interest in the regular curriculum. The other student is in the honors program. She loves the challenge of school and everything about it. Yet these two students with seemingly no common ground both *love* the discrete math problems. The first student may not try a traditional homework assignment, but will work all class and more finding circuits, paths, counting rabbits and more. The second student wants extra work in all areas and seems particularly interested in the various areas of discrete math we have covered. Two different students both touched in different ways by discrete mathematics.

Jeff Hoyle, LP '96, teaches in the Dartmouth (MA) Middle School.

When I first learned about discrete mathematics, it proved to be a wonderful vehicle to get slower general math and low level students involved. They loved the graph theory and fractals, as well as secret codes, fair division, and map coloring. These students loved these lessons because they were "fun" and gave them a great measure of success. I also had great success with these topics in my precalculus and calculus classes. There I was able to teach the same topics at a more mathematical and rigorous level. The reaction was the same ... the novelty was greatly appreciated. Now I am teaching future mathematics teachers and newly appointed teachers obtaining their masters degrees at Stony Brook University and at NYU. Most of them are unacquainted with discrete math, and absolutely are enthralled with it. A particular favorite has been doing logic problems using incompatibility graphs. I think that having spent two summers in this program has enriched my enthusiasm, knowledge, and joy in mathematics greatly. I have shared this experience with many hundreds of students, and now with at least 100 placed teachers in Stony Brook and NYU.

Elyse Magram, LP '90, teaches at Smithtown High School West in East Northport, NY.

I cannot think of another professional development experience that has changed my life so much. First, there is the material, but the program has gone far beyond that. I have a new set of colleagues, leaders with whom I can share ideas and get feedback, people who are *peers* as well as mentors and role models. I have gained self-confidence. I have grown from a teacher who attends conferences to a teacher who attends conferences as both a participant and a presenter. I was nervous at first, but now I really enjoy this experience. It all started with the LP! I hope I can give back as much as I have gained!

Judy Nesbit, LP '94, teaches at the Montclair (NJ) Kimberley Academy; she was recognized as NJ Non-Public School Teacher of the Year 1996.

The Leadership Program allowed me to bring discrete math to deaf education. It was there that I saw how visual discrete math could be — graphic representations of fractals, graph coloring, maps, paths and circuits, etc. I thought **this** is something **for** deaf kids — a visual representation of math concepts that translates well to American Sign Language, and can transfer logically to numerical or symbolic representation. I immediately incorporated it into my teaching with positive results! Since the time that I spent with the LP, I have earned my Ph.D. in Deaf Education with an emphasis in mathematics. I can honestly say that the LP has influenced this decision. Now, as a teacher educator, I regularly include the topics, materials, and activities I learned at the LP into my instruction, and include discrete math concepts, such as Fibonacci numbers and the Sierpinski Triangle, into math and logic "puzzles" which I co-create with a colleague as a regular feature in a publication for deaf and hard of hearing students ("World Around You"). I conclude by saying that if it were not for the LP in discrete math, I certainly would not be doing what I am doing today!

Claudia Pagliaro, LP '92, is an assistant professor at the University of Pittsburgh in the Education of Deaf and Hard of Hearing Students Program, Department of Instruction and Learning.

Beyond the large and new personal opportunities that have come to my life as a result of my participation in the LP, including a position as a staff and curriculum developer; the chance to be published; the opportunities to give talks and workshops to teachers in the U.S. and Europe; the development of a new high school curriculum in discrete mathematics implemented now in more than 20 schools; and the experience of doing a case study qualitative research project, I think the largest thing that has happened to me has been the widening and deepening of my understanding of what mathematics is as a living breathing discipline in our contemporary world; what it can do and what it is that a mathematician does. I have been fortunate to have met and interacted with mathematicians and researchers, and very important to me — women mathematicians — who have enabled me to feel the world of mathematics as more accessible than it ever could have been. This in turn I have been able to communicate to other teachers and to students, so that they can feel included in the world of mathematics rather than intimidated by it.

Susan Picker, LP '90, is a mathematics instruction specialist for the Office of the Superintendent of Manhattan (NY) High Schools.

I have been very excited to learn and present this relatively new area of mathematics — "discrete mathematics." Since attending the LP last summer, I have found examples of discrete math topics in many places — at workshops and in textbooks, etc. As I have worked with teachers and students this past year, I have developed a clearer understanding of how to explain this sometimes confusing area of math. I think the institute gave me this clarity of understanding and presentation. Although I have used my training this school year, I hope to use it even more as I offer a series of after-school problem-solving workshops for teachers, using LP activities, in my school district next school year.

Nancy Shields, LP '96, is a K–12 supervisor of mathematics at the Beeville (TX) Independent School District.

References

[1] Biehl, L. Charles, "Discrete Mathematics: A Fresh Start for Secondary Students", this volume.

[2] DeBellis, Valerie A., *Interactions between affect and cognition during mathematical problem solving: A two year case study of four elementary school children.* Doctoral dissertation, Rutgers University, 1996. Ann Arbor, Michigan: University Microfilm 96-30716.

[3] Malkevitch, Joseph, "Discrete Mathematics and the Public Perception of Mathematics", this volume.

[4] Picker, Susan, "Using Discrete Mathematics to Give Remedial Students a Second Chance", this volume.

[5] Rosenstein, Joseph G., "Discrete Mathematics in the Schools: An Opportunity to Revitalize School Mathematics", this volume.

[6] Rosenstein, Joseph G., "A Comprehensive View of Discrete Mathematics: Chapter 14 of the New Jersey Mathematics Curriculum Framework", this volume.

[7] Rosenstein, Joseph G., Deborah S. Franzblau, and Fred S. Roberts, *Discrete Mathematics in the Schools*, this volume.

[8] Scieszka, J. and L. Smith, *Math Curse*, Viking/Penguin Group, 1995.

DEPARTMENT OF MATHEMATICS, RUTGERS UNIVERSITY
E-mail address: `joer@dimacs.rutgers.edu`

CENTER FOR MATHEMATICS, SCIENCE, AND COMPUTER EDUCATION AND CENTER FOR DISCRETE MATHEMATICS AND THEORETICAL COMPUTER SCIENCE, RUTGERS UNIV.
E-mail address: `debellis@dimacs.rutgers.edu`

DIMACS Series in Discrete Mathematics
and Theoretical Computer Science
Volume 36, 1997

Computer Software for the Teaching of Discrete Mathematics in the Schools

Mario Vassallo and Anthony Ralston

1. Introduction

Despite the fact that the increasing emphasis on discrete mathematics has been motivated by the importance of computers, most of the software for teaching mathematics at the high school and college level, at least, has been oriented toward calculus. Although a number of software systems for discrete mathematics have been developed, most of them are meant to be used for research purposes rather than for teaching in or out of the classroom.

Can a software system meant for research purposes be suitable in education? We have surveyed three such systems, namely, *Mathematica/Combinatorica*, *GraphPack*, and *SetPlayer*. We found a number of good educational qualities. On the other hand, we also found them lacking in other vital characteristics. The reason is simple. The needs of a researcher in using a software system are quite different from the needs of the teacher and the student. At the college level, many professors might need to use a system both for teaching and for doing research. Should we ask for systems that could possibly run in a *teaching* mode and a *research* mode? Or should systems for teaching be totally separate from systems for research? We do not intend to answer these questions in this paper. Our aim is to look at possible classroom set-ups in which computer technology may be utilized and state what we should expect from a software system if it is to be used as a teaching and learning tool both in and out of the classroom. We shall also review the three software systems mentioned above and shall discuss how appropriate they may be if used as educational tools.

2. Computer Technology in the Classroom

Why should a teacher make use of computer technology in the classroom? The teacher should not use computer technology just because it is a modern

1991 *Mathematics Subject Classification.* Primary 00A05, 00A35, 68N99.

trend or just for the sake of using it or just because colleagues use it. The teacher should use computer technology because he/she truly believes that such usage will improve his/her teaching capabilities and the students will better understand the subject matter. The use of computer technology in the classroom requires, initially at least, a stronger commitment from the teacher. More hours of preparation need to be spent, especially during the first year when the teacher is learning how to use the technology in a manner that best fits his/her needs as well as the students' needs.

Here are two possible scenarios for using computer technology in the classroom:

Scenario 1: Instruction is conducted in a traditional classroom where the teacher will have a computer system at his/her disposal. A relatively large screen (preferably on top of or next to the chalkboard) will be needed so that the teacher can demonstrate subject-related material to the students. On their part, the students should be able to follow the demonstration without difficulty. In this mode, a number of computer systems would be available to students outside of the classroom to assist them in their work.

Scenario 2: Instruction is conducted in a laboratory environment with a number of computers linked together through a network. The teacher and students will all have a terminal at their disposal. A large screen similar to the one described in *Scenario 1* should also be available. The teacher should be able to control the mode of operation of the students' terminals, for example, by freezing the students' keyboards and making all monitors display the same information. Another possible mode is for the teacher to perform a number of operations and for the students to follow suit. The teacher may also decide to let the students work on their own by assigning tutorials or problems to solve. The students should be able to make use of the lab resources after class hours to assist them in their work.

Both scenarios have their advantages and disadvantages. *Scenario 1* sounds more practical. It is cheaper to implement. The essential situation in the classroom remains unchanged. However, the teaching content will change a lot and the whole presentation will probably need to be reworked. *Scenario 2* also requires a new style of teaching and student participation is essential. The teacher will be faced with additional problems in the classroom arising from the hardware and software being used. The flow of information will slow down in *Scenario 2* when the teacher decides to demonstrate how to perform some operations and the students have to try them out. Of course, combinations of these two scenarios as well as other variations are possible.

3. Desirable Qualities of Educational Software for the Teaching of Mathematics

When developing systems to be used in mathematics education as teaching tools, software designers should seriously consider the needs of the teacher and the student both in and outside of the classroom. The following is a list of characteristics that should be considered for such systems. The first four (A – D) are necessary for educational use, the fifth (E) is very desirable, and the remaining four (F – I) should be viewed as ideal qualities to be included as better systems are developed. This list is derived from our own experience with the good features and limitations of software for discrete mathematics.

(A) *Graphics Orientation*: The system should be heavily graphics oriented. It should make use of a graphics environment such as the *X Window system* or *MS-Windows*. The graphics user interface should consist of several types of windows including text windows, display windows, and menu windows. The user (teacher or student) should be able to interface with the system by using a keyboard, a mouse, and possibly other input devices. The user should be able to select choices simply by selecting menu options or icons, answer questions by checking options or typing a few characters, draw by indicating consecutive endpoints to be connected by straight lines, and paint or shade by moving the cursor over the screen. All of these features are particularly important for teachers who must be able to use the software in the classroom with minimum time lost giving instructions for using the software.

(B) *Ease of Use*: The system should be user-friendly so that a teacher can master its use in a short span of time. A teacher should not have to spend an unreasonable amount of time in using the system for class preparation. Also, the average student should be able to learn how to use the system without major problems. Error messages should be simple and meaningful. An extensive and sophisticated help system should be available.

(C) *Classroom Suitability*: In general, the teacher should be able to use the system in the classroom without wasting too much time in the issuing of commands. The system should be able to execute the software in such a way that the teacher can easily start, halt, speed up, and slow down the flow of information and backtrack to review a previous step. The software should be flexible enough to allow easy diversion from the programmed lesson to accommodate student questions.

(D) *Display of the Solution*: When producing a solution to a problem, the system should supply the user with three modes of operation. In the first mode, the whole solution will be displayed at one go. In the second mode, the solution will be displayed step-by-step. At every step, the display will pause until the user signals (through an input device) for the execution to resume. In the third mode, the solution will be displayed several steps at a time. The user will decide the number of steps to be displayed at a time. The user should be able to turn from one mode to another at will.

(E) *Completeness of the Subject*: The software should be able to solve common problems in the subject matter and cover other related topics.

(F) *Algorithm Animation*: The system should provide a mode in which it would take an algorithm (whether built-in or input by the user) and display a step-by-step animation of its execution. The user should be able to control the execution in the same manner as described in *Display of the Solution.* The user should also be able to backtrack to a previous step if the need arises.

(G) *Problem Mode*: The system should provide a mode in which a student may request problems on a specific topic. The system will then provide the student with a menu listing various degrees of difficulty. The student will choose the level of difficulty and the system will return with a set of problems to solve of the desired degree of difficulty. The student should be able to solve problems by typing in a solution. The software should be capable of checking the correctness of the student's solution. Meaningful error messages should be displayed when errors are detected and the student should be allowed to attempt a revised solution. If the student does not solve the problem correctly after a certain number of attempts, the system should have the ability to display a final or step-by-step solution. The system should also be able to save scripts of students' sessions so that the teacher can look at them to judge the progress of the students. At the end of a session, the system should return a brief synopsis of the student's performance (e.g., number of problems tried, number of correct solutions on the first attempt, number of correct solutions on second attempt, etc.)

(H) *Tutorials and Review Sessions*: If a student is unable to attend class on a particular day, he/she should be able to use the system outside of the classroom to learn about the topic that was earlier discussed in class. The system should also be able to assist a student in reviewing the subject matter for an upcoming test. The system should offer built-in tutorials and review sessions on each and every topic in the syllabus. These tutorials should be able to run at different levels of difficulty. It should also provide the teacher the option to customize his/her own tutorials and review sessions according to his/her students' needs. The system should contain a program that would assist the teacher in the customization of a tutorial or a review session.

(I) *Interfacing*: The system should be able to interface with other established software packages which provide the teacher with assistance in word processing and in keeping student grade records and statistics. Both teacher and student need to be able to make use of a mathematical typesetting system such as LaTeX. This will help the teacher in preparing quizzes, tests, and exams and in preparing class handouts and writing papers for publication. It will help the student when doing homework as well as when writing projects and papers.

Depending on their teaching style, some teachers will consider some of the above important and others less so. But we believe that each of these qualities will be considered desirable by many teachers. Of course, some of these qualities will be considerably more expensive than others to implement but all are feasible and should be considered in the development of any software for teaching discrete mathematics.

4. Existing Software Systems for Discrete Mathematics

Several software systems for Discrete Mathematics have been developed during the last few years. In this section we will survey three of these systems. We will describe what they do and discuss their suitability as teaching tools.

4.1. Mathematica and Combinatorica.

4.1.1. *Description.*

Mathematica (v2.0) is a general computer software system and language for mathematical computation and other applications **[8]**. It is written in C and may run under various operating systems, including *UNIX* and *MS-DOS*. *Mathematica* is made up of two parts: the *kernel*, which actually performs computations, and the *front end*, which handles interaction with the user. The front end makes use of a *textual* and *graphical* user interface. The *graphical* user interface uses `X Windows` or `Sun Windows` on *UNIX*-based machines and `MS Windows` on IBM-compatible PCs. There is also a version of *Mathematica* that runs on a Macintosh.

On computers with graphical user interfaces, *Mathematica* supports a special *notebook* interface. Notebooks are interactive documents, into which you can insert *Mathematica* input as well as ordinary text and graphics. A user interacts with notebooks by typing text and/or by using a mouse to indicate actions or choices.

The kernel and front end do not need to run on the same computer. They can be on separate computers connected through a network. The kernel can communicate with the front end by means of a high-level communication standard called *MathLink*.

Mathematica can be used

- as a calculator

  In[1]:= `5+7`
  Out[1]= *12*
  In[2]:= `Sqrt[9]`
  Out[2]= *3*

- to perform symbolic and algebraic operations; e.g.

  `Expand:`

  In[3]:= `(x + y)^2 + 9(2 + x)(x + y)`

  Out[3]= $9(2+x)(x+y)+(x+y)^2$

  In[4]:= `Expand[%]` [*% represents the last result*]

  Out[4]= $18x+10x^2+18y+11xy+y^2$

  `Factor:`

  In[5]:= `Factor[%]`

  Out[5]= $(x+y)(18+10x+y)$

  `Calculus:`

  In[6]:= `x^4 / (x^2 - 1)`

  Out[6]= $\frac{x^4}{-1+x^2}$

  In[7]:= `Integrate[%, x]`

  Out[7]= $x+\frac{x^3}{3}+\frac{log[-1+x]}{2}-\frac{log[1+x]}{2}$

  `Matrix Computations:`

  In[8]:= `m = Table[1/(i+j+1),{i, 3},{j, 3}]`

  Out[8]= $\{\{\frac{1}{3},\frac{1}{4},\frac{1}{5}\},\{\frac{1}{4},\frac{1}{5},\frac{1}{6}\},\{\frac{1}{5},\frac{1}{6},\frac{1}{7}\}\}$

  In[9]:= `Inverse[m]`

  Out[9]= $\{\{300,-900,630\},\{-900,2880,-2100\},\{630,-2100,1575\}\}$

- to create definitions

  In[10]:= `fac[n_] := n fac[n - 1]`

  In[11]:= `fac[1] = 1`

  Out[11]= 1

  In[12]:= `fac[20]`

  Out[12]= 2432902008176640000

- to trace steps in the solution process

  In[13]:= `Trace[fac[3]]`

  Out[13]= $\{fac[3], 3fac[3-1],$ [*fac(3) = 3 * fac(3-1)*]

  $\{\{3-1,-1+3,2\},$ [*3 - 1 = 2*]

  $fac[2], 2fac[2-1],$ [*fac(2) = 2 * fac(2-1)*]

  $\{\{2-1,-1+2,1\},$ [*2 - 1 = 1*]

  $fac[1], 1\},$ [*fac(1) = 1*]

  $2\ 1, 1\ 2, 2\},$ [*2 * 1 = 2*]

  $3\ 2, 2\ 3, 6\}$ [*3 * 2 = 6*]

- as a visualization system for functions and data

  In[14]:= `Plot[Sin[x^3], {x, -2, 2}]`

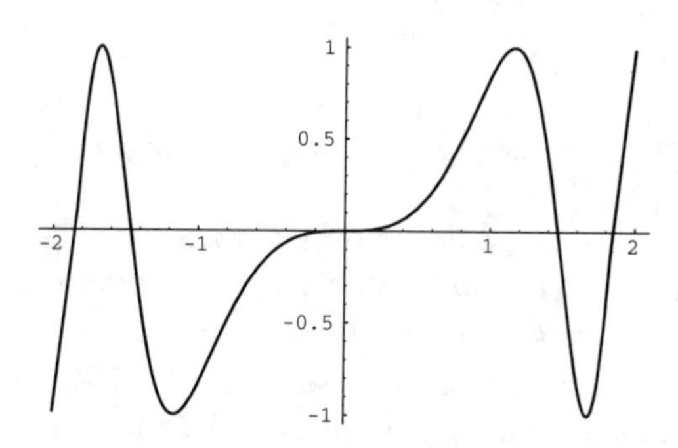

- to produce animated graphics or movies.

- to do set operations

  In[15]:= `Union[{1,2,3,5,7,11,13,17},{1,3,5,7,9,11,13}]`
  Out[15]:={1,2,3,5,7,9,11,13,17}
  In[16]:= `Intersect[{1,2,3,5,7,11,13,17}, {1,3,5,7,9,11,13}]`
  Out[16]:={1,3,5,7,11,13}

- as a high-level programming language in which you can create programs

  In[17]:=`n=17; While[(n=Floor[n/2])!=0, Print[n]]`
  8
  4
  2
  1

- to interface to other systems

  In[18]:= `(a^2 + b^2) / (x + y)^3`
  Out[18]= $\frac{a^2+b^2}{(x+y)^3}$
  In[19]:= `TeXForm[%]` [Conversion to TEX]
  Out[19]= TeXForm={{{a^2} + {b^2}} \over {(x+y)^3}}
  In[20]:= `FortranForm[%]` [Conversion to FORTRAN]
  *Out[20]= FortranForm = (a**2 + b**2) / (x + y)**3*

- to get information

  `?`*Name* [*show information on* `Name`]
  In[21]:= `?Log`
  Log[z] gives the natural logarithm of z (logarithm to base e).
  Log[b,z] gives the logarithm to base b.

- to handle files

  ≪filename *read in a file storing* `Mathematica` *input*
  !!filename *display the contents of a file*
  Save["*filename*", x_1, x_2, ...]
  save definitions for variables x_i in `filename`
  !*command* *execute an external OS command (available only on some systems)*

Mathematica can also be used as a software platform on which a user can run packages built for specific applications such as `Trigonometry`, `Statistics`, `Combinatorics`, `Vector Analysis`, and `Geometry`. One of these packages is called *Combinatorica* [7]. It is a computational environment for combinatorics and graph theory developed by Steven Skiena at SUNY at Stony Brook; it received a 1991 EDUCOM Higher Education Software Award for Distinguished Mathematics Software.

Combinatorica is made up of over 230 functions representing about 2500 lines of code. This means that, on average, a documented function is only eleven lines long – an impressive statistic.

Combinatorica offers a wide range of functions for:

- constructing combinatorial objects such as *permutations*, *subsets*, and *partitions*

  In[23]:= `DistinctPermutations[`$\{A, B, C\}$`]`
  Out[23]=$\{\{A, B, C\}, \{A, C, B\}, \{B, A, C\}, \{B, C, A\}, \{C, A, B\}, \{C, B, A\}\}$
  In[24]:= `KSubsets[{1,2,3,4,5},3]`
  Out[24]=$\{\{1, 2, 3\}, \{1, 2, 4\}, \{1, 2, 5\}, \{1, 3, 4\}, \{1, 3, 5\}, \{1, 4, 5\}, \{2, 3, 4\}, \{2, 3, 5\}, \{2, 4, 5\}, \{3, 4, 5\}\}$

- constructing a wide variety of graphs such as *cycles*, *trees*, *hypercubes*, and *random graphs*

  In[25]:= `ShowGraph[ K[5] ];`

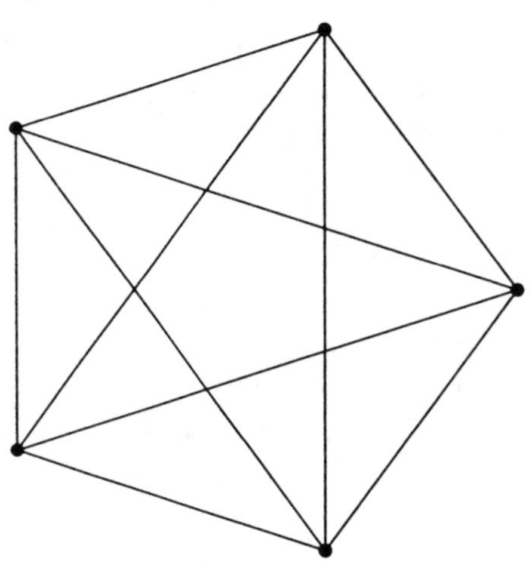

In[26]:= `ShowLabeledGraph[ RootedEmbedding[RandomTree[10],1]];`

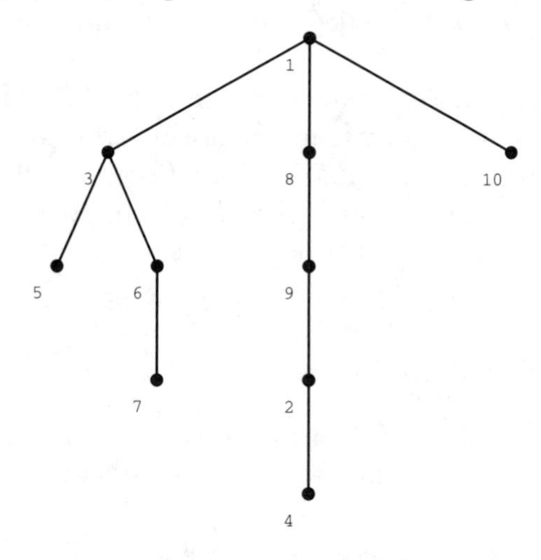

- constructing graphs from other graphs using *line graphs*, *graph products*, *joins* and *powers*
  In[27]:= `ShowGraph[ GraphProduct[ K[5], K[3] ] ];`

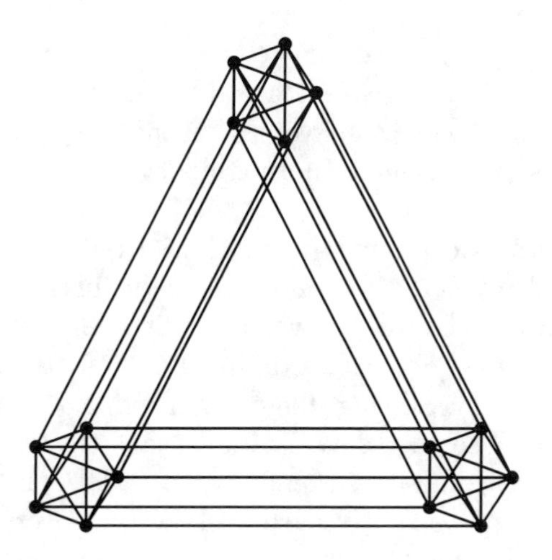

- testing graphs for certain properties (such as *bipartite*, *Eulerian*, and *planar*)
  In[28]:= `EulerianQ[ K[9] ]`
  *Out[28]=*True

Other functions compute invariants of a graph such as its chromatic number and diameter. *Combinatorica* also includes various graph algorithm implementations such as Dijkstra's Shortest Path Algorithm, Kruskal's Algorithm, The Traveling Salesman Problem, and a number of graph coloring algorithms.

Combinatorica reads and writes files in *SPREMB* format [**4**]. *SPREMB* is a system for developing graph algorithms that was designed at the University of Queensland, Australia. *SPREMB*'s graph editor, called GED, may be used in *Combinatorica* to edit graphs. In *SPREMB*, a graph is represented by n lines corresponding to the n vertices of the graph. Line i is of the form

$$ixyv_1v_2\cdots v_k$$

where

i	is the vertex label
x,y	are the coordinates of the vertex in the plane, $0 \le x, y \le 1$
v_j	are the vertices adjacent to vertex i

The following is the description of the graph K_5 in SPREMB format:

```
1  0.645842   0.974914  2  3  4  5
2  0.0728013  0.788722  1  3  4  5
3  0.0728013  0.186192  1  2  4  5
4  0.645842   0.        1  2  3  5
5  1.         0.487457  1  2  3  4
```

4.1.2. *Evaluation for Educational Use.*

Mathematica partially satisfies desirable qualities **A** (uses a graphics environment; user interfaces with system through keyboard and mouse), **C** (usage of script files), **H** (usage of notebooks), and **I** (interfacing with TEX, Fortran, and C).

Since *Mathematica* does not have a direct graphic interface, the graphics capabilities of *Combinatorica* are limited. For instance, the coloring of a graph is performed by labeling the vertices with non-negative integers representing different colors. For each graphical output, the system will create a window to display the resulting diagram. A window may also be used for animation purposes. The user may decide to save the windows for later use or to destroy them. No pull-down menus are available in standard mode.

Combinatorica is not easy to learn and to use. Users have to learn the syntax of a broad collection of commands. Typing in the commands is quite tedious. A solution to this is to create a script file. Like a batch file, a script is a sequence of commands that can be executed one after the other as many times as desired. Some syntax error messages are not very meaningful to the unsophisticated user. If incorrect data is entered to certain function calls, garbage is returned as a result. Help is available by issuing a command. A menu-driven *intelligent* help system would be desirable.

The standard way in which *Mathematica* works is to take any expression you give as input, evaluate the expression completely, and then return the result. For feedback on intermediate steps in the evaluation process, the user may use the command `Trace[`*expr*`]` (see example given above). This command returns a list which includes all the intermediate expressions involved in the evaluation of *expr*. Except in rather simple cases, however, the number of intermediate expressions generated in this way is very large, and the list returned by `Trace` is difficult to understand. The user may filter the expressions that `Trace` records by using `Trace[`*expr, form*`]`. This trace facility is rather primitive and would be of little use to a teacher trying to explain the solution of a problem in the classroom or to a student trying to understand the solution outside of class.

Notebook interfaces, when available, provide the user with various features that make the system easier to use. The amount of typing involved in entering commands can be reduced by typing in parts of names known to *Mathematica* and asking the interface to complete the name. Various parameters, particularly graphical ones, may be selected by using a mouse within pull-down menu windows. This interface, if further developed, could transform *Mathematica* into a better educational tool.

In its current state, *Combinatorica* would have a very limited use in the classroom. Occasionally, a teacher at the high school or college level may use this system through prepared scripts and notebooks. A student may use *Combinatorica* outside of class to access notebooks that review topics taught in the classroom, to learn through discovery, and to assist with homework assignments.

4.2. GraphPack.

4.2.1. *Description.*

GraphPack is a software system for manipulating graphs and sets [**5, 6**]. It was developed at RPI by a team led by Dr. Mukkai S. Krishnamoorthy. The system is written in C and runs under X Windows or Sun Windows. It supports a language called *LiLa* (short for LIttle LAnguage), a background environment program called *Kernel*, an *Implementation Selection Assistant* (ISA), a *Library* of functions, and a *User Interface.* It may be run in textual mode on any *UNIX*-based machine, or in graphical mode on *UNIX*-based machines with graphics capabilities (such as a Sun Sparcstation). To obtain a copy of *GraphPack*, send a request by electronic mail to `moorthy@cs.rpi.edu`.

LiLa is a programming language that was developed specifically for *GraphPack*. It is based on C with additional graph and set-theoretical primitives. These primitives are implemented as user-callable functions. The language makes use of all the standard C commands and contains the same features as C without the overhead of special declarations. A user may write *LiLa* procedures without specification of the data structures to be used for data storage. The language handles all of the input/output. The new commands are divided into several groups: *interpreter* commands, *set* commands, *list* commands, *stack* commands, *queue* commands, and *graph* commands. A help file is available that lists the commands and gives a concise description of each command.

The *Kernel* manages the data abstraction mechanism. It contains routines written in C for the primitive operations on the data structures. Different representations for a single data structure are available so that the most desirable format will be used for a particular basic operation application. It contains both a compiler and an interpreter for *LiLa*.

The *Information Selection Assistant* (ISA) is a module to manage the interface between the applications and the *Kernel* as well as selecting the implementation. The ISA may cooperate interactively with the user with queries to the user. The user decides whether to choose a particular implementation or leave the decision to the ISA. In batch mode, ISA will survey the code and decide what the correct implementation should be.

In the graphical mode, the *User Interface* consists of two windows that are used to access the routines and graphs. The `text` window runs the interpreter for LiLa. To use this window, the user has to type commands at the keyboard. The system responds by displaying the results. The system uses this window to display other messages to the user such as error messages and *LiLa* code generated during command execution. The second window is a `graphics` window. It contains the *canvas* on which the graphs are drawn. The user has to use a mouse to choose any one of the seven available pull-down menus and to select commands or options within a menu. To input a graph, a user may use the mouse to draw it within this window. Graphs may also be created manually or automatically in the `text` window, or loaded from a previously saved file. Operations in one window will result in changes in the other window.

The *Library* currently consists of a collection of programs that implement various graph algorithms and other logic commands. These programs may be used directly on given graphs to solve problems related to matching (bipartite graphs and general graphs), Minimal Spanning Trees, Graph Coloring, the Traveling Salesman Problem, and several others. The logic commands may be used to check boolean functions and relations between integers. All these programs can be called up from either window. The developers plan to add other programs that would generate certain special graphs (such as complete graphs, the Petersen graph, and the Tutte graph) and families of sets at the request of the user.

The following is a sample session to demonstrate how the *textual* mode may be used to create a graph and perform a number of operations:

<lila> `set(g1, make_graph(5));` [*make a graph of five vertices*]
Do you wish to name the vertices ? (y/n): `n`
Are the vertices of your graph weighted? (y/n): `n`
Choose one: directed or undirected graph (d/u): `u`
Enter the adjacency list for your graph below:
1: `2 3 4 5`

```
2: 1 3
3: 1 2 5
4: 1 5
5: 1 3 4
Does your graph contain weighted edges? (y/n): n

<lila> print(g1);
VERTICES : {1,2,3,4,5}
EDGES : {(1,2),(1,3),(1,4),(1,5),(2,3),(3,5),(4,5)}

<lila> print(cycle(g1));                [does g1 have a cycle?]
1                                       [yes]

<lila> euler(g1);
 Not Eulerian

<lila> erase({3,5}, edges(g1));
<lila> print(edges(g1));
{(1,2),(1,3),(1,4),(1,5),(2,3),(4,5)}

<lila> euler(g1);
 Eulerian

<lila> set(g3, make_graph(7));
Do you wish to name the vertices ? (y/n): n
Are the vertices of your graph weighted? (y/n): n
Choose one: directed or undirected graph (d/u): d
Enter the adjacency list for your graph below:
1: 2 3
2: 4 5
3: 6
4: 7
5:
6:
7:
Does your graph contain weighted edges? (y/n): n

<lila> print(g3);
VERTICES : {1,2,3,4,5,6,7}
EDGES : {(1,2),(1,3),(2,4),(2,5),(3,6),(4,7)}

<lila> print(bfs(g3));                  [breadth first search]
{(1,2),(1,3),(2,4),(2,5),(3,6),(4,7)}

<lila> bye
```

GraphPack offers the user a variety of set operations such as *union*, *intersection*, *difference*, *product*, and *power*. Set manipulation can only be performed in the *textual* mode. The following is a sample session to demonstrate how *GraphPack* may be used for set operations:

```
<lila> set(s1, make_set({1,2,3,5,7,11,13,17}));

<lila> print(s1);
{1,2,3,5,7,11,13,17}

<lila> set(s2, make_set({1,3,5,7,9,11,13}));
<lila> print(s2);
{1,3,5,7,9,11,13}

<lila> set(s3, union(s1,s2));

<lila> print(s3);
{1,2,3,5,7,9,11,13,17}

<lila> set(s4, intersection(s1,s2));

<lila> print(s4);
{1,3,5,7,11,13}

<lila> print(difference(s1,s2));
{2,17}

<lila> print(card(s4));
6

<lila> print(power(s4, 3));          [all subsets of s4 with 3 elements]
{{1,3,5},{1,3,7},{1,3,11},{1,3,13},{1,5,7},{1,5,11},
{1,5,13},{1,7,11},{1,7,13},{1,11,13},{3,5,7},{3,5,11},
{3,5,13},{3,7,11},{3,7,13},{3,11,13},{5,7,11},{5,7,13},
{5,11,13},{7,11,13}}

<lila> bye
```

4.2.2. *Evaluation for Educational Use.*

GraphPack partially satisfies desirable qualities **A** (uses *X Windows*; user interfaces with system through keyboard and/or mouse; in the `graphics` window, the user selects options in pull-down menus by clicking the mouse) and **C** (making use of the `graphics` window to enter commands).

The *graphical* mode of *GraphPack* has a number of very good features. Commands can be selected through pull-down menus. Graphs can be drawn by clicking and moving the mouse. The graphics window, however, cannot be used to do set operations.

The unsophisticated user will find that the *textual* mode of *GraphPack*, in its current state, is user-unfriendly and difficult to learn. Many commands are difficult to learn and a substantial amount of typing may be required. The user manual is very limited and the help files are basically a collection of syntactical expressions. A very good *help* system is desperately needed. There also are quite a few bugs that need to be fixed, and the output from several commands needs to be clearly explained.

The developers of *GraphPack* admit that their major objective was to provide theoretical computer scientists and mathematicians with a useful research tool. However, there is potential for it to be used in education. In fact, the developers want *GraphPack* to be thought similar to *Mathematica*.

GraphPack does not provide the user with a trace of the solution of an input problem. There is no *problem mode* for students to use.

This system can be used in conjunction with the mathematical subjects taught in high school and college level courses. At this level, students can use *GraphPack*'s full capabilities. Outside of the classroom, this software could be used by the students to assist in homework assignments or when reviewing concepts taught in the class. The graphical mode of the system can be used in the classroom as a teacher assistant or in a laboratory environment, provided that the text window is used very sparingly and the commands typed in are simple and short.

4.3. SetPlayer.

4.3.1. *Description.*

SetPlayer (release 4.0) is an interactive command-driven software system for set manipulation [**1**]. It was developed at RPI by D. Berque, R. Cecchini, M. Goldberg, and R. Rivenburgh. This system was designed to be used as an educational as well as a research tool for discrete mathematics and this is the major reason why it has more educationally-oriented features than the other two systems. It is written in C and runs under the *UNIX* operating system. It is available in two versions: a *textual* and *graphical* version. *SetPlayer* is available via anonymous ftp from `ftp.cs.rpi.edu`.

SetPlayer recognizes four data types: *integers*, *sets*, *tuples*, and *families*. These data types are also referred to as *entities*. In *SetPlayer* terminology,

- the term *set* stands for a finite set of integers;
- the term *tuple* stands for a finite sequence of integers;
- the term *family* stands for a finite collection of sets (a *set-family*) or a finite collection of tuples (a *tuple-family*). A *graph* may be represented by its edges as a set-family all of whose members have cardinality two. A *directed graph* is a family whose members are tuples of length 2.

The graphical version runs under the X window system and needs a terminal with graphics capabilities. It operates via four windows: the `help/history` window, the `database` window, the `display` window, and the `text` window.

The primary window is the text window. Using this window, users can define the entities that they wish to manipulate. The text window can be run in either the *menu mode* or the *command mode*.

To use the system in the *menu mode*, the user has to type in one of the various menu commands and then respond to a sequence of system prompts. It is possible to return to the preceding prompt (very helpful in case of an error) by typing a **!**. Once the user has responded to all the system prompts, *SetPlayer* will execute the command. The *menu mode* interface is very easy to use and is very suitable for novice users. This interface checks the user's input for certain common errors (such as giving the same name to different sets or families). The *menu mode* provides feedback which helps the user learn the syntax of the *command mode* version of the command being used.

The following sample session demonstrates how the *menu mode* is used to perform set operations.

% `setplayer`

Welcome to SetPlayer
Version 4.0
You are now in SetPlayer's Menu Mode.
(Type 'MENU' to see the Main Menu.)

Menu (1): **create**
*** SET CREATE ***
Enter the name of the SET to CREATE: **set1**
Enter the Elements of the New SET 'set1'
: **1..3,5,7,11,13,17** [1..3 denotes 1,2,3]
*** The SET 'set1' will be CREATED unless you enter '!':* <**CR**>
*** The SET 'set1' has been CREATED.*
→ *set1 = (1..3, 5, 7, 11, 13, 17)*
*** SET CREATE ***
Enter the name of the SET to CREATE: **set2**
Enter the Elements of the New SET 'set2'
: **1,3,5,7,9,11,13**
*** The SET 'set2' will be CREATED unless you enter '!':* <**CR**>
*** The SET 'set2' has been CREATED.*
→ *set2 = (1, 3, 5, 7, 9, 11, 13)*
*** SET CREATE ***
Enter the name of the SET to CREATE: **!**
Menu (3): **union**
*** SET UNION Operation ***
Enter the name to be given the Result SET: **set3**

Enter the names (up to 50) of the SETs to be UNIONED: **set1,set2**
The SETs will be UNIONED unless you enter '!': **<CR>**
** SET UNION is completed.
The Result SET is named 'set3'.
→ set3 = UNION(set1, set2)
** SET UNION Operation **
Enter the name to be given the Result SET: **!**

Menu (4): **intersect**
** SET INTERSECT Operation **
Enter the name to be given the Result SET: **set4**
Enter the names (up to 50) of SETs to be INTERSECTED: **set1,set2**
The SETs will be INTERSECTED unless you enter '!': **<CR>**
** SET INTERSECT is completed.
The Result SET is named 'set4'.
→ set4 = INTERSECT(set1, set2)
** SET INTERSECT Operation **
Enter the name to be given the Result SET: **!**

Menu (5): **print**
** ENTITY PRINT **
Enter the names of the ENTITIES you want to see : **set1,set2,set3,set4**
SET Name: set1
Cardinality = 8, (1..3, 5, 7, 11, 13, 17)
SET Name: set2
Cardinality = 7, (1, 3, 5, 7, 9, 11, 13)
SET Name: set3
Cardinality = 9, (1..3, 5, 7, 9, 11, 13, 17)
SET Name: set4
Cardinality = 6, (1, 3, 5, 7, 11, 13)
→ PRINT(set1..set4)

Menu (6): **exit**
Do you really wish to EXIT from SetPlayer, (Y)es or (N)o ? **y**
Do you wish to save the WORK-SPACE, (Y)es or (N)o ?
(Enter '!' to 'back-up'.)
: **n**
Do you wish to save the current HISTORY file, (Y)es or (N)o ?
(Enter '!' to 'back-up'.)
: **n**

The *command mode* interface is more flexible than the *menu mode* interface. Like the *menu mode*, this interface is case-sensitive, and allows the first three or more letters of a command to be used as an abbreviation of the command. In this mode, a user has to type in syntactically correct instructions without the computer prompts. For instance, the syntax for creating `set1` in *command mode* is:

Command(1): **set1 = (1..3, 5, 7, 11, 13, 17)**

In the *command mode*, a command may be split across several lines by terminating each line (except the last) with a back-slash. Any text which follows the characters /* is treated as a comment. When **!!** is entered, the previously executed command will be executed again. When **!n** is entered, the *n*th command from the current session is executed again. When **!string** is entered, the most recently entered command which began with *string* is executed again. The *command mode* is capable of handling nested expressions.

The display window allows users to view graphs of the entities created. Additionally, the display window allows users to manipulate the graphs that have been created.

The database window keeps track of all the integers, sets, tuples, families and groups that have been defined.

The history/help window has two functions. It enables users to review what procedures have been carried out and it offers a help utility. The help utility is simply program documentation. It is not an interactive type of help whereby users are guided to solutions by computer prompts. A dynamic type of help feature would be of great benefit to the user.

The user can *iconify*, *de-iconify*, *move*, and *re-size* any of the four major windows. These operations can be performed by using the mouse. Some of the windows may also be scrolled.

The textual version may be used on any *UNIX* based terminal. It makes use of just a text window. This window operates in the same manner as the text window in the graphical version.

4.3.2. *Evaluation for Educational Usefulness.*

SetPlayer satisfies desirable quality **B** (ease of use) and partially satisfies **A** (uses *X Windows*; the user interfaces with the system through keyboard and/or mouse) and **C** (commands may be issued without wasting too much time; `display` window is helpful).

The graphics orientation of *SetPlayer* can be improved. Commands have to be typed into the *text* window. No pull-down menus are available. The *menu mode* makes commands easy to compose. However, it is not convenient to use in the classroom. Step-by-step solutions are not available. To use the graphical version effectively, a user needs to have a very large monitor. We used 19 inch monitors and found it tedious to manage all four windows together.

It is a good idea to have a window for displaying help messages. However, the help facility of *SetPlayer* is restricted to the display of a help file. A sophisticated help system is badly needed. The *help/history* may also be used to review previously issued commands, but *SetPlayer* does not allow an actual script to be developed.

SetPlayer does not use characters for elements. For instance, it is not possible to create a set denoted as follows:

$$\texttt{Cars} = \{\textit{Chevrolet, Dodge, Ford}\}$$

Such a feature would be useful when working in an educational setting. It would be especially helpful when working with elementary school students who may be able to comprehend better the concept of sets with real objects rather than numbers.

SetPlayer is unable to produce mathematical symbols. A pull down window could be added to enable users to access useful mathematical operators (such as set operators). The user would click on the operator designation to make the corresponding symbol appear as part of the user's input in the display window.

In spite of these limitations, *SetPlayer* could be a helpful tool to use at the high school and college level, both by the teacher in the classroom and by the students outside of class. The graphical display of sets and their elements helps the user to understand better the problem being solved and the concepts involved. The database window helps the user remember the names of the entities being used and the members of defined sets and families.

5. Analysis and Conclusions

There are other software systems for discrete mathematics that we would like to test in the near future. Two of these which should be mentioned are *NETPAD* and *GraphLab*. *NETPAD* is a package developed by Nate Dean and Clyde Monma at Bell Communications Research (see [**3**]). It is an interactive system for the manipulation and analysis of networks. *GraphLab* was designed by Greg Shannon at Indiana University [**2**]. It is a visual and textual system for creating, editing, manipulating, and displaying graphs and for designing graph algorithms. We should also note that Steven Skiena, Mark Goldberg, Nate Dean, and Greg Shannon are developing a system (project LINK) which is intended to combine the capabilities of the systems each developed separately (see [**3**]).

At the pre-college level, teachers who are introduced to discrete mathematics seem to embrace it eagerly. Many teachers have introduced a number of topics in discrete mathematics into their syllabus. Suitable software systems will certainly accelerate this trend. However, much work still needs to be done in developing suitable educational software for discrete mathematics. We strongly recommend that the designing of sophisticated software systems for education should be undertaken by teams consisting of software engineers, educators and experts in the field of instruction (in our case, mathematicians). It is only through such endeavors that proper software systems will be developed that are satisfactory for in-class use.

Note: This paper was written in 1992. Since that time the essential picture presented in this paper has not changed substantially although some additional software for discrete mathematics has been developed (see Table 1 in [**3**]*). A Java version of GraphPak has recently been developed. To access it, try the URL:* `http://www.cs.rpi.edu/~moorthy` *and select "Graph Draw in Java".*

References

[1] D. Berque, R. Cecchini, M. Goldberg, and R. Rivenburgh, *The SetPlayer system: An overview and a user manual*, Technical Report 90-20, Rensselaer Polytechnic Institute, Troy NY, 1990.

[2] B. Birgisson and G. Shannon, *Graphview: An extensible interactive platform for manipulating and displaying graphs*, Technical Report 295, Computer Science Department, Indiana University, Bloomington IN, 1989.

[3] N. Dean and Y. Liu, "Discrete Mathematics Software for K-12 Education," this volume.

[4] P. Eades, I. Fogg, and D. Kelly, *Spremb: A system for developing graph algorithms*, Technical Report, Department of Computer Science, University of Queensland, St. Lucia, Queensland, Australia, 1988.

[5] M. Echeandia, *New functions for LILA interpreter and kernel for Graphpack*, Technical Report 88-25, Department of Computer Science, Rensselaer Polytechnic Institute, Troy NY, 1988.

[6] M. Krishnamoorthy, T. Spencer, M. Echeandia, A. Faulstich, G. Kyriazis, E. McCaughrin, C. Maroulis, and D. Pape, *Graphpack: a software system for computations of graphs and sets*, Technical Report 90-7, Rensselaer Polytechnic Institute, Troy NY, 1990.

[7] S. Skiena, *Implementing Discrete Mathematics: Combinatorics and Graph Theory with Mathematica*, Addison-Wesley, Redwood City CA, 1990.

[8] S. Wolfram, *Mathematica: A System for Doing Mathematics by Computer*, Second Edition. Addison-Wesley, Redwood City, California, 1991.

Lockheed Martin Federal Systems, Niagara Falls (NY)
E-mail address: `mario.vassallo@lmco.com`

Department of Computer Science, SUNY at Buffalo
Current address: Department of Computing, Imperial College, London
E-mail address: `ar9@doc.ic.ac.uk`

Selected Titles in This Series

(Continued from the front of this publication)

5 **Frank Hwang, Fred Roberts, and Clyde Monma, Editors,** Reliability of Computer and Communication Networks

4 **Peter Gritzmann and Bernd Sturmfels, Editors,** Applied Geometry and Discrete Mathematics, The Victor Klee Festschrift

3 **E. M. Clarke and R. P. Kurshan, Editors,** Computer-Aided Verification '90

2 **Joan Feigenbaum and Michael Merritt, Editors,** Distributed Computing and Cryptography

1 **William Cook and Paul D. Seymour, Editors,** Polyhedral Combinatorics